T0296488

LONDON MATHEMATICAL SOCIETY LECTURE NOTE SERIES

Managing Editor: Professor N.J. Hitchin, Mathematical Institute,
University of Oxford, 24–29 St Giles, Oxford OX1 3LB, United Kingdom

The titles below are available from booksellers, or, in case of difficulty, from Cambridge University Press.

46 *p*-adic Analysis: a short course on recent work, N. KOBLITZ
59 Applicable differential geometry, M. CRAMPIN & F.A.E. PIRANI
66 Several complex variables and complex manifolds II, M.J. FIELD
86 Topological topics, I.M. JAMES (ed)
88 FPF ring theory, C. FAITH & S. PAGE
90 Polytopes and symmetry, S.A. ROBERTSON
96 Diophantine equations over function fields, R.C. MASON
97 Varieties of constructive mathematics, D.S. BRIDGES & F. RICHMAN
99 Methods of differential geometry in algebraic topology, M. KAROUBI & C. LERUSTE
100 Stopping time techniques for analysts and probabilists, L. EGGHE
104 Elliptic structures on 3-manifolds, C.B. THOMAS
105 A local spectral theory for closed operators, I. ERDELYI & WANG SHENGWANG
107 Compactification of Siegel moduli schemes, C.-L. CHAI
109 Diophantine analysis, J. LOXTON & A. VAN DER POORTEN (eds)
113 Lectures on the asymptotic theory of ideals, D. REES
116 Representations of algebras, P.J. WEBB (ed)
119 Triangulated categories in the representation theory of finite-dimensional algebras, D. HAPPEL
121 Proceedings of *Groups - St Andrews 1985*, E. ROBERTSON & C. CAMPBELL (eds)
128 Descriptive set theory and the structure of sets of uniqueness, A.S. KECHRIS & A. LOUVEAU
130 Model theory and modules, M. PREST
131 Algebraic, extremal & metric combinatorics, M.-M. DEZA, P. FRANKL & I.G. ROSENBERG (eds)
138 Analysis at Urbana, II, E. BERKSON, T. PECK, & J. UHL (eds)
139 Advances in homotopy theory, S. SALAMON, B. STEER & W. SUTHERLAND (eds)
140 Geometric aspects of Banach spaces, E.M. PEINADOR & A. RODES (eds)
141 Surveys in combinatorics 1989, J. SIEMONS (ed)
144 Introduction to uniform spaces, I.M. JAMES
146 Cohen-Macaulay modules over Cohen-Macaulay rings, Y. YOSHINO
148 Helices and vector bundles, A.N. RUDAKOV *et al*
149 Solitons, nonlinear evolution equations and inverse scattering, M. ABLOWITZ & P. CLARKSON
150 Geometry of low-dimensional manifolds 1, S. DONALDSON & C.B. THOMAS (eds)
151 Geometry of low-dimensional manifolds 2, S. DONALDSON & C.B. THOMAS (eds)
152 Oligomorphic permutation groups, P. CAMERON
153 L-functions and arithmetic, J. COATES & M.J. TAYLOR (eds)
155 Classification theories of polarized varieties, TAKAO FUJITA
158 Geometry of Banach spaces, P.F.X. MÜLLER & W. SCHACHERMAYER (eds)
159 Groups St Andrews 1989 volume 1, C.M. CAMPBELL & E.F. ROBERTSON (eds)
160 Groups St Andrews 1989 volume 2, C.M. CAMPBELL & E.F. ROBERTSON (eds)
161 Lectures on block theory, BURKHARD KÜLSHAMMER
163 Topics in varieties of group representations, S.M. VOVSI
164 Quasi-symmetric designs, M.S. SHRIKANDE & S.S. SANE
166 Surveys in combinatorics, 1991, A.D. KEEDWELL (ed)
168 Representations of algebras, H. TACHIKAWA & S. BRENNER (eds)
169 Boolean function complexity, M.S. PATERSON (ed)
170 Manifolds with singularities and the Adams-Novikov spectral sequence, B. BOTVINNIK
171 Squares, A.R. RAJWADE
172 Algebraic varieties, GEORGE R. KEMPF
173 Discrete groups and geometry, W.J. HARVEY & C. MACLACHLAN (eds)
174 Lectures on mechanics, J.E. MARSDEN
175 Adams memorial symposium on algebraic topology 1, N. RAY & G. WALKER (eds)
176 Adams memorial symposium on algebraic topology 2, N. RAY & G. WALKER (eds)
177 Applications of categories in computer science, M. FOURMAN, P. JOHNSTONE & A. PITTS (eds)
178 Lower K- and L-theory, A. RANICKI
179 Complex projective geometry, G. ELLINGSRUD *et al*
180 Lectures on ergodic theory and Pesin theory on compact manifolds, M. POLLICOTT
181 Geometric group theory I, G.A. NIBLO & M.A. ROLLER (eds)
182 Geometric group theory II, G.A. NIBLO & M.A. ROLLER (eds)
183 Shintani zeta functions, A. YUKIE
184 Arithmetical functions, W. SCHWARZ & J. SPILKER
185 Representations of solvable groups, O. MANZ & T.R. WOLF
186 Complexity: knots, colourings and counting, D.J.A. WELSH
187 Surveys in combinatorics, 1993, K. WALKER (ed)
188 Local analysis for the odd order theorem, H. BENDER & G. GLAUBERMAN
189 Locally presentable and accessible categories, J. ADAMEK & J. ROSICKY
190 Polynomial invariants of finite groups, D.J. BENSON
191 Finite geometry and combinatorics, F. DE CLERCK *et al*
192 Symplectic geometry, D. SALAMON (ed)
194 Independent random variables and rearrangement invariant spaces, M. BRAVERMAN
195 Arithmetic of blowup algebras, WOLMER VASCONCELOS
196 Microlocal analysis for differential operators, A. GRIGIS & J. SJÖSTRAND
197 Two-dimensional homotopy and combinatorial group theory, C. HOG-ANGELONI *et al*

198 The algebraic characterization of geometric 4-manifolds, J.A. HILLMAN
199 Invariant potential theory in the unit ball of $\mathbf{C}^n$, MANFRED STOLL
200 The Grothendieck theory of dessins d'enfant, L. SCHNEPS (ed)
201 Singularities, JEAN-PAUL BRASSELET (ed)
202 The technique of pseudodifferential operators, H.O. CORDES
203 Hochschild cohomology of von Neumann algebras, A. SINCLAIR & R. SMITH
204 Combinatorial and geometric group theory, A.J. DUNCAN, N.D. GILBERT & J. HOWIE (eds)
205 Ergodic theory and its connections with harmonic analysis, K. PETERSEN & I. SALAMA (eds)
207 Groups of Lie type and their geometries, W.M. KANTOR & L. DI MARTINO (eds)
208 Vector bundles in algebraic geometry, N.J. HITCHIN, P. NEWSTEAD & W.M. OXBURY (eds)
209 Arithmetic of diagonal hypersurfaces over finite fields, F.Q. GOUVÊA & N. YUI
210 Hilbert C*-modules, E.C. LANCE
211 Groups 93 Galway / St Andrews I, C.M. CAMPBELL *et al* (eds)
212 Groups 93 Galway / St Andrews II, C.M. CAMPBELL *et al* (eds)
214 Generalised Euler-Jacobi inversion formula and asymptotics beyond all orders, V. KOWALENKO *et al*
215 Number theory 1992–93, S. DAVID (ed)
216 Stochastic partial differential equations, A. ETHERIDGE (ed)
217 Quadratic forms with applications to algebraic geometry and topology, A. PFISTER
218 Surveys in combinatorics, 1995, PETER ROWLINSON (ed)
220 Algebraic set theory, A. JOYAL & I. MOERDIJK
221 Harmonic approximation, S.J. GARDINER
222 Advances in linear logic, J.-Y. GIRARD, Y. LAFONT & L. REGNIER (eds)
223 Analytic semigroups and semilinear initial boundary value problems, KAZUAKI TAIRA
224 Computability, enumerability, unsolvability, S.B. COOPER, T.A. SLAMAN & S.S. WAINER (eds)
225 A mathematical introduction to string theory, S. ALBEVERIO, J. JOST, S. PAYCHA, S. SCARLATTI
226 Novikov conjectures, index theorems and rigidity I, S. FERRY, A. RANICKI & J. ROSENBERG (eds)
227 Novikov conjectures, index theorems and rigidity II, S. FERRY, A. RANICKI & J. ROSENBERG (eds)
228 Ergodic theory of $\mathbf{Z}^d$ actions, M. POLLICOTT & K. SCHMIDT (eds)
229 Ergodicity for infinite dimensional systems, G. DA PRATO & J. ZABCZYK
230 Prolegomena to a middlebrow arithmetic of curves of genus 2, J.W.S. CASSELS & E.V. FLYNN
231 Semigroup theory and its applications, K.H. HOFMANN & M.W. MISLOVE (eds)
232 The descriptive set theory of Polish group actions, H. BECKER & A.S. KECHRIS
233 Finite fields and applications, S. COHEN & H. NIEDERREITER (eds)
234 Introduction to subfactors, V. JONES & V.S. SUNDER
235 Number theory 1993–94, S. DAVID (ed)
236 The James forest, H. FETTER & B. GAMBOA DE BUEN
237 Sieve methods, exponential sums, and their applications in number theory, G.R.H. GREAVES *et al*
238 Representation theory and algebraic geometry, A. MARTSINKOVSKY & G. TODOROV (eds)
239 Clifford algebras and spinors, P. LOUNESTO
240 Stable groups, FRANK O. WAGNER
241 Surveys in combinatorics, 1997, R.A. BAILEY (ed)
242 Geometric Galois actions I, L. SCHNEPS & P. LOCHAK (eds)
243 Geometric Galois actions II, L. SCHNEPS & P. LOCHAK (eds)
244 Model theory of groups and automorphism groups, D. EVANS (ed)
245 Geometry, combinatorial designs and related structures, J.W.P. HIRSCHFELD *et al*
246 *p*-Automorphisms of finite *p*-groups, E.I. KHUKHRO
247 Analytic number theory, Y. MOTOHASHI (ed)
248 Tame topology and o-minimal structures, LOU VAN DEN DRIES
249 The atlas of finite groups: ten years on, ROBERT CURTIS & ROBERT WILSON (eds)
250 Characters and blocks of finite groups, G. NAVARRO
251 Gröbner bases and applications, B. BUCHBERGER & F. WINKLER (eds)
252 Geometry and cohomology in group theory, P. KROPHOLLER, G. NIBLO, R. STÖHR (eds)
253 The *q*-Schur algebra, S. DONKIN
254 Galois representations in arithmetic algebraic geometry, A.J. SCHOLL & R.L. TAYLOR (eds)
255 Symmetries and integrability of difference equations, P.A. CLARKSON & F.W. NIJHOFF (eds)
256 Aspects of Galois theory, HELMUT VÖLKLEIN *et al*
257 An introduction to noncommutative differential geometry and its physical applications 2ed, J. MADORE
258 Sets and proofs, S.B. COOPER & J. TRUSS (eds)
259 Models and computability, S.B. COOPER & J. TRUSS (eds)
260 Groups St Andrews 1997 in Bath, I, C.M. CAMPBELL *et al*
261 Groups St Andrews 1997 in Bath, II, C.M. CAMPBELL *et al*
263 Singularity theory, BILL BRUCE & DAVID MOND (eds)
264 New trends in algebraic geometry, K. HULEK, F. CATANESE, C. PETERS & M. REID (eds)
265 Elliptic curves in cryptography, I. BLAKE, G. SEROUSSI & N. SMART
267 Surveys in combinatorics, 1999, J.D. LAMB & D.A. PREECE (eds)
268 Spectral asymptotics in the semi-classical limit, M. DIMASSI & J. SJÖSTRAND
269 Ergodic theory and topological dynamics, M.B. BEKKA & M. MAYER
270 Analysis on Lie Groups, N.T. VAROPOULOS & S. MUSTAPHA
271 Singular perturbations of differential operators, S. ALBEVERIO & P. KURASOV
272 Character theory for the odd order function, T. PETERFALVI
273 Spectral theory and geometry, E.B. DAVIES & Y. SAFAROV (eds)
274 The Mandlebrot set, theme and variations, TAN LEI (ed)
276 Singularities of plane curves, E. CASAS-ALVERO
278 Global attractors in abstract parabolic problems, J.W. CHOLEWA & T. DLOTKO
279 Topics in symbolic dynamics and applications, F. BLANCHARD, A. MAASS & A. NOGUEIRA (eds)
281 Explicit birational geometry of 3-folds, ALESSIO CORTI & MILES REID (eds)

London Mathematical Society Lecture Note Series. 278

Global Attractors in Abstract Parabolic Problems

Jan W. Cholewa & Tomasz Dlotko
Silesian University, Poland

In cooperation with Nathaniel Chafee
Georgia Institute of Technology

CAMBRIDGE UNIVERSITY PRESS
Cambridge, New York, Melbourne, Madrid, Cape Town, Singapore, São Paulo

Cambridge University Press
The Edinburgh Building, Cambridge CB2 8RU, UK

Published in the United States of America by Cambridge University Press, New York

www.cambridge.org
Information on this title: www.cambridge.org/9780521794244

First published 2000

A catalogue record for this publication is available from the British Library

ISBN 978-0-521-79424-4 paperback

Transferred to digital printing 2007

Contents

Preface

The past forty years have witnessed an intensive study of problems in mathematical physics governed by dissipative equations and much progress has been achieved. Two already existing branches of mathematics have played a central role in these investigations: first, the qualitative theory of ordinary differential equations and - closely related to that - the theory of dynamical systems; second, the theory of partial differential equations. Thus, in this same connection, mathematicians have successfully applied finite dimensional concepts and techniques, suitably modified, to the study of semigroups generated in infinite dimensional spaces by evolutionary partial differential equations.

In recent years several authors have developed and exploited this combination of finite dimensional and infinite dimensional techniques. Specifically, we cite the monographs by J. K. Hale [HA 2], R. Temam [TE 1], A. V. Babin and M. I. Vishik [B-V 2], and O. A. Ladyzhenskaya [LA 3]. This present book is in that same vein. In it we shall set forth the theory of asymptotic behavior for dynamical systems corresponding to parabolic equations and - in that connection - we will expound the theory of *global attractors*.

An important notion in these developments is that of *sectorial operator*, an idea studied by A. Friedman [FR 1] and extensively exploited by D. Henry [HE 1] More recently, H. Amann [AM 5], A. Lunardi [LU 1] and H. Tanabe [TA 2] have employed sectorial operators in their own investigations, which have emphasized the union of finite dimensional and infinite dimensional methods.

Of course, a major theme in this present book will be the use of sectorial operators in the study of parabolic problems. Also, in this present book, we will employ deep results in the *theory of interpolation* reported by H. Triebel [TR].

This book is divided into nine chapters, a brief description of which is as follows.

Chapter 1 is a review of several topics necessary for our later work; the choice of these topics is to ensure that this book is suitably self contained. Specifically, Chapter 1 summarizes some basic facts and definitions from the theory of Sobolev spaces, the theory of elliptic operators and sectorial operators - whose negatives generate analytic semigroups - and the theory of stability for dynamical systems.

Chapter 2 is devoted to equations of the form

$$(i) \qquad \begin{cases} \dot{u} + Au = F(u), \ t > 0, \\ u(0) = u_0. \end{cases}$$

where A is a sectorial operator and where F is a nonlinear function. The chapter itself deals with the local solvability of (i) in fractional order spaces X^α - domains of fractional powers A^α of A.

In Chapter 3 we continue our study of equations (i). Specifically, in that chapter, we investigate the extendibility of local solutions to the whole half line $[0, +\infty)$, and we investigate the smoothing action of the semigroup generated by (i). In connection with all this we introduce a condition - Condition (A_2) in Chapter 3 - sufficient for the global solvability of equations (i). We have found that this condition is very useful in several applications.

Chapter 4 is devoted to the construction of global attractors for equations (i). A slight strengthening of the condition (A_2) just mentioned leads to an *a priori* estimate - asymptotically independent of initial data - for the global solution constructed in Chapter 3. This estimate guarantees the existence of a global attractor for equations (i).

Chapter 5 provides a bridge between the abstract results of Chapters 2, 3, 4 and the applications of those results to several specific problems. The purpose is to translate the abstract formulations of Chapters 2-4 into the language of higher order parabolic problems in the setting of appropriately selected Sobolev spaces. In particular, Condition (A_2) - formulated in Chapter 3 - becomes an admissibility condition for the global solvability of equations (i). That condition - rendered as Restriction I in Chapter 5 - pertains to the growth rate of the nonlinear term F appearing in equations (i) and is closely related to the Nirenberg-Gagliardo estimate reported in Proposition 1.2.2 of Chapter 1.

In Chapter 6 we apply the abstract theory developed in earlier chapters to various problems arising in the physical and engineering sciences. These examples are arranged in an order of increasing complication. We begin with a new approach to the *method of invariant regions* for second order parabolic systems. From this we proceed to the construction of a global attractor for the *Cahn-Hilliard equation*. For this construction we do not require the usual growth condition imposed on the nonlinear term. Next, we construct a global attractor for the 2-*dimensional Navier-Stokes equation*. Finally in Chapter 6, we discuss the problem of constructing a global attractor for the *Cauchy problem* governed by second order equations whose spatial variable x belongs to R^n. In this last application there are special difficulties caused by a lack of compactness in the resolvent operator. We overcome these difficulties through an appropriate use of weighted Sobolev spaces.

Chapter 7 contains a further description of the solutions to (i). Based on the *backward uniqueness property* we justify invertibility of flows corresponding to examples introduced in Chapter 6. We also extend the notion of an X^α solution to the case of all $\alpha \geq 0$, thus providing additional information about the regularity of solutions.

In Chapter 8 we discuss some extensions of results obtained in Chapters 2-4 to problems with non-Lipschitz nonlinearities. We also investigate there local stability of stationary solutions of the n-dimensional Navier-Stokes equations. Furthermore in this same chapter we study the smoothness - in a classical sense - of the solutions we obtained in Chapters 2-4 for parabolic problems. In the same spirit we extend to spaces of Hölder functions our earlier results concerning the existence of a global attractor in spaces of fractional order. Finally in Chapter 8, we briefly examine degenerate parabolic equations; here our methods are borrowed from the theory of *monotone operators.*

Chapter 9, which closes this monograph, comprises a list of notations used in this book, a compendium of definitions relevant to the main body of the text, an abstract version of the maximum principle, and an iteration proof of an *a priori* L^∞ estimate for second order equations used in Chapter 6. Finally in Chapter 9, we compare various definitions of solutions to parabolic problems which can be found in basic monographs.

At the very end of this book, following Chapter 9, there appears a list of references. The books and articles listed therein indicate the wide background of the subject treated in this present book. Also, throughout this present book we cite relevant works from that list.

Although we consider this book to be only an introduction to the study of dynamics, nevertheless we believe that it will be of interest to mathematicians working in the areas of dynamical systems, evolutionary partial differential equations and their applications in mathematical physics and biology. We also believe that this book will be a suitable text for graduate students preparing themselves for studies in this field. Thus, we have tried to include most of the important results and proofs necessary for understanding the material being presented and its possible extensions.

This present book differs in two essential respects from the other works cited earlier in this preface. First, the modern semigroup approach to studying evolutionary systems - an approach expounded by D. Henry [HE 1] - plays a central role in our own exposition. Second, our approach to investigating the asymptotics of an evolutionary system parallels the approach taken by A. M. Lyapunov in his studies of ordinary differential equations, namely, for a given initial value problem, one constructs a local solution, then proves its unique extendibility to the whole half line $[0, +\infty)$, and then proceeds to the asymptotics. That approach dictates the general scheme underlying a large portion of this present book.

Here, at this juncture in the preface, we want to indicate some of the specific features of that general scheme.

If, in equations (i) above, the nonlinear term F acting from some X^α into X - with $\alpha \in [0,1)$ - is Lipschitz continuous on bounded sets, then the local solvability of (i) on $X^{\alpha+\varepsilon}$ - with $\varepsilon > 0$, $\alpha + \varepsilon < 1$ - follows. Next, for most equations coming from mathematical physics, there is available an *a priori estimate*

$$(ii) \qquad \|u(t)\|_Y \leq c(\|u_0\|_{X^\alpha}),\ t \geq 0,$$

for all putative X^α solutions $u(t)$, where Y is some space including $D(A)$. Often the etimate (ii) is obtained with the aid of an *energy functional.*

If the growth of $F : X^\alpha \to X$ on local solutions $u(t)$ is *sublinear relative to an a priori estimate* (ii), i.e., if

$$(iii) \qquad \|F(u(t))\|_X \leq g(\|u(t)\|_Y)(1 + \|u(t)\|_{X^\alpha}),$$

where $g : R^+ \to R^+$ is nondecreasing, then each local $X^{\alpha+\varepsilon}$ solution ($\varepsilon > 0$, $\alpha + \varepsilon < 1$) can be uniquely extended to $[0, +\infty)$.

Finally, the *asymptotic independence* of the estimate (ii) relative to u_0, namely,

$$(iv) \qquad \limsup_{t \to +\infty} \|u(t)\|_Y \leq const.,$$

ensures, as we shall see, the existence of a *global attractor* $\mathcal{A}$ for the semigroup $\{T(t)\}$ generated by (i) on $X^{\alpha+\varepsilon}$.

Much of this present book concerns the development of ideas implicit in the general scheme described above. Two remarks are in order.

First, in Chapter 3 we shall present a sufficient condition for the global extendibility of solutions. This condition was originally set forth in [FR 1] and was later treated in [AM 1], [WA 1], and [C-D 1]. As observed in [C-D 3], a slight modification of this same sufficient condition automatically guarantees the existence of a global attractor (cf. Theorem 4.1.1 below). Thus our study of asymptotics easily fits into the general scheme outlined above: local solvability, global extendibility, and existence of a global attractor.

Second, if $Y \subset X$, then the conditions (ii), (iii), and (iv) above are together necessary and sufficient for (i) to generate a point dissipative semigroup $\{T(t)\}$ on $X^{\alpha+\varepsilon}$ ($\varepsilon > 0$, $\alpha + \varepsilon < 1$) with bounded sets yielding bounded orbits. Hence, for any sectorial equation (i) in which the operator A has a compact resolvent, the conditions (ii), (iii), and (iv) are together necessary and sufficient for the existence of a global attractor.

Acknowledgments. We the authors wish to express our gratitude to Professor Jack K. Hale who has stimulated our research for many years. Especially, we would like to thank Professor Nathaniel Chafee for his significant contribution in improving a large portion of the text. We owe him a lot for this friendly help. We are also grateful to our friends Professor Nicolas D. Alikakos, Doctor Alexandre N. de Carvalho, and Doctor Andrzej W. Turski for their encouragement and for many fruitful discussions concerning various subjects in this book. Furthermore, we want to acknowledge all the support we have received from the Polish State Committee for Scientific Research (KBN) and also the United States Information Agency (USIA) and the Council for International Exchange of Scholars (CIES) under the Fulbright Program. This sponsorship has given the first author a fruitful visit at the Center for Dynamical Systems and Nonlinear Studies at the Georgia Institute of Technology in Atlanta, Georgia, USA.

Finally, we would like to express our gratitude to the anonymous referees whose remarks and suggestions significantly improved the preliminary version of the manuscript. Also we would like to thank Doctor Alan Harvey and Mister Roger Astley, editors of mathematical sciences of the Cambridge University Press, for their professional assistance in the publication process.

Above all, for their patience and indulgence, the authors thank their families: Justyna and Tomasz Cholewa, and Maria and Paweł Dłotko. To them the authors lovingly dedicate this book.

CHAPTER 1

Preliminary Concepts

1.1. Elements of stability theory

A major portion of this book concerns the asymptotic behavior of solutions for nonlinear partial differential equations. Recent developments in the theory of dynamical systems show the unquestionable advantage of treating such solutions as an abstract flow on an appropriately selected phase space (e.g. [HE 1], [HA 2], [TE 1], [B-V 2]).

In this section we shall introduce some recently developed ideas in the theory of dynamical systems.[1] Our purpose here is to lay a foundation for analyzing and describing the long time dynamics of infinite dimensional differential equations.

1.1.1. Strongly continuous semigroups and stability of sets. We begin with the notion - basic for all our investigations - of a C^0 *semigroup* (*strongly continuous semigroup*).

Definition 1.1.1. *Let V be a metric space. A one parameter family $\{T(t)\}$ of maps $T(t): V \to V$, $t \geq 0$, is called a C^0 semigroup if*

(i) $T(0)$ is the identity map on V,
(ii) $T(t+s) = T(t)T(s)$ for all $t, s \geq 0$,
(iii) the function

$$[0,\infty) \times V \ni (t,x) \to T(t)x \in V$$

is continuous at each point $(t,x) \in [0,\infty) \times V$.

It is known that for the semigroups of bounded linear operators in a Banach space X the condition (iii) holds if and only if, at any element $x \in X$,

$$T(t)x \to x \quad \text{when} \quad t \to 0^+,$$

which is basically a consequence of the Banach *uniform boundedness property.*

[1] Recall that a list of notations used in the book and a compendium of the basic definitions are presented at the beginning of Chapter 9.

For purposes of convenience, we shall forthwith introduce several concepts closely related to Definition 1.1.1 above. These concepts will frequently appear in our further considerations.

- The semigroup $\{T(t)\}$ is said to be *compact* if $T(t) : V \to V$ is a compact map for each $t > 0$, i.e. each $T(t)$ takes bounded sets into precompact sets.
- The semigroup $\{T(t)\}$ is called *completely continuous* if it is compact and if for each bounded set $B \subset V$ and each number $t > 0$ the union $\bigcup_{s\in[0,t]} T(s)B$ is bounded in V.
- Let W_1, W_2 be two subsets of V. We say that W_2 is $\{T(t)\}$*-attracted* by W_1 if

$$d(T(t)W_2, W_1) \to 0 \quad \text{as} \quad t \to +\infty,$$

where, for each $t \geq 0$,

$$d(T(t)W_2, W_1) := \sup_{w_2\in T(t)W_2} \inf_{w_1\in W_1} dist_V(w_2, w_1).$$

Of course, it is true that the quantity $d(T(t)W_2, W_1)$ enters into the construction of the Hausdorff distance between the two sets $T(t)W_2$ and W_1. This is a matter we will not pursue here. However, we note that, roughly speaking, the number $d(T(t)W_2, W_1)$ measures how much the set $T(t)W_2$ lies outside the set W_1.
- Given any two subsets $W_1, W_2 \subset V$, we say that W_1 *absorbs* W_2 under $\{T(t)\}$ if there exists a number $t_0 \geq 0$ such that $T(t)W_2 \subset W_1$ for all $t \geq t_0$. Notice that W_1 attracts W_2 if and only if each open neighborhood $\mathcal{N}_{W_1}$ of W_1 in V absorbs W_2.
- An element $v \in V$ is called an *equilibrium point* for $\{T(t)\}$ if $T(t)v = v$ for all $t \geq 0$. Extending this notion, we say that a set $\mathcal{A} \subset V$ is $\{T(t)\}$*-invariant* if $T(t)\mathcal{A} = \mathcal{A}$ for all $t \geq 0$. Also, we will call $\mathcal{A} \subset V$ *positively* $\{T(t)\}$*-invariant* if $T(t)\mathcal{A} \subset \mathcal{A}$ for all $t \geq 0$.
- For any set $B \subset V$ the two sets $\gamma^+(B)$ and $\omega(B)$ defined by

$$\gamma^+(B) := \bigcup_{t\geq 0} T(t)B,$$

$$\omega(B) := \bigcap_{s\geq 0} cl_V \bigcup_{t\geq s} T(t)B$$

are called, respectively, the *positive orbit* and the *ω-limit set* of B. Thus, $\omega(B)$ consists of all points $v \in V$ for which there exist positive numbers $t_n \nearrow +\infty$ and points $v_n \in B$ with $T(t_n)v_n \to v$.
- For any point $v \in V$ we let S_v^- denote the set of all functions $\phi : (-\infty, 0] \to V$ such that $\phi(0) = v$ and such that $T(t)\phi(s) = \phi(t + s)$ whenever $-\infty < s \leq -t \leq 0$. Here we allow the possibility that S_v^- is empty, that S_v^- consists of exactly one element ϕ, or that S_v^- consists of more than one element ϕ. By a *negative orbit* through a given point $v \in V$ we mean any

set

$$\gamma_\phi^-(v) = \bigcup_{t\geq 0}\{\phi(-t)\},$$

where $\phi \in S_v^-$. Of course, there may or may not exist a nonempty negative orbit.

- For each point $v \in V$, a *complete orbit* through v is any set

$$\gamma_\phi(v) := \gamma^+(v) \cup \gamma_\phi^-(v),$$

where $\phi \in S_v^-$. Since we allow a negative orbit to be empty, we remark that a complete orbit $\gamma_\phi(v)$ is invariant if and only if its component $\gamma_\phi^-(v)$ is nonempty.

One of our principal preoccupations in this book will be the *stability properties* of invariant sets. This necessitates the following definition.

Definition 1.1.2. *Let $\mathcal{A} \subset V$ be nonempty and $\{T(t)\}$-invariant. We say that:*

(i) $\mathcal{A}$ is ***stable*** *if and only if for each open neighborhood U of $\mathcal{A}$ there exists an open neighborhood W of $\mathcal{A}$ such that $T(t)W \subset U$ for all $t \geq 0$;*
(ii) $\mathcal{A}$ is ***asymptotically stable*** *if and only if $\mathcal{A}$ is stable and attracts each point lying in some open neighborhood of $\mathcal{A}$;*
(iii) $\mathcal{A}$ is ***uniformly asymptotically stable*** *if and only if $\mathcal{A}$ is stable and attracts some open neighborhood of itself.*

When $\mathcal{A}$ is compact, in order to conclude that $\mathcal{A}$ is uniformly asymptotically stable, it suffices to show that $\mathcal{A}$ attracts some one of its open neighborhoods. More precisely:

Observation 1.1.1. *Let $\{T(t)\}$ be a C^0 semigroup in a metric space V. If $\mathcal{A}$ is compact and $\{T(t)\}$-invariant and if $\mathcal{A}$ attracts at least one of its own open neighborhoods, then $\mathcal{A}$ is stable.*

Proof. For the sake of argument suppose that $\mathcal{A}$ is not stable. Then there exists a neighborhood $\mathcal{N}_\mathcal{A}$ of $\mathcal{A}$ having the property that to each neighborhood W of $\mathcal{A}$ there correspond a point $w \in W$ and a number $t_w \geq 0$ such that $T(t_w)w \notin \mathcal{N}_\mathcal{A}$. From this and from the compactness of $\mathcal{A}$ there follows the existence of two sequences $t_n \to t_0 \in [0, +\infty]$ and $w_n \to w_0 \in \mathcal{A}$ such that

$$T(t_n)w_n \notin \mathcal{N}_\mathcal{A} \text{ for all } n \in N. \tag{1.1.1}$$

In the case that $t_0 = +\infty$, (1.1.1) implies that there is no integer $n_0 > 0$ such that the bounded set $\{w_n,\ n \geq n_0\}$ is attracted to $\mathcal{A}$. This is absurd.

In the case that $t_0 < +\infty$, we have $T(t_n)w_n \to T(t_0)w_0$ and, since $\mathcal{A}$ is invariant, we must conclude that $T(t_0)w_0 \in \mathcal{A}$. Hence, the neighborhood $\mathcal{N}_\mathcal{A}$ contains all but finitely many elements of the set $\{T(t_n)w_n\}$. In view of (1.1.1) this also is absurd.

Thus in fact, $\mathcal{A}$ is stable. □

Compact, uniformly asymptotically stable subsets of the *phase space* are usually called *local attractors*. That is:

Definition 1.1.3. *A set $\mathcal{A} \subset V$ is called a* **local attractor** *for the semigroup $\{T(t)\}$ on V if and only if*

(i) *$\mathcal{A}$ is nonempty, compact, and invariant with respect to $\{T(t)\}$;*
(ii) *$\mathcal{A}$ attracts some open neighborhood $\mathcal{N}_{\mathcal{A}}$ of $\mathcal{A}$.*

For many infinite dimensional problems the notion of local attractor is, from a certain point of view, inadequate. Specifically, later in this book we will be seeking a characterization of long time behavior in terms of the dynamics of a suitably selected finite dimensional system. Such a characterization becomes possible if one can show the existence of a *global attractor* and then delineate its phase portrait. The notion of global attractor involves stability properties stronger than those formulated in Definition 1.1.3. Specifically:

Definition 1.1.4. *By a* **global attractor** *for $\{T(t)\}$ we mean a nonempty, compact, $\{T(t)\}$-invariant set $\mathcal{A} \subset V$ which attracts every bounded subset of V.*

Observation 1.1.2. *The global attractor - if it exists - is unique and also is maximal in the class of bounded invariant subsets of V.*

Proof. Indeed, let $\mathcal{A}, \mathcal{A}_1$ be compact invariant subsets of the phase space V, and suppose that $\mathcal{A}, \mathcal{A}_1$ each attract all the bounded subsets of V. Then, keeping in mind that $\mathcal{A}, \mathcal{A}_1$ are both bounded, we have

$$d(\mathcal{A}, \mathcal{A}_1) = d(T(t)\mathcal{A}, \mathcal{A}_1) \to 0 \quad \text{as} \quad t \to +\infty,$$
$$d(\mathcal{A}_1, \mathcal{A}) = d(T(t)\mathcal{A}_1, \mathcal{A}) \to 0 \quad \text{as} \quad t \to +\infty.$$

Consequently,

$$d(\mathcal{A}, \mathcal{A}_1) = d(\mathcal{A}_1, \mathcal{A}) = 0.$$

But both $\mathcal{A}$, $\mathcal{A}_1$ are closed. Therefore, $\mathcal{A} = \mathcal{A}_1$. One can also see that each bounded invariant subset of V must be contained in the global attractor. □

Observation 1.1.3. *The global attractor $\mathcal{A}$ is minimal in the class of all those closed bounded sets B in V which attract bounded sets.*

Proof. Indeed, consider any closed bounded set $B \subset V$ which attracts all bounded subsets of V. Then, arguing in a manner similar to the proof above for Observation 1.1.2, we obtain $d(\mathcal{A}, B) = 0$. From this it follows that $\mathcal{A} \subset B$. □

Before rendering our next observation we want to recall the notion of connectedness. Two sets $K, L \subset V$ are said to be separated if and only if

$$L \cap cl_V K = K \cap cl_V L = \emptyset. \tag{1.1.2}$$

A set $S \subset V$ is called *connected* if and only if S cannot be decomposed into two separated sets K, L.

Observation 1.1.4. *The global attractor $\mathcal{A}$ is connected if and only if there exists a connected bounded set $B \subset V$ such that $\mathcal{A} \subset B$.*

Proof. If $\mathcal{A}$ is connected, then, taking $B = \mathcal{A}$, we trivially have a bounded connected set $B \subset V$ such that $\mathcal{A} \subset B$.

The converse assertion requires more argumentation. Specifically, suppose that B is a bounded connected subset V which contains $\mathcal{A}$. From the continuity of $T(t)$ it follows that the image $T(t)B$, $t \geq 0$, is connected.

By hypothesis, $\mathcal{A}$ is the global attractor for $\{T(t)\}$. Hence, $\mathcal{A}$ attracts every bounded set in V. It follows that, for each open neighborhood $\mathcal{N}_{\mathcal{A}}$ of $\mathcal{A}$, there exists a number $t_{\mathcal{N}_{\mathcal{A}}} > 0$ such that

$$\mathcal{A} \subset T(t)B \subset \mathcal{N}_{\mathcal{A}} \quad \text{for all} \quad t \geq t_{\mathcal{N}_{\mathcal{A}}}. \tag{1.1.3}$$

For the sake of argument, suppose that $\mathcal{A}$ is not connected, i.e., $\mathcal{A} = K \cup L$ where $K, L \subset V$ are nonempty and separated. We know that $\mathcal{A}$ is closed and we know that K, L satisfy (1.1.2). Hence, the sets K, L must each be closed. Thus, $\mathcal{A}$ decomposes into a sum of two disjoint nonempty closed sets K, L.

Since K, L are disjoint and closed, there must exist two open sets $\mathcal{U}_K, \mathcal{U}_L \subset V$ such that

$$\mathcal{U}_K \cap \mathcal{U}_L = \emptyset, \ K \subset \mathcal{U}_K, \ L \subset \mathcal{U}_L. \tag{1.1.4}$$

Let $\mathcal{N}_{\mathcal{A}}$ be the particular open neighborhood of $\mathcal{A}$ given by $\mathcal{N}_{\mathcal{A}} := \mathcal{U}_K \cup \mathcal{U}_L$. We know that $T(t)B$ is connected and that $\mathcal{U}_K$, $\mathcal{U}_L$ are separated. Therefore, with the aid of (1.1.3) we have either $\mathcal{A} = K \cup L \subset T(t)B \subset \mathcal{U}_K$ or $\mathcal{A} = K \cup L \subset T(t)B \subset \mathcal{U}_L$. But this last statement contradicts (1.1.4). The proof is complete. □

We remark that, if a metric space V is connected, then the global attractor $\mathcal{A}$ for a C^0 semigroup $T(t) : V \to V$, $t \geq 0$, is connected. However, this same statement is not necessarily true for discrete semigroups (see [GO-SA]).

Our goal now is to establish conditions guaranteeing the *existence* of the global attractor $\mathcal{A}$. To that end, we will introduce a class of semigroups having the property that, for each compact invariant set $B \subset V$, asymptotic stability is equivalent to uniform asymptotic stability.

1.1.2. Existence of a global attractor. In that which follows we will set forth conditions - formulated in [HA 2] - which guarantee the existence of a global attractor. These conditions relate to the notions of *dissipativeness* and *asymptotic smoothness* for $\{T(t)\}$.

Definition 1.1.5. *The semigroup $\{T(t)\}$ is called* **point dissipative** *if and only if there exists a nonempty, bounded set $B \subset V$ which attracts every point in V. The semigroup $\{T(t)\}$ is called* **bounded dissipative** *if and only if there exists a nonempty, bounded set $B \subset V$ which attracts every bounded subset of V.*

Definition 1.1.6. *The semigroup $\{T(t)\}$ is called* **asymptotically smooth** *if and only if each nonempty, closed, bounded, positively invariant set $W \subset V$ contains a nonempty, compact subset C which attracts W.*

Clearly, if the semigroup $\{T(t)\}$ has a global attractor $\mathcal{A}$ in V, then $\{T(t)\}$ must be dissipative in the sense of Definition 1.1.5. Furthermore, under the same conditions, $\{T(t)\}$ is of necessity asymptotically smooth. Indeed, consider any nonempty, closed, bounded set $W \subset V$ such that

$$T(t)W \subset W \quad \text{for } t \geq 0. \tag{1.1.5}$$

Certainly, $\mathcal{A} \cap W$ is a closed subset of the compact set $\mathcal{A}$. Hence, $\mathcal{A} \cap W$ is compact. Moreover, since $\mathcal{A}$ attracts bounded sets, condition (1.1.5) ensures that $\mathcal{A} \cap W$ is nonempty and attracts W. Thus, as asserted, $\{T(t)\}$ is asymptotically smooth. We summarize these remarks as follows.

Observation 1.1.5. *If $\{T(t)\}$ is a C^0 semigroup on a metric space V and if $\{T(t)\}$ has a global attractor $\mathcal{A}$, then $\{T(t)\}$ is bounded dissipative and asymptotically smooth.*

At this moment we want to establish several important properties of *ω-limit sets for bounded sets* in the case that the semigroup $\{T(t)\}$ is asymptotically smooth (see [HA 2, Lemma 3.2.1]) .

Proposition 1.1.1. *Let $\{T(t)\}$ be a C^0 semigroup acting on a metric space V. If $\{T(t)\}$ is asymptotically smooth, if B is a nonempty subset of V, and if, for some number $t_B \geq 0$, the set*

$$\bigcup_{s \geq t_B} T(s)B$$

is bounded, then $\omega(B)$ is nonempty, compact, and invariant. Furthermore, $\omega(B)$ attracts B.

Proof. We start with the proof that $\omega(B)$ is compact. Since $\bigcup_{s \geq t_B} T(s)B$ is positively invariant and the maps $T(t)$ $(t \geq 0)$ are continuous, we have

$$T(t) cl_V \bigcup_{s \geq t_B} T(s)B \subset cl_V \bigcup_{s \geq t_B} T(s)B, \; t \geq 0.$$

Since the semigroup is asymptotically smooth there exists a nonempty, compact set $C \subset cl_V \bigcup_{s \geq t_B} T(s)B$ attracting $cl_V \bigcup_{s \geq t_B} T(s)B$. Hence C attracts B and, therefore,

$$\forall_{\substack{\mathcal{N}_C \text{ bounded, open} \\ \text{neighborhood of } C}} \exists_{\substack{\mathcal{N}'_C \text{ bounded, open} \\ \text{neighborhood of } C}} \exists_{t_{\mathcal{N}_C} \geq 0} \tag{1.1.6}$$

$$cl_V \bigcup_{t \geq t_{\mathcal{N}_C}} T(t)B \subset cl_V \mathcal{N}'_C \subset \mathcal{N}_C.$$

Condition (1.1.6) ensures that $\omega(B)$ is contained in each open neighborhood of C, which (since C is closed) implies the inclusion $\omega(B) \subset C$. Furthermore, using (1.1.6) and the compactness of C, one may show the existence of sequences $t_n \nearrow +\infty$ and $\{v_n\} \subset B$ for which $\{T(t_n)v_n\}$ is convergent. The set $\omega(B)$ is thus nonempty and, since $\omega(B)$ is closed and C is compact, $\omega(B)$ must be compact.

Next we shall prove that

$$d\big(T(t)B, \omega(B)\big) \to 0 \quad \text{when} \quad t \to +\infty. \tag{1.1.7}$$

Suppose that (1.1.7) is violated, i.e. there are an $\varepsilon > 0$, a sequence $\{t_n\}$ increasing to infinity and a sequence $\{v_n\} \subset B$ such that

$$\inf_{y \in \omega(B)} dist_V\big(T(t_n)v_n, y\big) > \varepsilon, \; n \in N. \tag{1.1.8}$$

Then $\{T(t_n)v_n\}$ cannot have a convergent subsequence; otherwise such a limit point would belong to $\omega(B)$ and (1.1.8) would not be true. However, from (1.1.6) it follows that for any bounded, open neighborhood $\mathcal{N}_C$ of C almost all elements of $\{T(t_n)y_n\}$ are contained in $\mathcal{N}_C$. Since C is compact this allows one to choose from $\{T(t_n)y_n\}$ a convergent subsequence, which contradicts (1.1.8).

To prove that $\omega(B)$ is invariant note first, from the definition of $\omega(B)$ and continuity of $T(t)$, that the set $\omega(B)$ is positively invariant. Consider further any point $v_0 \in \omega(B)$. Then $T(t_n)v_n \to v_0$ in V for some sequences $t_n \nearrow +\infty$ and $\{v_n\} \subset B$. Fix $t \geq 0$ and define

$$w_n := T(t_n - t)v_n, \; n \geq n_t.$$

Certainly

$$T(t)w_n \to v_0, \tag{1.1.9}$$

whereas, by (1.1.7) and the compactness of $\omega(B)$, there exists a subsequence $\{w_{n'}\}$ of $\{w_n\}$ convergent to some $w_0 \in \omega(B)$. This and (1.1.9) ensure that $T(t)w_0 = v_0$ which proves the inclusion $\omega(B) \subset T(t)\big(\omega(B)\big)$. The proof is complete. □

Corollary 1.1.1. *(see* [LA 3, Proposition 2.2]*) Let $\{T(t)\}$ be a C^0 semigroup on a metric space V, and suppose that $\{T(t)\}$ has a global attractor $\mathcal{A}$. Then*

(i) $\mathcal{A}$ is the union of the ω-limit sets of all bounded subsets of V,
(ii) $\mathcal{A}$ is the union of the ω-limit sets of all compact subsets of V,
(iii) $\mathcal{A}$ is the union of all bounded, invariant complete orbits through $v \in V$,
(iv) $\mathcal{A}$ is the union of all precompact, invariant complete orbits through $v \in V$.

Proof. First note from Observation 1.1.5 that $\{T(t)\}$ must be asymptotically smooth. Since $\mathcal{A}$ attracts bounded sets, for each bounded set $B \subset V$ there is a $t_B \geq 0$ such that $T(t_B)\big(\gamma^+(B)\big)$ is bounded. Therefore $\omega(B)$ is compact, invariant and, consequently, $\omega(B) \subset \mathcal{A}$. Moreover, $\omega(\mathcal{A}) = \mathcal{A}$, and that completes the proof of (i). Similar reasoning establishes (ii). To prove (iii) (and (iv)), note first that, for each $v \in \mathcal{A}$, a negative orbit through v is nonempty. Hence, for each $v \in \mathcal{A}$ there exists an invariant, bounded, and precompact complete orbit through v. Therefore, $\mathcal{A}$ is contained in the union of such orbits. Obviously, each bounded, invariant complete orbit must lie within $\mathcal{A}$. Thus conditions (iii) (and (iv)) are established. □

In Definition 1.1.6 above we introduced the notion of asymptotically smooth semigroup and thus, in effect, singled out such semigroups for special attention. Our motive for doing this is embodied in the next observation.

Observation 1.1.6. *Let $\{T(t)\}$ be a C^0 semigroup in a metric space V and let $\mathcal{A} \subset V$ be a nonempty, compact, invariant set. If $\{T(t)\}$ is asymptotically smooth, then the following two conditions are equivalent:*

(a) $\mathcal{A}$ is asymptotically stable;
(b) $\mathcal{A}$ is uniformly asymptotically stable.

Proof. From Definition 1.1.2 it trivially follows that $(b) \Longrightarrow (a)$. Consequently, we need only prove that $(a) \Longrightarrow (b)$.

By hypothesis, $\mathcal{A}$ is asymptotically stable. Hence, there exists at least one bounded open neighborhood U of $\mathcal{A}$ such that $\mathcal{A}$ attracts points of U. Consider another open neighborhood U' of $\mathcal{A}$ such that $cl_V U' \subset \mathcal{U}$. Since $\mathcal{A}$ is stable, there must exist an open neighborhood $\mathcal{U}''$ of $\mathcal{A}$ such that

$$W := cl_V \bigcup_{t \geq 0} T(t)U'' \subset clU' \subset U.$$

Clearly W is nonempty, closed, bounded, and positively invariant. Therefore $\omega(W) \subset W$ and, by virtue of Proposition 1.1.1, $\omega(W)$ is compact and invariant. Also, $\omega(W)$ attracts W. Hence, in particular,

$$\omega(W) \text{ attracts } U''. \tag{1.1.10}$$

Also, since $\mathcal{A}$ is a compact, invariant subset of U'',

$$\mathcal{A} \subset \omega(W). \tag{1.1.11}$$

It is clear from the preceding considerations that $\mathcal{A}$ attracts points of $\omega(W)$. Thus, the continuity of $\{T(t)\}$ and the stability of $\mathcal{A}$ yield the statement that, for each open neighborhood $\mathcal{N}_\mathcal{A}$ of $\mathcal{A}$,

$$\exists_{\substack{\mathcal{N}'_\mathcal{A} \text{ an open} \\ \text{neighborhood of } \mathcal{A}}} \forall_{v \in \omega(W)} \exists_{t_v \geq 0} \exists_{\varepsilon_v > 0} \quad T(t_v)B_V(v, \varepsilon_v) \subset \mathcal{N}'_\mathcal{A} \subset \gamma^+(\mathcal{N}'_\mathcal{A}) \subset \mathcal{N}_\mathcal{A}, \tag{1.1.12}$$

where $B_V(v, \varepsilon_v)$ denotes the open ball in V centered at v and having radius ε_v. Let $\mathcal{N}_\mathcal{A}$ be fixed. From the compactness of $\omega(W)$ there follows

$$\exists_{k \in N} \exists_{v_1, \ldots, v_k \in \omega(W)} \; \omega(W) \subset B_V(v_1, \varepsilon_{v_1}) \cup \ldots \cup B_V(v_k, \varepsilon_{v_k}). \tag{1.1.13}$$

Introduce

$$t_{\max} := \max\{t_{v_1}, \ldots, t_{v_k}\}.$$

The assertions (1.1.13), (1.1.12) and the invariance of $\omega(W)$ imply

$$\omega(W) = T(t_{\max})\big(\omega(W)\big) \subset T(t_{\max})\left(\bigcup_{j=1}^{k} B_V(v_j, \varepsilon_{v_j})\right) = \bigcup_{j=1}^{k} T(t_{\max})B_V(v_j, \varepsilon_{v_j})$$

$$= \bigcup_{j=1}^{k} T(t_{\max} - t_{v_j})T(t_{v_j})B_V(v_j, \varepsilon_{v_j}) \subset \gamma^+(\mathcal{N}'_{\mathcal{A}}) \subset \mathcal{N}_{\mathcal{A}}.$$

All of this implies that $\omega(W)$ is contained in each open neighborhood of $\mathcal{A}$. Since $\mathcal{A}$ is compact, we obtain $\omega(W) \subset \mathcal{A}$ which, by virtue of (1.1.11), leads us to the equality $\omega(W) = \mathcal{A}$. Taking into account (1.1.10) we see that $\mathcal{A}$ attracts its open neighborhood U''. Thus, $\mathcal{A}$ is uniformly asymptotically stable and the proof is complete. □

The theorem stated below, due to J. K. Hale [HA 2], gives sufficient conditions for the existence of a global attractor.

Theorem 1.1.2. *Let $\{T(t)\}$ be a C^0 semigroup on a metric space V. If $\{T(t)\}$ is point dissipative, asymptotically smooth, and keeps orbits of bounded sets bounded, then $\{T(t)\}$ has a global attractor in V.*

Proof. The proof proceeds in two steps.
Step 1. We shall first show the existence of a bounded set $\mathcal{O} \subset V$ such that each compact set $C \subset V$ has an open neighborhood $\mathcal{N}_C$ which is absorbed by $\mathcal{O}$.

By hypothesis there exists a nonempty bounded set $W_0 \subset V$ which attracts points of V. Let $\mathcal{N}_{W_0}$ be any bounded open neighborhood of W_0. Using the continuity of $\{T(t)\}$ and the fact that $\mathcal{N}_{W_0}$ absorbs points of V we conclude that

$$\text{(1.1.14)} \qquad \forall_{v\in V}\ \exists_{\tau_v \geq 0}\ \exists_{B_V(v,\varepsilon_v)}\ T(\tau_v)B_V(v, \varepsilon_v) \subset \mathcal{N}_{W_0},$$

where $B_V(v, \varepsilon_v) \subset V$ denotes the open ball in V centered at v and having radius ε_v. Next, choose $t_{\mathcal{N}_{W_0}} \geq 0$ such that

$$\mathcal{O} := \bigcup_{t \geq t_{\mathcal{N}_{W_0}}} T(t)(\mathcal{N}_{W_0})$$

is bounded. By our assumptions, $\mathcal{O}$ is positively invariant and absorbs points of V. Moreover, from (1.1.14) we observe that

$$\text{(1.1.15)} \qquad \forall_{v\in V}\ \exists_{t_v := \tau_v + t_{\mathcal{N}_{W_0}} \geq 0}\ \exists_{B_V(v,\varepsilon_v)}\ T(t)B_V(v, \varepsilon_v) \subset \mathcal{O} \text{ for } t \geq t_v.$$

Consider now any compact set $C \subset V$. Certainly $C \subset \bigcup_{v\in C} B_V(v, \varepsilon_v)$ so that, from the compactness of C,

$$\text{(1.1.16)} \qquad C \subset B_V(v_1, \varepsilon_{v_1}) \cup \ldots \cup B_V(v_k, \varepsilon_{v_k}) =: \mathcal{N}_C$$

for some $k \in N$ and $v_1, \ldots, v_k \in C$. With the aid of (1.1.15), (1.1.16) we obtain

$$T(t)C \subset T(t)\mathcal{N}_C = \bigcup_{j=1}^{k} T(t)\left(B_V(v_j, \varepsilon_{v_j})\right) \subset \mathcal{O} \text{ for } t \geq \max\{t_{v_1}, \ldots, t_{v_k}\}.$$

Step 2. We shall now construct a compact invariant set $\mathcal{A}$ which attracts each bounded subset of V.

Let $B \subset V$ be a bounded set. By our assumptions Proposition 1.1.1 ensures that $\omega(B)$ is compact and attracts B, i.e.

$$\forall_{\substack{\mathcal{N}_{\omega(B)} \text{ an open} \\ \text{neighborhood of } \omega(B)}} \exists_{t_B \geq 0} \; T(t)B \subset \mathcal{N}_{\omega(B)} \text{ for } t \geq t_B. \tag{1.1.17}$$

However, as shown in Step 1, there exists some open neighborhood $\mathcal{N}_{\omega(B)}$ of $\omega(B)$ absorbed by $\mathcal{O}$. From this result and condition (1.1.17) we obtain

$$\forall_{\substack{B \subset V \\ B \text{ bounded}}} \exists_{\tau_B \geq 0} \; T(t)B \subset \mathcal{O} \text{ for } t \geq \tau_B. \tag{1.1.18}$$

Let $\mathcal{A} := \omega(\mathcal{O})$. Using Proposition 1.1.1 again we find that $\mathcal{A}$ is compact, is invariant and attracts $\mathcal{O}$. Furthermore $\mathcal{A}$ attracts bounded subsets of V since, as shown in (1.1.18), bounded sets are absorbed by $\mathcal{O}$. The proof is complete. □

Remark 1.1.1. This theorem was introduced in [HA 2] with the comment that the result is somewhat surprising inasmuch as, in the presence of a weak form of dissipation, orbits of bounded sets converge uniformly to some compact set. With regard to applications, this last comment is particularly significant. Indeed, it is much easier to obtain pointwise estimates on trajectories than it is to establish the existence of an absorbing set.

Remark 1.1.2. As a result of Theorem 1.1.2 and Observation 1.1.5 we conclude that *if $\{T(t)\}$ is a C^0 semigroup on a metric space V and if, under $\{T(t)\}$, the orbits of bounded sets are bounded, then $\{T(t)\}$ has a global attractor in V if and only if $\{T(t)\}$ is point dissipative and asymptotically smooth.* In general the boundedness of orbits for bounded sets is not necessary for the existence of a global attractor, although this is true for a large number of systems (e.g. compact semigroups corresponding to sectorial equations discussed in Chapter 3 have this property). From the point of view of Theorem 1.1.2, such an assumption may be weakened, since what we actually used in the proof was the property that

$$\forall_{\substack{B \subset V \\ B \text{ bounded}}} \exists_{t_B \geq 0} \bigcup_{t \geq t_B} T(t)B \text{ is bounded in V.} \tag{1.1.19}$$

On the basis of this remark we have (see [LA 3, Theorem 3.4])

Corollary 1.1.3. *A C^0 semigroup $\{T(t)\}$ on a metric space V has a global attractor if and only if $\{T(t)\}$ is point dissipative, is asymptotically smooth, and satisfies the condition* (1.1.19).

Remark 1.1.3. Looking again at the proof of Theorem 1.1.2 let us also note that instead of the point dissipativeness of $\{T(t)\}$ we might assume a weaker condition:

$$\exists_{\substack{B \subset V \\ B \text{ bounded}}} \forall_{u_0 \in V} \exists_{t_{u_0} \geq 0} \; T(t_{u_0})u_0 \in B \tag{1.1.20}$$

(i.e. $\gamma^+(u_0) \cap B \neq \emptyset$ for each $u_0 \in V$). Corollary 1.1.3 then becomes

Corollary 1.1.4. *A C^0 semigroup on a metric space V has a global attractor if and only if it is asymptotically smooth and satisfies the conditions* (1.1.19) *and* (1.1.20).

1.1.3. The role of Lyapunov function. The idea of a *Lyapunov function* proves to be very useful in the study of long time behavior for solutions of evolutionary equations. There are several distinct - logically inequivalent - definitions for the notion of Lyapunov function. The following definition is particularly appropriate for some problems to be treated in Chapter 6 below.

Definition 1.1.7. *Let $\{T(t)\}$ be a C^0 semigroup in a metric space V. A functional $\mathcal{L} : V \to R$ is said to be a* ***Lyapunov function*** *for $\{T(t)\}$ if*

(i) $\mathcal{L}$ is continuous on V and bounded below,
(ii) for each $v \in V$ function $(0, +\infty) \ni t \to \mathcal{L}(T(t)v) \in R$ is nonincreasing,
(iii) for any $v \in V$ we have

$$\mathcal{L}(T(t)v) = const. \text{ for all } t \geq 0 \text{ implies that } T(t)v = v \text{ for all } t \geq 0.$$

Proposition 1.1.2. *Let $\{T(t)\}$ be an asymptotically smooth C^0 semigroup on a metric space V and suppose that $\{T(t)\}$ admits a Lyapunov function $\mathcal{L}$. Also, suppose that for each point $v \in V$ the orbit $\gamma^+(v)$ is bounded in V. Then for each point $v \in V$ the corresponding ω-limit set $\omega(v)$ lies within the set $\mathcal{E}$ of all equilibria for $\{T(t)\}$:*

$$\mathcal{E} := \{w \in V : \ T(t)w = w \text{ for all } t \geq 0\}. \tag{1.1.21}$$

If, moreover, the set $\mathcal{E}$ is bounded in V, then $\{T(t)\}$ is point dissipative in V.

Proof. Let $\mathcal{L}$ be a Lyapunov function for $\{T(t)\}$. Based on Proposition 1.1.1 our assumptions ensure that the ω-limit set $\omega(v)$ of any element $v \in V$ is nonempty and attracts v. For fixed $v \in V$, let us choose any $y \in \omega(v)$ and a sequence $t_n \to +\infty$, such that

$$\lim_{n\to\infty} T(t_n)v = y.$$

Since $\mathcal{L}$ is bounded below we have, for some $\beta \in R$,

$$\inf_{t\geq 1} \mathcal{L}(T(t)v) = \beta. \tag{1.1.22}$$

Using the above condition and the property that $\mathcal{L}(T(t)v)$ is a nonincreasing function of t, we justify the existence of a limit

$$\lim_{t\to+\infty} \mathcal{L}(T(t)v) = \beta.$$

From the continuity of $\mathcal{L}$ and the semigroup properties of $\{T(t)\}$ we also obtain

$$\begin{aligned} \mathcal{L}(T(t)y) &= \mathcal{L}\left(T(t) \lim_{n\to\infty} T(t_n)v\right) = \lim_{n\to\infty} \mathcal{L}\big(T(t)T(t_n)v\big) \\ &= \lim_{n\to\infty} \mathcal{L}(T(t+t_n)v) = \beta \text{ for each } t \geq 0. \end{aligned}$$

This proves that y is an equilibrium point and therefore the union $\bigcup_{v\in V}\omega(v)$ is contained in $\mathcal{E}$. Recalling that $\bigcup_{v\in V}\omega(v)$ attracts points of V it is now clear that $\{T(t)\}$ is point dissipative whenever $\mathcal{E}$ is a bounded subset of V. The proof is complete. □

The following result is an immediate consequence of Proposition 1.1.2 and Corollary 1.1.3.

Corollary 1.1.5. *Let $\{T(t)\}$ be an asymptotically smooth C^0 semigroup on a metric space V, and suppose that $\{T(t)\}$ satisfies the condition* (1.1.19). *Furthermore, suppose that all the equilibria of $\{T(t)\}$ lie in a bounded subset of V. If $\{T(t)\}$ has a Lyapunov function $\mathcal{L}$, then $\{T(t)\}$ has a global attractor.*

1.1.4. Compact semigroups. An important class of semigroups appearing in applications - especially when those applications involve parabolic problems - is the class of compact semigroups. To show that Theorem 1.1.2 is applicable to the study of such semigroups, we shall first prove that compact semigroups are asymptotically smooth and fulfill the requirement (1.1.19) (see [H-R 1, Theorem 1.2]).

Lemma 1.1.1. *If $\{T(t)\}$ is a compact C^0 semigroup on a metric space V, then $\{T(t)\}$ is asymptotically smooth.*

Proof. In the spirit of Definition 1.1.6, we consider any nonempty, closed, bounded, positively invariant set $B \subset V$. Since $\gamma^+(B) \subset B$ and since $\{T(t)\}$ is compact, the set $cl_V T(1)\gamma^+(B)$ is compact. But $\omega(B)$ is a closed subset of $cl_V T(1)\gamma^+(B)$. Consequently, $\omega(B)$ is compact.

We also see that $\omega(B) \subset cl_V\gamma^+(B) \subset cl_V B = B$. Hence it only remains for us to show that $d\big(T(t)B, \omega(B)\big) \to 0$ as $t \to +\infty$.

For the sake of argument, suppose that there are an $\varepsilon > 0$ and a sequence $t_n \to +\infty$ such that

$$d\big(T(t_n)B, \omega(B)\big) > \varepsilon,\ n \in N. \tag{1.1.23}$$

Then there must exist a sequence $\{y_n\} \subset B$ for which the sequence $\{T(t_n)y_n\}$ does not have a convergent subsequence. However, almost all elements of the sequence $\{T(t_n)y_n\}$ are contained in the compact set $cl_V T(1)\gamma^+(B)$ and this contradicts (1.1.23). Finally, recalling the *Cantor condition* in metric spaces - namely, the intersection of a decreasing family of nonempty compact sets is nonempty - we conclude that $\omega(B)$ is empty only if B is empty. The proof is complete. □

Lemma 1.1.2. *If $\{T(t)\}$ is a compact C^0 semigroup on a metric space V, then*

$$\forall_{\substack{B\subset V\\ B \text{ bounded}}}\ \forall_{\tau_2>\tau_1>0}\ \bigcup_{t\in[\tau_1,\tau_2]} T(t)B \ \textit{ is bounded in } V. \tag{1.1.24}$$

Proof. If (1.1.24) is not true one can choose a nonempty, bounded set $B \subset V$, two numbers $\tau_2 > \tau_1 > 0$, and two sequences $\{v_n\} \subset B$, $\{t_n\} \subset [\tau_1, \tau_2]$ such that

$$y_n := dist_V(T(t_n)v_n, T(t_0)v_0) > n,\ n = 1, 2, \dots . \tag{1.1.25}$$

However, there is a subsequence $\{t_{n'}\}$ such that $t_{n'} \to \tau^* \in [\tau_1, \tau_2]$. Furthermore, by the compactness of $T(t)$, $t > 0$, there is a subsequence $\{v_{n''}\}$, for which $T(\frac{\tau^*}{2})v_{n''} \to v^* \in cl_V T(\frac{\tau^*}{2})B$. Since the semigroup is continuous with respect to both its arguments we obtain

$$T(t_{n''})v_{n''} = T\left(t_{n''} - \frac{\tau^*}{2}\right) T\left(\frac{\tau^*}{2}\right) v_{n''} \to T\left(\frac{\tau^*}{2}\right) v^*.$$

But this last result implies that $\{T(t_{n''})v_{n''}\}$ is bounded, which contradicts (1.1.25). The proof is complete. □

Corollary 1.1.6. *Let $\{T(t)\}$ be a C^0 semigroup on a metric space V. If $\{T(t)\}$ is compact and point dissipative, then $\{T(t)\}$ has a global attractor in V.*

Proof. In view of Corollary 1.1.3 it suffices to verify the condition (1.1.19).

Let W_0 be a bounded subset of V which attracts points and let $\mathcal{N}_{W_0}$ be any bounded, open neighborhood of W_0. We first show that

$$\forall_{\substack{B \subset V \\ B \text{ bounded}}} \exists_{\tau_B \geq 2}\, T(\tau_B)B \subset \bigcup_{t \in [1, \tau_B]} T(t)\mathcal{N}_{W_0}. \tag{1.1.26}$$

Indeed, as in (1.1.14) we have

$$\forall_{v \in V} \exists_{\tau_v \geq 0} \exists_{B_V(v, \varepsilon_v)}\, T(\tau_v)B_V(v, \varepsilon_v) \subset \mathcal{N}_{W_0}, \tag{1.1.27}$$

where $B_V(v, \varepsilon_v) \subset V$ is an open ball centered at v and having radius ε_v. Since B is bounded and since $T(1)$ is compact, we can choose finitely many such balls so that

$$cl_V T(1)B \subset B_V(v_1, \varepsilon_{v_1}) \cup \ldots \cup B_V(v_{k(B)}, \varepsilon_{v_{k(B)}}). \tag{1.1.28}$$

For $\tau_B := 2 + \tau_{max} = 2 + \max\{\tau_{v_j} : j = 1, 2, \ldots\}$, the conditions (1.1.28) and (1.1.27) lead to the inclusion

$$\begin{aligned} T(\tau_B)B &= T(\tau_{max} + 2)B \subset T(\tau_{max} + 1) cl_V T(1)B \\ &\subset T(\tau_{max} + 1) \bigcup_{j=1}^{k(B)} B_V(v_j, \varepsilon_{v_j}) \subset \bigcup_{j=1}^{k(B)} T(\tau_{max} + 1 - \tau_{v_j}) T(\tau_{v_j}) B_V(v_j, \varepsilon_{v_j}) \\ &\subset \bigcup_{j=1}^{k(B)} T(\tau_{max} + 1 - \tau_{v_j})\mathcal{N}_{W_0} \subset \bigcup_{t \in [1, \tau_B]} T(t)\mathcal{N}_{W_0}. \end{aligned}$$

As a consequence of Lemma 1.1.2, the set $\mathcal{O} = \bigcup_{t \in [1,2]} T(t)\mathcal{N}_{W_0}$ is bounded in V. Thus, applying (1.1.26) to $\mathcal{O}$, we obtain

$$\bigcup_{t \in [\tau_{\mathcal{O}}+1, \tau_{\mathcal{O}}+2]} T(t)\mathcal{N}_{W_0} = T(\tau_{\mathcal{O}})\mathcal{O} \subset \bigcup_{t \in [1, \tau_{\mathcal{O}}]} T(t)\mathcal{N}_{W_0}. \tag{1.1.29}$$

Next, from (1.1.29) there follows

$$\bigcup_{t \in [1, \tau_{\mathcal{O}}+2]} T(t)\mathcal{N}_{W_0} = \bigcup_{t \in [1, \tau_{\mathcal{O}}+1]} T(t)\mathcal{N}_{W_0} \cup \bigcup_{t \in [\tau_{\mathcal{O}}+1, \tau_{\mathcal{O}}+2]} T(t)\mathcal{N}_{W_0} \subset \bigcup_{t \in [1, \tau_{\mathcal{O}}+1]} T(t)\mathcal{N}_{W_0}.$$

This last inclusion further ensures that

$$\bigcup_{t\geq 1} T(t)\mathcal{N}_{W_0} \subset \bigcup_{t\in[1,\tau_{\mathcal{O}}+1]} T(t)\mathcal{N}_{W_0},$$

and now (1.1.26) yields the condition

$$\forall_{\substack{B\subset V\\ B \text{ bounded}}} \exists_{\tau_B\geq 2} \bigcup_{t\geq\tau_B} T(t)B \subset \bigcup_{t\geq 1} T(t)\mathcal{N}_{W_0} \subset \bigcup_{t\in[1,\tau_{\mathcal{O}}+1]} T(t)\mathcal{N}_{W_0}.$$

By virtue of Lemma 1.1.2, the set $\bigcup_{t\in[1,\tau_{\mathcal{O}}+1]} T(t)\mathcal{N}_{W_0}$ is bounded. The proof is complete. □

The next statement relates the notion of a Lyapunov function to Corollary 1.1.6.

Corollary 1.1.7. *Let $\{T(t)\}$ be a compact C^0 semigroup on a metric space V. Furthermore, suppose that all the equilibria of $\{T(t)\}$ lie in a bounded subset of V. If there exists a Lyapunov function for $\{T(t)\}$, then $\{T(t)\}$ has a global attractor.*

Remark 1.1.4. Note that, for the case in which the semigroup $\{T(t)\}$ is compact, Proposition 1.1.2 (and hence also Corollary 1.1.7) can be proved without the assumption that the Lyapunov function is bounded below. Indeed, the only place in the proof of Proposition 1.1.2 where we specifically used this assumption was formula (1.1.22). But, if the semigroup $\{T(t)\}$ is known to be compact and if the orbit $\gamma^+(v)$ is known to be bounded, then clearly the image $T(1)\gamma^+(v)$ is precompact in V and (1.1.22) immediately follows. Of course, the preceding argument in no way refers to an assumed boundedness of $\mathcal{L}$.

In addition to all this, we want to point out that the examples which we shall present in Chapter 6 yield semigroups which - in a certain natural way - are compact. In fact, as we shall see, even if the semigroup $\{T(t)\}$ is generically noncompact - as, for example, in the case of Cauchy problems considered on the whole of R^n - we are able to slightly alter the original phase space so as to bring about the required compactness of $\{T(t)\}$.

From the point of view of applications the usefulness of the procedure just indicated is clear. Indeed, when the semigroup is compact, we have better control over the behavior of the given system.

In general, we shall not again use Theorem 1.1.2 (or Corollary 1.1.5), which explicitly requires asymptotic smoothness for $\{T(t)\}$. Rather, we will employ Corollaries 1.1.6 and 1.1.7, which require compactness for $\{T(t)\}$.

Remarks on completely continuous semigroups. We want to formulate some assertions concerning the special case in which the semigroup $\{T(t)\}$ is *completely continuous for* $t > 0$ in the sense of [HA 2, p. 36], which means

(1.1.30)

the semigroup is compact and for each bounded set $B \subset V$ and each number $\tau > 0$ the set $\bigcup_{t\in[0,\tau]} T(t)B$ is bounded in V.

For brevity's sake we shall henceforth refer to such semigroups $\{T(t)\}$ as *completely continuous*, omitting explicit reference to the inequality $t > 0$.

Note that the second requirement in (1.1.30) can be significantly weakened inasmuch as, for a compact C^0 semigroup $\{T(t)\}$, Theorem 1.1.6 implies that the boundedness of $\bigcup_{t\in[0,\tau]} T(t)B$ for a single number $\tau > 0$ is equivalent to the boundedness of $\bigcup_{t\in[0,\tau]} T(t)B$ for every number $\tau > 0$.

Regarding complete continuity, we also have

Corollary 1.1.8. *If $\{T(t)\}$ is a C^0 semigroup on a metric space V and if $\{T(t)\}$ is completely continuous and point dissipative on V, then $\{T(t)\}$ has a global attractor $\mathcal{A}$ in V.*

Remark 1.1.5. Note that in both Corollaries 1.1.6 and 1.1.8 the assumption of point dissipativeness for $\{T(t)\}$ can be replaced by the weaker condition (1.1.20) (see Remark 1.1.3)

Asymptotic compactness of $\{T(t)\}$. It is sometimes convenient to have available criteria for asymptotic smoothness distinct from those stipulated in the definition. With this in mind, we introduce the following condition (see [LA 3, Chapter 3]):

(1.1.31)

for any nonempty set B having the property that there exists a number $t_B \geq 0$ such that $\bigcup_{t\geq t_B} T(t)B$ is bounded, each sequence $\{T(t_n)v_n\}$, where $t_n \nearrow +\infty$ and $\{v_n\} \subset B$, has a convergent subsequence.

We shall say that $\{T(t)\}$ is *asymptotically compact* in V if (1.1.31) holds.

We want to establish

Proposition 1.1.3. *Let $\{T(t)\}$ be a C^0 semigroup on a metric space V. Then $\{T(t)\}$ is asymptotically smooth if and only if $\{T(t)\}$ is asymptotically compact.*

Proof. Suppose that $\{T(t)\}$ is asymptotically smooth. Let $B \subset V$ be nonempty and have the property that the union $\bigcup_{t\geq t_B} T(t)B$ is bounded for some $t_B \geq 0$. By Proposition 1.1.1, $\omega(B)$ is nonempty, is compact, and attracts B. This allows us to choose a convergent subsequence from any sequence of the sort $\{T(t_n)v_n\}$ with $t_n \nearrow +\infty$ and $\{v_n\} \subset B$, and thus we have (1.1.31).

Now suppose that $\{T(t)\}$ is *asymptotically compact.* Consider any nonempty, closed, bounded, and positively invariant set $B \subset V$. We need to prove that $\omega(B)$ is nonempty, attracts B, is invariant, and is compact.

By virtue of (1.1.31), there exists an element $v \in B$ such that $\omega(v) \neq \emptyset$. Trivially, $\omega(v) \subset \omega(B)$. Hence, $\omega(B) \neq \emptyset$.

Clearly, $\omega(B) \subset B$. If - for the sake of argument - $\omega(B)$ does not attract B, then, in accordance with (1.1.7) and (1.1.8), one can choose sequences $\{v_n\} \subset B$,

$t_n \nearrow \infty$ such that $\{T(t_n)v_n\}$ has no convergent subsequence. This contradicts (1.1.31). Thus, $\omega(B)$ attracts B.

To prove that $\omega(B)$ is invariant we need only establish the inclusions $\omega(B) \subset T(t)\omega(B)$, $t \geq 0$. Choose any number $t > 0$, element $v \in \omega(B)$, sequence $t_n \nearrow +\infty$, and sequence $\{v_n\} \subset B$. Assume that $T(t_n)v_n \to v$. By virtue of (1.1.31) there are subsequences $t_{n'} \nearrow +\infty$ and $\{v_{n'}\} \subset B$ such that $T(t_{n'} - t)v_{n'}$ converges to some element $w \in \omega(B)$. Therefore $T(t)w = v$ and we have established that $\omega(B)$ is invariant.

To prove that $\omega(B)$ is compact, we note that $\omega(B)$ is bounded and that $T(t)\omega(B) = \omega(B)$ for all $t \geq 0$. From (1.1.31) it immediately follows that $\omega(B)$ is compact, which completes the proof. □

1.2. Inequalities. Elliptic operators

1.2.1. Elementary inequalities. Here we shall set forth several important inequalities which in later chapters we shall use for performing various estimates.

Lemma 1.2.1. *(Cauchy inequality) Let $a, b \in R$ and let $\varepsilon > 0$. Then*

$$ab \leq \frac{\varepsilon}{2}a^2 + \frac{1}{2\varepsilon}b^2. \tag{1.2.1}$$

Proof. For the proof it suffices to note that

$$0 \leq \left(\sqrt{\varepsilon}a - \frac{1}{\sqrt{\varepsilon}}b\right)^2 = \varepsilon a^2 + \frac{1}{\varepsilon}b^2 - 2ab,$$

from which (1.2.1) follows. □

Lemma 1.2.2. *(Young inequality) Let $\varepsilon > 0$, $a, b \geq 0$, $p, q > 1$, and $\frac{1}{p} + \frac{1}{q} = 1$. Then*

$$ab \leq \varepsilon\frac{a^p}{p} + \frac{1}{\varepsilon^{\frac{q}{p}}}\frac{b^q}{q}. \tag{1.2.2}$$

Proof. The case $ab = 0$ is obvious. Hence we can assume that $ab > 0$. Recalling that the exponential function is convex, we obtain

$$ab = e^{\ln ab} = e^{\frac{1}{p}\ln a^p + \frac{1}{q}\ln b^q} \leq \frac{1}{p}e^{\ln a^p} + \frac{1}{q}e^{\ln b^q} = \frac{1}{p}a^p + \frac{1}{q}b^q.$$

From this there easily follows

$$ab = \left(\varepsilon^{\frac{1}{p}}a\right)\left(\frac{b}{\varepsilon^{\frac{1}{p}}}\right) \leq \frac{1}{p}\left(\varepsilon^{\frac{1}{p}}a\right)^p + \frac{1}{q}\left(\frac{b}{\varepsilon^{\frac{1}{p}}}\right)^q = \varepsilon\frac{a^p}{p} + \frac{1}{\varepsilon^{\frac{q}{p}}}\frac{b^q}{q},$$

which establishes (1.2.2). □

In the literature inequality (1.2.2) often appears in a slightly different form. Thus, we render here

Corollary 1.2.1. *Let $a, b \geq 0$, $\varepsilon > 0$, $m > 1$. Then*

$$ab \leq \frac{1}{m}\varepsilon^m a^m + \frac{m-1}{m}\varepsilon^{-\frac{m}{m-1}} b^{\frac{m}{m-1}}.$$

Lemma 1.2.3. *(Hölder inequality) Let $u, v : \Omega \to R$ be Lebesgue measurable functions, $p, q > 1$, and $\frac{1}{p} + \frac{1}{q} = 1$. Then*

$$\|uv\|_{L^1(\Omega)} \leq \|u\|_{L^p(\Omega)} \|v\|_{L^q(\Omega)}. \tag{1.2.3}$$

Proof. The case when the right side of (1.2.3) is infinite or equal to zero is trivial. Otherwise, using the Young inequality with $\varepsilon := \frac{\|v\|_{L^q(\Omega)}}{\|u\|^{p-1}_{L^p(\Omega)}}$, we get

$$\int_\Omega |uv| dx \leq \int_\Omega \left(\varepsilon \frac{|u|^p}{p} + \frac{1}{\varepsilon^{\frac{q}{p}}} \frac{|v|^q}{q} \right) dx = \varepsilon \frac{\|u\|^p_{L^p(\Omega)}}{p} + \frac{1}{\varepsilon^{\frac{q}{p}}} \frac{\|v\|^q_{L^q(\Omega)}}{q}$$

$$= \frac{\|v\|_{L^q(\Omega)}}{\|u\|^{p-1}_{L^p(\Omega)}} \frac{\|u\|^p_{L^p(\Omega)}}{p} + \left(\frac{\|u\|^{p-1}_{L^p(\Omega)}}{\|v\|_{L^q(\Omega)}} \right)^{\frac{q}{p}} \frac{\|v\|^q_{L^q(\Omega)}}{q} = \left(\frac{1}{p} + \frac{1}{q} \right) \|u\|_{L^p(\Omega)} \|v\|_{L^q(\Omega)}.$$

Therefore, (1.2.3) is proved. □

Lemma 1.2.4. *(Bernoulli inequality) Let $a, b > 0$, $\rho > \nu \geq 0$, and consider a continuous function $y : [0, \tau_0) \to [0, +\infty)$ differentiable on $(0, \tau_0)$ - where $0 < \tau_0 \leq +\infty$ - and satisfying*

$$y'(t) \leq a y^\nu(t) - b y^\rho(t), \quad \textit{for all} \quad t \in (0, \tau_0). \tag{1.2.4}$$

Then

$$\sup_{t \in [0, \tau_0)} \{y(t)\} \leq \max\left\{ y(0), \left(\frac{a}{b} \right)^{\frac{1}{\rho - \nu}} \right\}. \tag{1.2.5}$$

Furthermore, if $\tau_0 = \infty$, then

$$\limsup_{t \to +\infty} y(t) \leq \left(\frac{a}{b} \right)^{\frac{1}{\rho - \nu}}. \tag{1.2.6}$$

Proof. The function $h(y) = a y^\nu - b y^\rho$ appearing on the right hand side of (1.2.4) is negative at each y larger than $\left(\frac{a}{b}\right)^{\frac{1}{\rho-\nu}}$, which number is the unique positive root of the equation $h(y) = 0$. Therefore, given (1.2.4), $y(t)$ is fully controlled by the initial value $y(0)$ and the quantity $\left(\frac{a}{b}\right)^{\frac{1}{\rho-\nu}}$, as is stated in the conditions (1.2.5) and (1.2.6). □

Lemma 1.2.5. *Let $y \in C^0([0, \infty)) \cap C^1((0, \infty))$, let f be nonnegative and continuous on $[0, \infty)$, and suppose that $\lim_{t \to +\infty} f(t) = M$. If, for some $a > 0$, y satisfies*

$$y'(t) \leq -a y(t) + f(t), \; t \in (0, +\infty), \tag{1.2.7}$$

then

$$\limsup_{t\to+\infty} y(t) \le \frac{M}{a}.$$

Proof. With our assumptions f is bounded on $[0,+\infty)$. Fix $\varepsilon > 0$ and a time $t_0 > 0$, such that $f(t) \le M+\varepsilon$ for $t \ge t_0$. Solving the differential inequality (1.2.7) for $t \ge t_0$, we find

$$\begin{aligned} y(t) &\le y(0)e^{-at} + \left(\int_0^{t_0} + \int_{t_0}^{t}\right) f(s)e^{as}dse^{-at} \\ &\le y(0)e^{-at} + \sup_{s\in[0,+\infty)} f(s)\,\frac{e^{a(t_0-t)} - e^{-at}}{a} + (M+\varepsilon)\frac{1-e^{-a(t-t_0)}}{a} \to \frac{M+\varepsilon}{a} \end{aligned}$$

when $t \to +\infty$. Since ε was arbitrary, the proof is complete. □

Lemma 1.2.6. *(Asymptotic Gronwall inequality) Let the differentiable function* $y : (0,+\infty) \to [0,+\infty)$ *be integrable on* $(0,+\infty)$ *and satisfy the inequality*

$$y'(\tau) \le const.y(\tau), \ \textit{for} \ \ \tau > 0. \tag{1.2.8}$$

Then

$$\lim_{t\to+\infty} y(t) = 0. \tag{1.2.9}$$

Proof. When *const.* = 0, (1.2.9) is immediate. Fixing $t > 0$ we obtain from (1.2.8) that

$$\frac{d}{d\tau}\left(y(\tau)e^{const.(t-\tau)}\right) \le 0, \quad \text{for} \ \ \tau > 0. \tag{1.2.10}$$

For *const.* < 0 the condition (1.2.9) follows by integrating (1.2.10). Consider the case *const.* > 0. Fixing $r > 0$, $s \in [t,t+r]$ and integrating (1.2.10) over the interval $[s,t+r]$, we get

$$y(t+r)e^{-const.r} \le y(s)e^{const.(t-s)} \le y(s). \tag{1.2.11}$$

Since inequality (1.2.11) is valid for $s \in [t,t+r]$ we can integrate it over $s \in [t,t+r]$, so that

$$y(t+r)re^{-const.r} = \int_t^{t+r} y(t+r)e^{-const.r}ds \le \int_t^{t+r} y(s)ds. \tag{1.2.12}$$

As a consequence of the finite value of the integral $\int_0^\infty y(s)ds$, the upper limit $\limsup_{t\to+\infty} \int_t^{t+r} y(s)ds$ is zero, so that (1.2.9) follows. □

Lemma 1.2.7. *(Uniform Gronwall inequality) Let the differentiable function* $y : (0,\tau_0) \to R$ *satisfy the conditions*

$$y'(\tau) \le const.y(\tau), \ \textit{for} \ \ \tau \in (0,\tau_0), \tag{1.2.13}$$

and

$$\exists_{r\in(0,\tau_0)}\, \exists_{const._1>0}\, \forall_{t\in(0,\tau_0-r)} \int_t^{t+r} y(s)ds \le const._1. \tag{1.2.14}$$

Then

$$y(t+r) \leq \frac{const._1}{r} e^{const.r}, \quad for \ \ t \in (0, \tau_0 - r). \tag{1.2.15}$$

Proof. In order to show (1.2.15), we shall repeat the proof of Lemma 1.2.6 until (1.2.12) is established. Then, applying (1.2.14) to the right side of (1.2.12) we obtain the condition

$$y(t+r)re^{-const.r} \leq \int_t^{t+r} y(s)ds \leq const._1, \tag{1.2.16}$$

which is a counterpart of (1.2.15). □

The preceding considerations can be generalized to the following stronger version of Lemma 1.2.7 (see [TE 1, p. 89]).

Lemma 1.2.8. *(Generalized Gronwall inequality) Let $g, h, y : [\tau_0, +\infty) \to (0, +\infty)$ be continuous functions, continuously differentiable on $(\tau_0, +\infty)$. Assume further that for some $r > 0$ and all $t \geq \tau_0$, the following conditions hold:*

$$\begin{cases} y'(t) \leq g(t)y(t) + h(t), \\ \int_t^{t+r} g(s)ds \leq const._1, \ \int_t^{t+r} h(s)ds \leq const._2, \\ \int_t^{t+r} y(s)ds \leq const._3. \end{cases} \tag{1.2.17}$$

Then y fulfills the uniform estimate

$$y(t+r) \leq \left(\frac{const._3}{r} + const._2 \right) e^{const._1}, \ t \geq \tau_0. \tag{1.2.18}$$

Lemma 1.2.9. *(Volterra type inequality) Let $\alpha, \beta \in [0,1)$, $a \geq 0$, $b > 0$; and let $y : [0, \tau) \to [0, +\infty)$ be a continuous function satisfying the inequality*

$$y(t) \leq \frac{a}{t^\alpha} + b \int_0^t \frac{1}{(t-s)^\beta} y(s)ds, \quad for \ \ t \in (0, \tau). \tag{1.2.19}$$

Then

$$\sup_{t \in [0,\tau)} t^\alpha y(t) \leq a \ const.(b, \alpha, \beta, \tau), \tag{1.2.20}$$

where $const.(b, \alpha, \beta, \tau)$ is a continuous function increasing with respect to τ.

Proof. The proof proceeds in three steps.

Step 1. Choose $t^* = \min\{\frac{\tau}{2}, T^*\}$, where

$$\frac{bT^{*1-\beta}}{2^{1-\alpha-\beta}} \left(\frac{1}{1-\alpha} + \frac{1}{1-\beta} \right) = \frac{1}{2} \tag{1.2.21}$$

and introduce the following linear transformation Φ, acting on the continuous functions $f : (0, \tau) \to [0, +\infty)$:

$$f \stackrel{\Phi}{\to} \Phi(f)(t) := b \int_0^t \frac{1}{(t-s)^\beta} f(s)ds.$$

Clearly, we can write (1.2.19) in the form

$$y(t) \le \frac{a}{t^\alpha} + \Phi(y)(t), \ t \in (0, t^*). \tag{1.2.22}$$

Since Φ is linear and since $f \le g$ implies $\Phi f \le \Phi g$ for all functions f, g in the domain of Φ, the inequality (1.2.22) yields

$$\Phi(y)(t) \le a\Phi\left(\frac{1}{t^\alpha}\right) + \Phi^2(y)(t).$$

Consequently,

$$y(t) \le \frac{a}{t^\alpha} + a\Phi\left(\frac{1}{t^\alpha}\right) + \Phi^2(y)(t), \ t \in (0, t^*). \tag{1.2.23}$$

Proceeding by induction we obtain that for all $n \in N$

$$y(t) \le \frac{a}{t^\alpha} + a\sum_{k=1}^{n-1} \Phi^k\left(\frac{1}{t^\alpha}\right) + \Phi^n(y)(t), \ t \in (0, t^*). \tag{1.2.24}$$

Furthermore, we observe that

$$\Phi(y)(t) \le \|y\|_{L^\infty(0,t^*)} b \int_0^t \frac{1}{(t-s)^\beta} ds = \|y\|_{L^\infty(0,t^*)} \frac{bt^{1-\beta}}{1-\beta}. \tag{1.2.25}$$

The inequalities (1.2.21) and (1.2.25) together yield the estimate

$$\|\Phi(y)\|_{L^\infty(0,t^*)} < const. \|y\|_{L^\infty(0,t^*)}$$

with

$$const. := \frac{bt^{*1-\beta}}{1-\beta} < 1.$$

Hence,

$$\|\Phi^n(y)\|_{L^\infty(0,t^*)} \le const.^n \|y\|_{L^\infty(0,t^*)} \to 0 \quad \text{when} \quad n \to \infty. \tag{1.2.26}$$

Next, using (1.2.21) again, we obtain the estimate

$$\begin{aligned}\Phi\left(\frac{1}{t^\alpha}\right) &= \left(\int_0^{\frac{t}{2}} + \int_{\frac{t}{2}}^t\right) b\frac{1}{(t-s)^\beta}\frac{1}{s^\alpha} ds \le \frac{b}{(\frac{t}{2})^\beta}\frac{(\frac{t}{2})^{1-\alpha}}{1-\alpha} + \frac{b}{(\frac{t}{2})^\alpha}\frac{(\frac{t}{2})^{1-\beta}}{1-\beta} \\ &= \frac{1}{t^\alpha}\frac{bt^{1-\beta}}{2^{1-\alpha-\beta}}\left(\frac{1}{1-\alpha} + \frac{1}{1-\beta}\right) \le \frac{1}{2}\frac{1}{t^\alpha}, \ t \in (0, t^*).\end{aligned}$$

By induction,

$$\begin{aligned}\Phi^{k+1}\left(\frac{1}{t^\alpha}\right) &= \left(\int_0^{\frac{t}{2}} + \int_{\frac{t}{2}}^t\right) b\frac{1}{(t-s)^\beta}\left[\Phi^k\left(\frac{1}{t^\alpha}\right)\right] ds \\ &\le \left(\int_0^{\frac{t}{2}} + \int_{\frac{t}{2}}^t\right) b\frac{1}{(t-s)^\beta}\left[\frac{1}{2^k}\frac{1}{s^\alpha}\right] ds \le \frac{1}{2^{k+1}}\frac{1}{t^\alpha}, \ t \in (0, t^*).\end{aligned} \tag{1.2.27}$$

From (1.2.27) there follows

$$\sum_{k=1}^{n-1} \Phi^k\left(\frac{1}{t^\alpha}\right) \leq \sum_{k=1}^{n-1} \frac{1}{2^k}\frac{1}{t^\alpha} \leq \frac{1}{t^\alpha}, \quad t \in (0, t^*). \tag{1.2.28}$$

With the aid of (1.2.26) and (1.2.28) we can pass to the limit as $n \to \infty$ in (1.2.24). The result is

$$y(t) \leq \frac{a}{t^\alpha} + a\sum_{k=1}^{\infty} \Phi^k\left(\frac{1}{t^\alpha}\right) \leq \frac{2a}{t^\alpha}, \quad t \in (0, t^*). \tag{1.2.29}$$

This proves that

$$\sup_{t\in[0,t^*]} t^\alpha y(t) \leq 2a. \tag{1.2.30}$$

Step 2. Define

$$h(r) = \left(2\left(\frac{\tau}{t^*}\right)^\alpha + \frac{\tau}{t^*}r\right), \quad r \in R,$$

and fix $\delta \in (0, \tau - t^*)$ so that

$$\frac{b\delta^{1-\beta}}{1-\beta} \leq \frac{1}{2}. \tag{1.2.31}$$

We shall prove that, for each $t_0 \in [t^*, \tau - \delta)$, the following implication holds:

$$\text{if } \sup_{t\in[0,t_0]} t^\alpha y(t) \leq ar, \text{ then } \sup_{t\in[0,t_0+\delta]} t^\alpha y(t) \leq ah(r). \tag{1.2.32}$$

Indeed, if $t_0 \geq t^*$, if $\sup_{t\in[0,t_0]} t^\alpha y(t) \leq ar$, and if $t \in [t_0, t_0+\delta]$, then from (1.2.19) there follows

$$\begin{aligned} y(t) &\leq \frac{a}{t^\alpha} + b\left(\int_0^{t_0} + \int_{t_0}^{t}\right)\frac{1}{(t-s)^\beta}y(s)ds \\ &\leq \frac{a}{t^\alpha} + abr\int_0^{t_0}\frac{1}{(t-s)^\beta s^\alpha}ds + b\sup_{s\in[t_0,t_0+\delta]} y(s)\int_{t_0}^{t}\frac{1}{(t-s)^\beta}ds, \end{aligned} \tag{1.2.33}$$

where - keeping in mind that $t^* \le t_0 \le t \le t_0 + \delta < \tau$ and that (1.2.21) holds -

(1.2.34)

$$\begin{aligned}
\int_0^{t_0} \frac{1}{(t-s)^\beta s^\alpha} ds &= \left(\int_0^{\frac{t_0}{2}} + \int_{\frac{t_0}{2}}^{t_0}\right) \frac{1}{(t-s)^\beta s^\alpha} ds \\
&\le \frac{1}{(t-\frac{t_0}{2})^\beta} \frac{1}{1-\alpha} \left(\frac{t_0}{2}\right)^{1-\alpha} + \frac{1}{(\frac{t_0}{2})^\alpha} \frac{1}{1-\beta} \left(\left(t-\frac{t_0}{2}\right)^{1-\beta} - (t-t_0)^{1-\beta}\right) \\
&\le \frac{1}{(\frac{t_0}{2})^\beta} \frac{1}{1-\alpha} \left(\frac{t_0}{2}\right)^{1-\alpha} + \frac{1}{(\frac{t_0}{2})^\alpha} \frac{1}{1-\beta} \left(t-\frac{t_0}{2} - (t-t_0)\right)^{1-\beta} \\
&= \left(\frac{1}{1-\alpha} + \frac{1}{1-\beta}\right) \left(\frac{t_0}{2}\right)^{1-\alpha-\beta} \le \frac{1}{2b} \frac{t_0^{1-\alpha-\beta}}{t^{*1-\beta}} \\
&\le \frac{1}{2b} \frac{t_0^{1-\alpha-\beta}}{t^{*1-\beta}} \left(\frac{t_0}{t^*}\right)^\beta = \frac{1}{2b} \frac{\tau^{1-\alpha}}{t^*}.
\end{aligned}$$

Combining (1.2.31), (1.2.33), and (1.2.34), we obtain

$$y(t) \le \frac{a}{t^\alpha} + \frac{1}{2} a \frac{\tau^{1-\alpha}}{t^*} r + \frac{1}{2} \sup_{s \in [t_0, t_0+\delta]} y(s), \ t \in [t_0, t_0 + \delta].$$

Next, recalling that $t^* \le t_0$, we see that

$$\sup_{t \in [t_0, t_0+\delta]} y(t) \le \frac{2a}{t_0^\alpha} + a \frac{\tau^{1-\alpha}}{t^*} r \le a \left(\frac{2}{t^{*\alpha}} + \frac{\tau^{1-\alpha}}{t^*} r\right).$$

Since $t \le t_0 + \delta < \tau$ and since $\sup fg \le \sup f \sup g$ we have the estimate

$$\sup_{t \in [t_0, t_0+\delta]} t^\alpha y(t) \le a \left(2\left(\frac{\tau}{t^*}\right)^\alpha + \frac{\tau}{t^*} r\right) = ah(r),$$

and so finally

$$\sup_{t \in [0, t_0+\delta]} t^\alpha y(t) \le a \max\{r, h(r)\} = ah(r).$$

Step 3. Given τ, t^*, and δ as above, there exist numbers $\theta \in [0, \delta)$, $k \in N$, such that $\tau - t^* = k\delta + \theta$. Hence, from (1.2.30) - with the aid of (1.2.32) - we have

$$\sup_{t \in [0, \tau-\delta]} t^\alpha y(t) \le a(\underbrace{h \circ \ldots \circ h}_{k \text{ times}})(2).$$

Therefore, for any sequence $t_n \to t^{*-}$, we have

$$\sup_{t \in [0, t_n-\delta]} t^\alpha y(t) \le a(\underbrace{h \circ \ldots \circ h}_{k \text{ times}})(2), \ n \in N.$$

Again we apply (1.2.32) and now we obtain

$$\sup_{t \in [0, t_n]} t^\alpha y(t) \le a(\underbrace{h \circ \ldots \circ h}_{k+1 \text{ times}})(2), \ n \in N.$$

This completes the proof of Lemma 1.2.9. □

1.2.2. Embedding theorems. In the modern approach to partial differential equations different function spaces, most often of the Sobolev and Hölder type, are constantly used. We need to compare the norms of such spaces and these connections are expressed as *embedding theorems.* We start with formulation of the classical version of such a theorem valid for $W^{m,p}$ spaces of natural order m. For the proof of it we refer to [FR 1, p. 29], [AD, p. 97] but we shall follow the presentation of [HE 2].

For $m \in N$ and $p \in [1, \infty]$ we consider the Sobolev space $W^{m,p}(\Omega)$. It is convenient to introduce the *net smoothness number* characterizing the smoothness properties of the elements of $W^{m,p}(\Omega)$:

$$\text{(1.2.35)} \qquad \textit{net smoothness}\ (f) := m - \frac{n}{p} \quad \text{for all} \quad f \in W^{m,p}(\Omega).$$

This quantity is decisive for relations between different Sobolev spaces. Let $\Omega \subset R^n$ be a domain lying on one side of its boundary $\partial\Omega$ with $\partial\Omega$ being a Lipschitz surface (see Section 9.1 for the *extension property*). We shall recall now the classical Sobolev embedding theorem.

Sobolev embeddings for natural order spaces. *Let $k, m \in N$, $1 \le p, q \le +\infty$ and Ω be as above. Then the following continuous inclusions take place:*

$$\text{(1.2.36)} \qquad W^{m,p}(\Omega) \subset \begin{cases} W^{k,q}(\Omega) & \text{if } m - \frac{n}{p} \ge k - \frac{n}{q},\ q \ge p \\ & (\text{unless } m - \frac{n}{p} = k,\ q = \infty), \\ C^{k+\mu}(\overline{\Omega}) & \text{if } 0 \le \mu \le m - k - \frac{n}{p} < 1 \\ & (\text{unless } m - \frac{n}{p} \in N). \end{cases}$$

Moreover, for bounded Ω the above inclusions are compact in the case of strict inequalities; $m - \frac{n}{p} > k - \frac{n}{q}$ or $m - \frac{n}{p} > k + \mu$, respectively.

Further, the generalization of the embedding theorem to fractional order spaces $W^{s,p}$ with real $s \ge 0$ will be described. Let us start with some definitions.

We shall consider below bounded domains Ω having the *extension property* as described in Section 9.1 and recall the definition of the $W^{s,p}$ spaces with $0 \le s \in R$.

For $\Omega = R^n$, let [TR, Definition 2.3.1(d)]

$$\text{(1.2.37)} \qquad W^{s,p}(R^n) = \begin{cases} H^s_p(R^n), & s = 0, 1, 2, \dots, \\ B^s_{p,p}(R^n), & 0 < s \notin N, \end{cases}$$

where $B^s_{p,p}(R^n)$ are Besov spaces and $H^s_p(R^n)$ are the spaces of Bessel potentials (see [TR, Definition 2.3.1]). It is worth noting that $H^s_p(R^n) \ne B^s_{p,p}(R^n)$ unless $p = 2$ (see [TR, Remark 2.3.3/4]).

For an arbitrary domain $\Omega \subset R^n$ (bounded or not), $-\infty < s < +\infty$, $1 < p < +\infty$, the space $B^s_{p,p}(\Omega)$ consists of (see [TR, 4.2.1]) restrictions $\phi_{|\Omega}$ (in the sense of distributions) of elements of $B^s_{p,p}(R^n)$ and is normed by

$$\|f\|_{B^s_{p,p}(\Omega)} = \inf_{\substack{\phi_{|\Omega} = f \\ \phi \in B^s_{p,p}(R^n)}} \|\phi\|_{B^s_{p,p}(R^n)}.$$

Simultaneously, $H_p^s(\Omega)$ consists of restrictions $\phi_{|\Omega}$ of elements $\phi \in H_p^s(R^n)$ and is normed by

$$\|f\|_{H_p^s(\Omega)} = \inf_{\substack{\phi_{|\Omega}=f \\ \phi \in H_p^s(R^n)}} \|\phi\|_{H_p^s(R^n)}.$$

We extend naturally (1.2.37) to

$$W^{s,p}(\Omega) = \begin{cases} H_p^s(\Omega), & s = 0,1,2,\ldots, \\ B_{p,p}^s(\Omega), & 0 < s \notin N, \end{cases} \tag{1.2.38}$$

pointing out as before that $H_p^s(\Omega) \neq B_{p,p}^s(\Omega)$ unless $p = 2$ (see [TR, Theorem 4.6.1(b)]).

Next we shall recall the following two consequences of the *interpolation theory* (see [TR, 2.4.2]).

For $-\infty < s_0, s_1 < +\infty,\ 1 < p_0, p_1 < +\infty,\ 0 < \theta < 1,\ s_0 \neq s_1$,

$$s = (1-\theta)s_0 + \theta s_1, \quad \frac{1}{p} = \frac{1-\theta}{p_0} + \frac{\theta}{p_1} \tag{1.2.39}$$

we have from [TR]

$$[H_{p_0}^{s_0}(R^n), H_{p_1}^{s_1}(R^n)]_{\theta,p} = H_p^s(R^n). \tag{1.2.40}$$

Also, by [TR, Theorem 2.4.1(c)], for the same range of parameters,

$$(B_{p_0,p_0}^{s_0}(R^n), B_{p_1,p_1}^{s_1}(R^n))_{\theta,p} = B_{p,p}^s(R^n),\ 0 < \theta < 1, \tag{1.2.41}$$

where s, p are given by (1.2.39).

Sobolev embeddings for fractional order spaces. Basing on [TR, 4.6.1] we recall the generalization of the Sobolev embedding theorem to fractional order spaces.

Proposition 1.2.1. *For $0 \le t \le s < +\infty$, $1 < q \le p < +\infty$ and $\Omega = R^n$ or Ω bounded having the extension property (see Section 9.1) the following continuous embeddings are valid:*

$$W^{s,q}(\Omega) \subset W^{t,p}(\Omega) \quad \textit{whenever} \quad s - \frac{n}{q} \ge t - \frac{n}{p}. \tag{1.2.42}$$

When Ω is bounded and $t < s$, the condition $q \le p$ can be neglected.

Embeddings in $C^k(\overline{\Omega})$ ($W^{k,\infty}(\Omega)$). Embeddings of the Sobolev spaces in $C^k(\overline{\Omega})$ or $W^{k,\infty}(\Omega)$ are of special importance for our further studies. Using the result of [TR, Theorem 4.6.1(e)], we claim an embedding

$$W^{s,q}(\Omega) \subset C^{k+\mu}(\overline{\Omega}), \tag{1.2.43}$$

valid whenever $s - \frac{n}{q} \ge k + \mu$ ($s \in R$, $q \in (1,+\infty)$, $k \in N$, $\mu \in (0,1)$), and we require strict inequality $s - \frac{n}{q} > k$ when $\mu = 0$. Here $\Omega \subset R^n$ is assumed to be a Lipschitz domain. We remark that (1.2.43) holds also for $\Omega = R^n$.

Agmon inequalities. Finally we recall the interpolation estimate due to S. Agmon (see [AG], [TE 1, p. 50]), valid for bounded domains $\Omega \subset R^n$ with C^n smooth boundary $\partial\Omega$:

(1.2.44)
$$\begin{cases} \|\phi\|_{L^\infty(\Omega)} \leq const. \|\phi\|^{\frac{1}{2}}_{H^{\frac{n}{2}-1}(\Omega)} \|\phi\|^{\frac{1}{2}}_{H^{\frac{n}{2}+1}(\Omega)}, & \phi \in H^{\frac{n}{2}+1}(\Omega), n \text{ even}, \\ \|\phi\|_{L^\infty(\Omega)} \leq const. \|\phi\|^{\frac{1}{2}}_{H^{\frac{n-1}{2}}(\Omega)} \|\phi\|^{\frac{1}{2}}_{H^{\frac{n+1}{2}}(\Omega)}, & \phi \in H^{\frac{n+1}{2}}(\Omega), n \text{ odd}. \end{cases}$$

1.2.3. Nirenberg-Gagliardo type inequalities. In many estimates of the present book the leading role is played by the *interpolation inequality* that was stated by L. Nirenberg and E. Gagliardo in 1959. The proof of it can be found in [FR 1, p. 27], [TA 2, p. 111], also in [HE 1, p. 37] or directly in [NI 1]. We follow the formulation of [FR 1].

Theorem 1.2.2. *(Nirenberg-Gagliardo inequality) Let $m \in N$, $\Omega \subset R^n$ be a domain having the C^m smooth boundary $\partial\Omega$. Let $v \in W^{m,r}(\Omega) \cap L^q(\Omega)$ with $1 \leq r, q \leq +\infty$. Then, for any integer j, $0 \leq j < m$, and any number $\theta \in [\frac{j}{m}, 1)$, define*

$$\frac{1}{p} = \frac{j}{n} + \theta\left(\frac{1}{r} - \frac{m}{n}\right) + (1-\theta)\frac{1}{q}. \tag{1.2.45}$$

If $m - j - \frac{n}{r}$ is not a nonnegative integer, then

$$\|D^j u\|_{L^p(\Omega)} \leq const. \|u\|^{\theta}_{W^{m,r}(\Omega)} \|u\|^{1-\theta}_{L^q(\Omega)}, \tag{1.2.46}$$

where D^j denotes any partial derivative of order j and const. depends on Ω, r, q, m, j, θ. If $m - j - \frac{n}{r}$ is a nonnegative integer, then (1.2.46) holds with $\theta = \frac{j}{m}$.

Extension of Nirenberg-Gagliardo inequality to fractional spaces. Such an extension of Theorem 1.2.2 follows from the deep interpolation results given in [TR] and was formulated in [AM 1].

Proposition 1.2.2. *Let $1 < p, p_0, p_1 < +\infty$, $0 \leq s_0 < s_1 < +\infty$, $\theta \in (0,1)$ and s, p be given by (1.2.39); then*

$$\forall_{v \in W^{s_1,p_1}(\Omega) \cap W^{s_0,p_0}(\Omega)} \ \|v\|_{W^{s^-,p}(\Omega)} \leq c_\theta \|v\|^{\theta}_{W^{s_1,p_1}(\Omega)} \|v\|^{1-\theta}_{W^{s_0,p_0}(\Omega)}, \tag{1.2.47}$$

where

(1.2.48)
$$s^- \begin{cases} = s & \text{if } s_0, s, s_1 \in N \text{ or } s_0, s, s_1 \in R \setminus N, \\ < s \text{ and arbitrarily close to } s & \text{otherwise}. \end{cases}$$

This proposition is valid for an arbitrary domain Ω in R^n having the extension property (see Section 9.1); in particular for $\Omega = R^n$.

Proof. First, the case $\Omega = R^n$ will be considered. We shall recall the following inclusions (see [TR, 2.3.3]):

$$(1.2.49) \quad H_p^{s+\varepsilon}(R^n),\ W^{s+\varepsilon,p}(R^n) \subset B^s_{p,p}(R^n) \subset H_p^{s-\varepsilon}(R^n),\ W^{s-\varepsilon,p}(R^n),$$

valid for any $-\infty < s < +\infty$, $1 < p < +\infty$ and $\varepsilon > 0$.

When $s_0, s, s_1 \in N$ and $H^{s_0}_{p_0}(R^n), H^s_p(R^n), H^{s_1}_{p_1}(R^n)$ are Bessel spaces, the estimate (1.2.47) follows directly from (1.2.40) and the *moments inequality* [TR, Theorem 1.3.3(g)], also for all $s_0, s, s_1 \in R \setminus N$ we will stay inside the scale of Besov spaces and (1.2.47) follows from (1.2.41) and [TR, Theorem 1.3.3(g)].

For arbitrary $-\infty < s_0 < s_1 < +\infty$ (1.2.41) implies

$$(1.2.50) \quad \forall_{v\in B^{s_1}_{p_1,p_1}(R^n)\cap B^{s_0}_{p_0,p_0}(R^n)} \ \|v\|_{B^s_{p,p}(R^n)} \le c\|v\|^{\theta}_{B^{s_1}_{p_1,p_1}(R^n)}\|v\|^{1-\theta}_{B^{s_0}_{p_0,p_0}(R^n)}.$$

For $0 < s_0 < s_1 < +\infty$ we can choose $\varepsilon > 0$ arbitrary small and such that $s_0 - \varepsilon$, $s_1 - \varepsilon$ and $s - \varepsilon = (1-\theta)(s_0 - \varepsilon) + \theta(s_1 - \varepsilon)$ are not natural numbers, so that by (1.2.50), for p, s given by (1.2.39)

$$\forall_{v\in W^{s-\varepsilon,p_0}(R^n)\cap W^{s_1-\varepsilon,p_1}(R^n)} \ \|v\|_{W^{s-\varepsilon,p}(R^n)} \le c\|v\|^{\theta}_{W^{s_1-\varepsilon,p_1}(R^n)}\|v\|^{1-\theta}_{W^{s_0-\varepsilon,p_0}(R^n)}.$$

Using (1.2.49) to estimate the right hand side of the inequality above we obtain

$$\forall_{v\in W^{s_0,p_0}(R^n)\cap W^{s_1,p_1}(R^n)} \ \|v\|_{W^{s-\varepsilon,p}(R^n)} \le c'\|v\|^{\theta}_{W^{s_1,p_1}(R^n)}\|v\|^{1-\theta}_{W^{s_0,p_0}(R^n)}.$$

Since ε was arbitrarily close to 0 this means precisely that

$$\forall_{v\in W^{s_0,p_0}(R^n)\cap W^{s_1,p_1}(R^n)} \ \|v\|_{W^{s^-,p}(R^n)} \le c'\|v\|^{\theta}_{W^{s_1,p_1}(R^n)}\|v\|^{1-\theta}_{W^{s_0,p_0}(R^n)},$$

where $s^- < s$ is arbitrarily close to s.

If $s_0 = 0$, choosing again $\varepsilon > 0$ arbitrarily small and such that $s_1 - \varepsilon$ and $s - \varepsilon$ are not natural numbers we find by (1.2.41) for p, s given by (1.2.39), that

$$\forall_{v\in B^{-\varepsilon}_{p_0,p_0}(R^n)\cap W^{s_1-\varepsilon,p_1}(R^n)} \ \|v\|_{W^{s-\varepsilon,p}(R^n)} \le c\|v\|^{\theta}_{W^{s_1-\varepsilon,p_1}(R^n)}\|v\|^{1-\theta}_{B^{-\varepsilon}_{p_0,p_0}(R^n)}$$

(the case $s_0 = 0$ needs to be distinguished because the space $W^{-\varepsilon,p_0}(R^n)$ was not defined), and applying (1.2.49) on the right hand side, we get

$$(1.2.51) \quad \forall_{v\in W^{0,p_0}(R^n)\cap W^{s_1,p_1}(R^n)} \ \|v\|_{W^{s^-,p}(R^n)} \le c'\|v\|^{\theta}_{W^{s_1,p_1}(R^n)}\|v\|^{1-\theta}_{W^{0,p_0}(R^n)}$$

as before. Now the estimate (1.2.47) will be extended from R^n to an arbitrary domain $\Omega \subset R^n$ having the *extension property* (Definition 9.1.1). The proof is complete. □

Remark 1.2.1. As a consequence of the Sobolev type inclusion reported in Proposition 1.2.1,

$$(1.2.52) \quad \begin{gathered} W^{s,q}(\Omega) \subset W^{t,p}(\Omega), \\ s - \frac{n}{q} \ge t - \frac{n}{p},\ s,t \in R,\ 1 < q \le p < +\infty, \end{gathered}$$

the main estimate of Proposition 1.2.2 will be extended to

$$(1.2.53) \quad \|v\|_{W^{t,p}(\Omega)} \le c\|v\|^{\theta}_{W^{s_1,p_1}(\Omega)}\|v\|^{1-\theta}_{W^{s_0,p_0}(\Omega)}$$

where we need to assume that

$$t - \frac{n}{p} < (\leq) (1-\theta)\left(s_0 - \frac{n}{p_0}\right) + \theta\left(s_1 - \frac{n}{p_1}\right) \tag{1.2.54}$$

and

$$\frac{1}{p} \leq \frac{1-\theta}{p_0} + \frac{\theta}{p_1} \tag{1.2.55}$$

(here $\theta \in [0,1]$, $0 \leq s_0 < s_1 < +\infty$, $1 < p, p_0, p_1 < +\infty$). Note that the condition $t \leq s = (1-\theta)s_0 + \theta s_1$, needed for validity of the Sobolev embedding, follows from (1.2.54), (1.2.55).

For $0 < \theta < 1$ the equality sign in (1.2.54) is permitted under the additional restrictions described in (1.2.48).

For $\theta = 1$ the estimate (1.2.53) is a direct consequence of Proposition 1.2.1, provided that $t \leq s_1$, $t - \frac{n}{p} \leq s_1 - \frac{n}{p_1}$ and $1 < p_1 \leq p < +\infty$. Note that if Ω is bounded weaker assumptions can be taken, as described in Proposition 1.2.1. The case $\theta = 0$ in (1.2.52) is quite similar and follows from Proposition 1.2.1 with $t \leq s_0$, $t - \frac{n}{p} \leq s_0 - \frac{n}{p_0}$, $1 < p_0 \leq p < +\infty$ (and, as before, with appropriately weakened assumptions for bounded Ω).

1.2.4. Properties of uniformly strongly elliptic operators. The primary problem in the semigroup approach to differential equations is to show that the operator appearing in the main part is closable and, what is even much more important, to determine explicitly the domain of its closed linear extension. This usually requires advanced studies of elliptic operators leading to suitable Calderón-Zygmund type estimates.

Example considerations of this type are contained in the following lemma.

Lemma 1.2.10. *The differential Laplace operator $\Delta : L^2(R^n) \to L^2(R^n)$ considered on $C_0^\infty(R^n)$ is closable in $L^2(R^n)$ and the domain of its closed linear extension is equal to $H^2(R^n)$.*

Proof. A key step in this proof is to show for $\phi \in C_0^\infty(R^n)$ the estimate of the type

$$const.\|\phi\|_{H^2(R^n)} \leq \|\phi\|_{L^2(R^n)} + \|\Delta\phi\|_{L^2(R^n)} \leq const.'\|\phi\|_{H^2(R^n)}. \tag{1.2.56}$$

The right inequality in (1.2.56) is a simple consequence of the definition of the $H^2(R^n)$ norm. Indeed, for $\phi \in C_0^\infty(R^n)$ we have

$$\|\phi\|_{L^2(R^n)} + \|\Delta\phi\|_{L^2(R^n)} \leq \|\phi\|_{L^2(R^n)} + \sum_{i=1}^{n} \left\|\frac{\partial^2\phi}{\partial x_i^2}\right\|_{L^2(R^n)} \leq \|\phi\|_{H^2(R^n)},$$

which shows that right hand side inequality in (1.2.56) holds with *const.'* $= 1$.

Consider next the Fourier transform F, which is known to be an isomorphism on the space $\mathcal{S}(R^n) \subset C^\infty(R^n)$ of complex valued functions *rapidly decreasing at infinity*, and denote by F^{-1} its inverse on $\mathcal{S}(R^n)$. For $\phi \in C_0^\infty(R^n)$, $r,s = 1,\dots,n$, we have $F\frac{\partial^2\phi}{\partial x_r \partial x_s} = -\xi_r\xi_s F\phi$ and $F[(I-\Delta)\phi] = (1+|\xi|^2)F\phi$. Since in addition, F preserves the $L^2(R^n)$ norm, we obtain

(1.2.57)
$$\begin{aligned}\left\|\frac{\partial^2\phi}{\partial x_r \partial x_s}\right\|_{L^2(R^n)} &= \left\|F\frac{\partial^2\phi}{\partial x_r \partial x_s}\right\|_{L^2(R^n)} = \|(-\xi_r\xi_s)F\phi\|_{L^2(R^n)}\\ &\leq \|(1+|\xi|^2)F\phi\|_{L^2(R^n)} = \|F(I-\Delta)\phi\|_{L^2(R^n)} = \|(I-\Delta)\phi\|_{L^2(R^n)}.\end{aligned}$$

For $r = 1,\dots,n$, using the elementary inequality $|\xi_r| \leq (1+|\xi|^2)$ and estimating as in (1.2.57), we get similarly

(1.2.58)
$$\begin{aligned}\left\|\frac{\partial\phi}{\partial x_r}\right\|_{L^2(R^n)} &= \left\|F\frac{\partial\phi}{\partial x_r}\right\|_{L^2(R^n)} = \|i\xi_r F\phi\|_{L^2(R^n)}\\ &\leq \|(1+|\xi|^2)F\phi\|_{L^2(R^n)} = \|F(I-\Delta)\phi\|_{L^2(R^n)} = \|(I-\Delta)\phi\|_{L^2(R^n)}.\end{aligned}$$

Collecting (1.2.57) and (1.2.58) we obtain the crucial estimate

$$\|\phi\|_{H^2(R^n)} \leq (1+n+n^2)\|(I-\Delta)\phi\|_{L^2(R^n)}, \quad \phi \in C_0^\infty(R^n), \tag{1.2.59}$$

which proves the left inequality in (1.2.56) with *const.* $= \frac{1}{1+n+n^2}$.

Denote by G the graph of $\Delta : C_0^\infty(R^n) \to L^2(R^n)$ in $L^2(R^n) \times L^2(R^n)$. Based on (1.2.56) it is now possible to show that $cl_{L^2(R^n)\times L^2(R^n)}G$ is a graph of some closed linear extension of $\Delta : C_0^\infty(R^n) \to L^2(R^n)$ and to determine its domain.

To verify that $cl_{L^2(R^n)\times L^2(R^n)}G$ is a graph of a linear operator take $\{\phi_n\} \subset C_0^\infty(R^n)$ for which there exist $L^2(R^n)$ limits $\phi_n \to 0$ and $\Delta\phi_n \to \chi$. From the left inequality in (1.2.56) $\{\phi_n\}$ is a Cauchy sequence in $H^2(R^n)$ so that $\|\phi_n\|_{H^2(R^n)} \to 0$. But the right estimate in (1.2.56) ensures then that $\|\Delta\phi_n\|_{L^2(R^n)} \to 0$. Therefore $\chi = 0$ which shows the operator $\Delta : C_0^\infty(R^n) \to L^2(R^n)$ is closable.

Consider further $(g_1, g_2) \in cl_{L^2(R^n)\times L^2(R^n)}G$ and take $\{\phi_n\} \subset C_0^\infty(R^n)$ such that $(\phi_n, \Delta\phi_n) \in G$ and $\phi_n \to g_1$, $\Delta\phi_n \to g_2$ in $L^2(R^n)$. As a consequence of (1.2.56), $\{\phi_n\}$ is then a Cauchy sequence in $H^2(R^n)$. Therefore, g_1 belongs to $H^2(R^n)$, which shows that the domain of the closed extension of $\Delta : C_0^\infty(R^n) \to L^2(R^n)$ is contained in $H^2(R^n)$.

To prove that the converse inclusion holds, take $g \in H^2(R^n)$ and a sequence $\{\phi_n\} \subset C_0^\infty(R^n)$ for which $\phi_n \to g$ in $H^2(R^n)$. From the estimate (1.2.56) it is then seen that $\{\Delta\phi_n\}$ is a Cauchy sequence in $L^2(R^n)$, and hence there exists $g_0 \in L^2(R^n)$ such that

$$(\phi_n, \Delta\phi_n) \overset{L^2(R^n)\times L^2(R^n)}{\longrightarrow} (g, g_0) \in cl_{L^2(R^n)\times L^2(R^n)}G.$$

The domain of a closed linear extension of $\Delta : C_0^\infty(R^n) \to L^2(R^n)$ is thus equal to $H^2(R^n)$ and the proof is complete. □

Remark 1.2.2. Let Ω be a bounded domain in R^n. Conditions (1.2.59), (1.2.56) ensure validity of the estimate

$$\frac{1}{1+n+n^2}\|\phi\|_{H^2(\Omega)} \leq \|\phi\|_{L^2(\Omega)} + \|\Delta\phi\|_{L^2(\Omega)} \leq \|\phi\|_{H^2(\Omega)}, \ \phi \in C_0^\infty(\Omega).$$

As seen from the inequality above, the operator

$$\Delta : C_0^\infty(\Omega) \to L^2(\Omega)$$

is closable in $L^2(\Omega)$ and, via density arguments, the domain of its *smallest closed extension* (see [YO 1, p. 78]) is equal to $H_0^2(\Omega)$.

Regular elliptic boundary value problems. Consider now the linear $2m$-th order differential operator

$$A = \sum_{|\sigma|\leq 2m} a_\sigma(x) D^\sigma \tag{1.2.60}$$

whose (complex or real) coefficients a_σ are uniformly continuous in $\overline{\Omega}$ and $\Omega \subset R^n$, $n \geq 2$, is a bounded domain with the boundary $\partial\Omega$ of the class C^{2m}. Assume that the following *uniform strong ellipticity condition* holds:

$$\exists_{c>0}\ \forall_{x\in\Omega}\ \forall_{\xi\in R^n}\ (-1)^m Re\left[\sum_{|\sigma|=2m} a_\sigma(x)\xi^\sigma\right] \geq c|\xi|^{2m}. \tag{1.2.61}$$

Then, as shown in [FR 1, Theorem 18.1], for $u \in H^{2m}(\Omega) \cap H_0^m(\Omega)$ we have

$$const.\|u\|_{H^{2m}(\Omega)} \leq \|Au\|_{L^2(\Omega)} + \|u\|_{L^2(\Omega)} \leq const.'\|u\|_{H^{2m}(\Omega)},$$

which guarantees that

$$A : H^{2m}(\Omega) \cap H_0^m(\Omega) \to L^2(\Omega)$$

is a closed operator in $L^2(\Omega)$. Furthermore,

$$\text{the range of } A + \lambda I \text{ is the whole of } L^2(\Omega) \tag{1.2.62}$$

provided that $\lambda > \lambda_0$ and λ_0 is chosen sufficiently large (see [FR 1, Theorem 18.2]).

This may be generalized for A acting in $L^p(\Omega)$, $p \in (1, +\infty)$, under more general boundary conditions (A and Ω as above). Take the set $\{B_j, j = 1, \ldots, m\}$ of m_j-th order ($m_j \in \{0, 1, \ldots, 2m-1\}$) boundary operators $B_j = \sum_{|\sigma|\leq m_j} b_\sigma^j(x) D^\sigma$, where $b_\sigma^j \in C^{2m-m_j}(\partial\Omega)$ for $|\sigma| \leq m_j$, $m_i \neq m_j$ for $i \neq j$, $i, j = 1, \ldots, m$, and consider

$$A : \{\phi \in C^{2m}(\overline{\Omega}) : \ B_1\phi_{|\partial\Omega} = \ldots = B_m\phi_{|\partial\Omega} = 0\} \to L^p(\Omega).$$

Since A is uniformly strongly elliptic, the extended *roots condition* holds (see [TA 2, Section 5.2]), that is:

EXTENDED ROOTS CONDITION. *Let $x \in \partial\Omega$, $N(x)$ denote the outward unit normal vector to $\partial\Omega$ at x and H_x be the hyperplane tangent to $\partial\Omega$ at x. The extended roots condition is satisfied, if for arbitrary $x \in \partial\Omega$, any nonzero $\xi \in H_x$ and each complex λ lying on the ray $\arg\lambda = \theta$ where $\theta \in [\frac{\pi}{2}, \frac{3\pi}{2}]$ is arbitrary, the polynomial $p(z) =$*

$\sum_{|\sigma|=2m} a_\sigma(x)(\xi+zN(x))^\sigma-(-1)^m\lambda$ *has exactly* m *roots* $z_1^+(x,\xi,\lambda),\dots,z_m^+(x,\xi,\lambda)$ *with positive imaginary parts.*

Assume further that the *strong complementary condition* is satisfied, i.e.:

STRONG COMPLEMENTARY CONDITION. *Let* $x\in\partial\Omega$, $N(x)$ *denote the outward unit normal vector to* $\partial\Omega$ *at* x *and* H_x *be the hyperplane tangent to* $\partial\Omega$ *at* x. *The strong complementary condition is satisfied, if for each* $x\in\partial\Omega$, *any vector* $\xi\in H_x$ *and each complex* λ *lying on the ray* $\arg\lambda=\theta$ *where* $(\xi,\lambda)\neq(0,0)$ *and* $\theta\in[\frac{\pi}{2},\frac{3\pi}{2}]$ *is arbitrary, the polynomials* $P_j(z)$ $(j=1,\dots,m)$, $P_j(z)=\sum_{|\sigma|=m_j} b^j_\sigma(x)(\xi+zN(x))^\sigma$, *are linearly independent modulo the polynomial* $Q(z)=(z-z_1^+(x,\xi,\lambda))\dots(z-z_m^+(x,\xi,\lambda))$, *where the roots* $z_i^+(x,\xi,\lambda)$ *were defined in the roots condition.*

Taking $\xi=0$ in the preceeding condition one obtains the *normality condition* (see [TA 2, Section 5.2]).

NORMALITY CONDITION. *If* $i\neq j$ *then* $m_i\neq m_j$ *and*

$$\sum_{|\sigma|=m_j} b^j_\sigma(x)\,(N(x))^\sigma\neq 0,\ x\in\partial\Omega,\ j=1,\dots,m,$$

where $N(x)$ *is the outward unit normal vector to* $\partial\Omega$ *at* x.

As mentioned above we connect with the order of operators A, B_j, $j=1,\dots,m$, the following *smoothness condition.*

SMOOTHNESS CONDITION. *We assume that*

- $a_\sigma\in C(\overline{\Omega})$ *for* $|\sigma|\le 2m$,
- $b^j_\sigma\in C^{2m-m_j}(\partial\Omega)$ *for* $|\sigma|\le m_j$, $j=1,\dots,m$.

Definition 1.2.1. *If the triple* $(A,\{B_j\},\Omega)$ *is such that* A *is a uniformly strongly elliptic operator,* $\partial\Omega$ *and coefficients of* A, B_j $(j=1,\dots,m)$ *are smooth in the sense introduced above and, moreover, the strong complementary condition holds, then* $(A,\{B_j\},\Omega)$ *is called a* ***regular elliptic boundary value problem.***

Remark 1.2.3. The notion of the regular elliptic boundary value problem in Definition 1.2.1 is stronger than the original such notion appearing e.g. in [FR 1] or [TR]. Avoiding detailed explanation (see [TA 1, Section 5.2]) we only remark that our interest here is not only the estimate (1.2.63) (which holds for an elliptic operator A under the usual *roots, normality* and *complementing* conditions; [FR 1], [TR]) but formulas (1.2.65) and (1.2.66) which require A to be *uniformly strongly elliptic* and to satisfy the *strong complementary condition.*

As a particular consequence of the above assumptions on the triple $(A,\{B_j\},\Omega)$ the elements of $\{\phi\in C^{2m}(\overline{\Omega}):\ B_1\phi_{|\partial\Omega}=\dots=B_m\phi_{|\partial\Omega}=0\}$ fulfill the estimate (see [FR 1, Theorem 19.1])

$$\text{(1.2.63)}\quad const.\|u\|_{W^{2m,p}(\Omega)}\le\|Au\|_{L^p(\Omega)}+\|u\|_{L^p(\Omega)}\le const.'\|u\|_{W^{2m,p}(\Omega)}.$$

This ensures that A considered on the domain $D(A) = W^{2m,p}_{\{B_j\}}(\Omega)$, where

(1.2.64)
$$W^{2m,p}_{\{B_j\}}(\Omega) := cl_{W^{2m,p}(\Omega)}\{\phi \in C^{2m}(\overline{\Omega}) : B_1\phi_{|\partial\Omega} = \ldots = B_m\phi_{|\partial\Omega} = 0\},$$

is a closed operator in $L^p(\Omega)$. Furthermore, as follows from [FR 1, Theorem 19.4], there exist *const.* > 0 and $\Lambda_0 > 0$ such that for each complex λ with $Re\,\lambda < -\Lambda_0$

$$(1.2.65) \qquad (\lambda I - A) \text{ takes } W^{2m,p}_{\{B_j\}}(\Omega) \text{ onto } L^p(\Omega)$$

and

$$(1.2.66) \quad \sum_{j=0}^{2m} |\lambda|^{\frac{(2m-j)}{2m}} \|u\|_{W^{j,p}(\Omega)} \le const. \|(\lambda I - A)u\|_{L^p(\Omega)}, \ u \in W^{2m,p}_{\{B_j\}}(\Omega).$$

See [TR, 5.3.4] for the generalization of (1.2.63), also [TR, 5.5.1] for the description of properties of A considered on the fractional order spaces. In the latter case the spaces $H^s_{p,\{B_j\}}(\Omega)$, $B^s_{p,p\{B_j\}}(\Omega)$, $s > 0$, $p \in (1, +\infty)$ may be introduced by

$$(1.2.67) \qquad H^s_{p,\{B_j\}}(\Omega) = \{\phi \in H^s_p(\Omega) : \forall_{i \in \{j:\ m_j < s - \frac{1}{p}\}} \ B_i\phi_{|\partial\Omega} = 0\},$$

$$(1.2.68) \qquad B^s_{p,p,\{B_j\}}(\Omega) = \{\phi \in B^s_{p,p}(\Omega) : \forall_{i \in \{j:\ m_j < s - \frac{1}{p}\}} \ B_i\phi_{|\partial\Omega} = 0\}$$

(see [TR, §4.3.3]), which allows us to generalize (1.2.64) to the formula

$$(1.2.69) \qquad W^{s,p}_{\{B_j\}}(\Omega) = \{\phi \in W^{s,p}(\Omega) : \forall_{i \in \{j:\ m_j < s - \frac{1}{p}\}} \ B_i\phi_{|\partial\Omega} = 0\}.$$

Compactness of the resolvent. Condition (1.2.66) ensures that $\lambda I - A$ is an invertible operator in $L^p(\Omega)$ for each complex λ with $Re\,\lambda < -\Lambda_0$, and $(\lambda I - A)^{-1}$ is bounded. As a consequence of (1.2.65) the inverse $(\lambda I - A)^{-1}$ is seen to be defined on the whole of $L^p(\Omega)$, so that the half plane $\{\lambda \in C : Re\,\lambda < \Lambda_0\}$ is contained in the resolvent set $\rho(A)$.

From the point of view of further applications to the problems in bounded subdomains of R^n an important property of the elliptic operators is the *compactness of the resolvent.*

Definition 1.2.2. *Let X be a Banach space and $A : D(A) \to X$ a linear operator in X. We say that A has* **compact resolvent** *if $(\lambda I - A)^{-1} : X \to X$ is a compact map for each $\lambda \in \rho(A)$.*

Let us consider $\lambda_1, \lambda_2 \in \rho(A)$ and the operators $(\lambda_1 I - A)^{-1}$, $(\lambda_2 I - A)^{-1}$. The *resolvent equation* (see [YO 1, p. 211]) then reads

$$(1.2.70) \qquad (\lambda_1 I - A)^{-1} = \left[(\lambda_2 - \lambda_1)(\lambda_1 I - A)^{-1} + I\right](\lambda_2 I - A)^{-1},$$

which shows that:

Observation 1.2.1. *The resolvent of A is compact if and only if $(\lambda I - A)^{-1}$ is compact for some $\lambda \in \rho(A)$.*

Considering again the regular elliptic boundary value problem $(A, \{B_j\}, \Omega)$ we have from (1.2.66) the estimate

$$\|(\lambda I - A)^{-1} v\|_{W^{2m,p}(\Omega)} \leq const. \|v\|_{L^p(\Omega)}, \; Re\,\lambda < \Lambda_0.$$

Therefore, as a consequence of the Rellich-Kondrachov theorem (see [AD, Theorem 6.2], Section 1.2.2), we get the compactness of $(\lambda I - A)^{-1}$ in $L^p(\Omega)$ for $Re\,\lambda < \Lambda_0$. From this and Observation 1.2.1 we thus obtain the following conclusion.

Proposition 1.2.3. *For $(A, \{B_j\}, \Omega)$ being a regular elliptic boundary value problem introduced in Definition* 1.2.1 *the resolvent of A is compact.*

1.3. Sectorial operators

The notion of *sectorial operator* is of an unquestionable importance in contemporary studies of differential equations. A large number of phenomena in the applied sciences are described by systems of equations with a sectorial operator in the main part. The problems falling into this class may be treated within the same theoretical approach, very useful for the description of evolution and asymptotic behavior of these systems.

For the purpose of the following definition let us denote by $S_{a,\phi}$, where $a \in R$ and $\phi \in (0, \frac{\pi}{2})$, a sector of the complex plane given by the relation (see [HE 1, Definition 1.3.1])

$$S_{a,\phi} := \{\lambda \in C : \; \phi \leq |arg(\lambda - a)| \leq \pi, \; \lambda \neq a\}. \tag{1.3.1}$$

Definition 1.3.1. *Consider a linear, closed and densely defined operator $A : X \supset D(A) \to X$ acting in a Banach space X. Then A is a* **sectorial operator** *in X if and only if there exist $a \in R$, $\phi \in (0, \frac{\pi}{2})$ and $M > 0$ such that*

$$\text{the resolvent set } \rho(A) \text{ contains the sector } S_{a,\phi}, \tag{1.3.2}$$

$$\|(\lambda I - A)^{-1}\|_{\mathcal{L}(X,X)} \leq \frac{M}{|\lambda - a|}, \quad \text{for each } \lambda \in S_{a,\phi}. \tag{1.3.3}$$

Remark 1.3.1. Choose an arbitrary $\omega \in R$. It is easy to see that A is sectorial if and only if $A_\omega := A + \omega I$ is sectorial. Indeed, via a translation argument, $S_{a,\phi} \subset \rho(A)$ and (1.3.3) holds if and only if

$$S_{a+\omega,\phi} = \{\lambda' \in C : \; \phi \leq |arg\big(\lambda' - (a + \omega)\big)| \leq \pi, \; \lambda' \neq a + \omega\} \subset \rho(A_\omega)$$

and, simultaneously,

$$\|(\lambda' I - A_\omega)^{-1}\|_{\mathcal{L}(X,X)} \leq \frac{M}{|\lambda' - (a + \omega)|}, \quad \text{for each } \lambda' \in S_{a+\omega,\phi}.$$

In addition, we also have

$$Re\,\sigma(A_\omega) \geq a + \omega,$$

so that it is always possible to choose $\omega > 0$ for which $Re\,\sigma(A_\omega) > 0$ and consider A_ω as a *positive operator.*

Equivalent sectoriality conditions. Referring to Definition 1.3.1, let us list some equivalent conditions for A to be sectorial.

Proposition 1.3.1. *Let $A : X \supset D(A) \to X$ be a linear closed and densely defined operator in a Banach space X and consider the operators $A_\omega = A + \omega I$ with $\omega \in R$. Then the following conditions are equivalent:*

(a) A_ω is sectorial in X for some $\omega \in R$,
(b) A_ω is sectorial in X for each $\omega \in R$,
(c) There exist $k, \omega \in R$ such that the resolvent set $\rho(A_\omega)$ of A_ω contains a half plane $\{\lambda \in C:\ Re\,\lambda \le k\}$ and

$$\|\lambda(\lambda I - A_\omega)^{-1}\|_{\mathcal{L}(X,X)} \le M \quad for \quad Re\,\lambda \le k. \tag{1.3.4}$$

Proof. Certainly, (a) and (b) are equivalent as shown in Remark 1.3.1. Next, the implication $(a) \Longrightarrow (c)$ is formal. Indeed if A_{ω_0} fulfills conditions of Definition 1.3.1 then repeating the considerations of Remark 1.3.1 we find that $\mathcal{S}_{0,\phi} \subset \rho(A_{\omega_0} - aI)$ and $\|\lambda'\big(\lambda' - (A_{\omega_0} - aI)\big)^{-1}\|_{\mathcal{L}(X,X)} \le M$ for all $\lambda' \in \mathcal{S}_{0,\phi}$, so that (c) is satisfied with $\omega = \omega_0 - a$ and arbitrary $k < 0$.

Let us now proceed to the proof of $(c) \Longrightarrow (a)$, which is essential here (see [LU 1, Proposition 2.1.11]). First let us recall that if A is a closed linear operator in a nontrivial Banach space X then (see [ML, Chapter 3, Theorem 4.1])

$$\lambda_0 \in \rho(A) \text{ implies that } \mathcal{B}_{\lambda_0} \subset \rho(A), \tag{1.3.5}$$

where $\mathcal{B}_{\lambda_0}$ is the following open ball on a complex plane:

$$\mathcal{B}_{\lambda_0} := \Big\{\lambda \in C:\ |\lambda - \lambda_0| < \frac{1}{\|(\lambda_0 I - A)^{-1}\|_{\mathcal{L}(X,X)}}\Big\}. \tag{1.3.6}$$

Furthermore,

$$(\lambda I - A)^{-1} = \sum_{n=0}^{\infty}(\lambda_0 - \lambda)^n \left[(\lambda_0 I - A)^{-1}\right]^{n+1} \quad \text{for } \lambda \in \mathcal{B}_{\lambda_0}. \tag{1.3.7}$$

We may assume without loss of generality that $k < 0$ in (1.3.4), so that

$$\{\lambda \in C:\ Re\,\lambda \le k\} \subset \rho(A_\omega) \tag{1.3.8}$$

and

$$\|(\lambda I - A_\omega)^{-1}\|_{\mathcal{L}(X,X)} \le \frac{M}{|\lambda|} \le \frac{M}{|\lambda - k|} \quad \text{for } Re\,\lambda \le k, \lambda \ne k. \tag{1.3.9}$$

Define

$$\mathcal{S}_{k,\arctan(2M)} = \{\lambda \in C:\ \arctan(2M) \le |arg\big(\lambda - k\big)\big| \le \pi,\ \lambda \ne k\}$$

and take $\lambda \in \mathcal{S}^{<}_{k,\arctan(2M)} := \mathcal{S}_{k,\arctan(2M)} - \{\lambda \in C : \ Re\,\lambda \leq k\}$. Then

$$|Re(\lambda - k)| = \frac{|Im\lambda|}{|\tan(arg(\lambda - k))|} \leq \frac{|Im\lambda|}{2M} \tag{1.3.10}$$

so that, from (1.3.5) and assumption (c),

$$\lambda \in \mathcal{B}\left(k + iIm\lambda, \frac{|k + iIm\lambda|}{M}\right) \subset \rho(A_\omega) \ \text{ for all } \lambda \in \mathcal{S}^{<}_{k,\arctan(2M)}. \tag{1.3.11}$$

Connecting (1.3.7), (1.3.9), (1.3.10) and the simple inequality

$$|Im(\lambda - k)|^2 \left(1 + \frac{1}{(2M)^2}\right) \geq |\lambda - k|^2$$

following from (1.3.10), we obtain

(1.3.12)

$$\begin{aligned}
&\|(\lambda I - A_\omega)^{-1}\|_{\mathcal{L}(X,X)} \\
&\leq \sum_{n=0}^{\infty} |\lambda - (k + iIm\lambda)|^n \left\| \left((k + iIm\lambda)I - A_\omega\right)^{-1} \right\|^{n+1}_{\mathcal{L}(X,X)} \\
&\leq \sum_{n=0}^{\infty} |Re\,\lambda - k|^n \frac{M^{n+1}}{|k + iIm\lambda|^{n+1}} = \frac{M}{|k + iIm\lambda|} \sum_{n=0}^{\infty} \left[\frac{|Re\,\lambda - k|}{|k + iIm\lambda|} M\right]^n \\
&\leq \frac{M}{|Im\lambda|} \sum_{n=0}^{\infty} \left[\frac{|Re\,\lambda - k|}{|Im\lambda|} M\right]^n \leq \frac{M}{|Im\lambda|} \sum_{n=0}^{\infty} \left[\frac{1}{2M} M\right]^n \\
&= \frac{2M}{|Im\lambda|} \leq \frac{2\sqrt{1 + \frac{1}{(2M)^2}}\,M}{|\lambda - k|} \quad \text{for } \ \lambda \in \mathcal{S}^{<}_{k,\arctan(2M)}.
\end{aligned}$$

Based on assumption (1.3.8) and (1.3.11), it is thus seen that $\mathcal{S}_{k,\arctan(2M)} \subset \rho(A_\omega)$ whereas (1.3.9), (1.3.12) imply that

$$\|(\lambda I - A_\omega)^{-1}\|_{\mathcal{L}(X,X)} \leq \frac{2\sqrt{1 + \frac{1}{(2M)^2}}\,M}{|\lambda - k|} \quad \text{for } \ \lambda \in \mathcal{S}_{k,\arctan(2M)}. \tag{1.3.13}$$

The proof is complete. □

The well known result from semigroup theory is that sectorial operators coincide with the negative generators of *analytic semigroups.*

Definition 1.3.2. *Let X be a Banach space and $\{T(t)\}$ a C^0 semigroup in X. We say that $\{T(t)\}$ is an* **analytic semigroup**, *if there exist a sector of the complex plane*

$$\mathcal{S} = \{z \in C : \ \phi_1 < arg\,z < \phi_2\}$$

($\phi_1 < 0 < \phi_2$), and a family of bounded, linear operators $T(z) : X \to X$, $z \in \mathcal{S}$, coinciding with $T(t)$ for $t \geq 0$, and such that

- *$z \to T(z)$ is analytic in $\mathcal{S}$,*
- *$\lim_{z \to 0, z \in \mathcal{S}} T(z)x = x$ for every $x \in X$,*

- $T(z_1 + z_2) = T(z_1)T(z_2)$ *for* $z_1, z_2 \in \mathcal{S}$.

Then we have

Theorem 1.3.1. *Let* X *be a Banach space. Then a densely defined linear operator* A *is a negative generator of an analytic semigroup* $\{T(t)\}$ *of bounded linear operators* $T(t) : X \to X$, $t \geq 0$, *if and only if* A *is a sectorial operator in* X.

The proof that minus a sectorial operator generates an analytic semigroup may be found in [HE 1, Theorem 1.3.4]. For the converse implication note first that $\{T(t)\}$ is an analytic semigroup if and only if it is strongly differentiable for $t > 0$ and

$$\exists_{C_1>0}\, \exists_{\omega_1>0}\, \|AT(t)\|_{\mathcal{L}(X,X)} \leq \frac{C_1}{t} e^{\omega_1 t}, \ t > 0$$

(see [PAZ 2, p. 65]). Therefore, for some $\omega_2 \in R$, $\{e^{\omega_2 t}T(t)\}$ fulfills the assumptions of [PAZ 2, Theorem 5.2 (d)]. Consequently, by [PAZ 2, Theorem 5.2], there exists $\omega \in R$ such that $A + \omega I$ is a sectorial operator, which ensures that A is sectorial itself (see Proposition 1.3.1).

The above theorem follows also from [FA, Theorem 4.2.1].

Remark 1.3.2. In the context of Definitions 1.1.1, 1.3.1 and 1.3.2 the denseness assumption in Theorem 1.3.1 is superfluous. However, some authors consider semigroups with generators not densely defined (see [A-T], [LU 1]). In that case examples are known (see [SIN]) of semigroups that are analytic (in a weaker sense than in Definition 1.3.2) for $t > 0$ but *not* strongly continuous.

From now on the notation $\{e^{-At}\}$ will be used for the analytic semigroup corresponding to the infinitesimal generator $-A$.

1.3.1. Two basic estimates. Especially interesting in further applications will be the case $Re\, \sigma(A) > 0$, that is when elements of the spectrum have positive real parts. Then the estimate of the $\|\cdot\|_{\mathcal{L}(X,X)}$ norm of the semigroup operators shows that the process decays at infinity. In consequence, the powers A^α of A may be defined, which will be of particular importance for further considerations of an abstract differential equation in X. We thus cite the following result (see [HE 1, Theorem 1.3.4]).

Theorem 1.3.2. *Let* A *be a sectorial operator in a Banach space* X *such that* $Re\, \sigma(A) > a > 0$. *Then, for some positive constants* c_0, c_1,

$$\|e^{-At}\|_{\mathcal{L}(X,X)} \leq c_0 e^{-at}, \ t \geq 0, \tag{1.3.14}$$

$$\|Ae^{-At}\|_{\mathcal{L}(X,X)} \leq \frac{c_1}{t} e^{-at}, \ t > 0. \tag{1.3.15}$$

In the considerations concerning power operators A^α we also use results coming from various sources, like the original papers by H. Komatsu [KO 1], [KO 2] or the key monograph by H. Triebel [TR]. The remark below justifies that condition (1.3.43) formulated further in Subsection 1.3.3 allows us to use the results mentioned.

Remark 1.3.3. As a consequence of the Hille-Yosida theorem (e.g. [PAZ 2, Theorem 5.3, p. 20], [TR, §1.13.1]), the condition (1.3.14) implies in particular that the interval $(-a, +\infty)$ is contained in $\rho(-A)$ and

$$\|(-\lambda I - A)^{-1}\|_{\mathcal{L}(X,X)} \leq \frac{M}{\lambda + a}, \; \lambda > -a.$$

Since

$$\frac{M}{\lambda + a} \leq \frac{M}{\lambda + 1}, \; a \geq 1, \; \lambda \geq 0,$$
$$\frac{M}{\lambda + a} < \frac{M}{a(\lambda + 1)}, \; a \in (0,1), \; \lambda \geq 0,$$

we obtain the estimate

$$\|(-\lambda I - A)^{-1}\|_{\mathcal{L}(X,X)} \leq \max\left\{M, \frac{M}{a}\right\} \frac{1}{\lambda + 1}, \; \lambda \geq 0. \tag{1.3.16}$$

Consequently, setting $t = -\lambda$, we then have

$$\begin{cases} (-\infty, 0] \subset (-\infty, a) \subset \rho(A) \text{ and} \\ \|(tI - A)^{-1}\|_{\mathcal{L}(X,X)} \leq \frac{const.}{1+|t|}, \; t \in (-\infty, 0]; \end{cases} \tag{1.3.17}$$

i.e. A is a *positive operator* in the sense of [TR, §1.14.1]. Furthermore, (1.3.14) and (1.3.15) ensure together that A fulfills the requirements of Definition 1.3.1 with $a = 0$, so that

$$\begin{cases} \textit{the resolvent set } \rho(A) \textit{ contains the sector } S_{0,\phi}, \\ \textit{and } \|(\lambda I - A)^{-1}\|_{\mathcal{L}(X,X)} \leq \frac{M}{|\lambda|}, \textit{ for } \lambda \in S_{0,\phi}, \end{cases} \tag{1.3.18}$$

where $\phi \in (0, \frac{\pi}{2})$ and $M > 0$ are fixed (see [KO 1, Theorem 12.2], also [FR 1, Problem (5), p. 108]).

The above remark and [HE 1, Excercise 7, Chapter I] yield the following conclusion.

Corollary 1.3.3. *Consider a linear, closed and densely defined operator $A : X \supset D(A) \to X$ acting in a Banach space X. Then A is a sectorial operator with $Re(\sigma(A)) > 0$ if and only if both* (1.3.17) *and* (1.3.18) *are satisfied.*

1.3.2. Examples of sectorial operators. We start with two simple examples.

Example 1.3.1. Each bounded linear operator defined on a Banach space X is sectorial.

Proof. If A is a bounded linear operator in X then $\{\lambda \in C : |\lambda| > \|A\|_{\mathcal{L}(X,X)}\} \subset \rho(A)$ and $(\lambda I - A)^{-1} = \sum_{n=0}^{\infty} \frac{A^n}{\lambda^{n+1}}$ (see [ML, Lemma 4.2]). In particular the half plane $\{\lambda \in C : Re\,\lambda \leq -2\|A\|_{\mathcal{L}(X,X)}\}$ is contained in $\rho(A)$ and

$$\|\lambda(\lambda I - A)^{-1}\|_{\mathcal{L}(X,X)} \leq \sum_{n=0}^{\infty} \left(\frac{\|A\|_{\mathcal{L}(X,X)}}{|\lambda|}\right)^n \leq \sum_{n=0}^{\infty} \left(\frac{1}{2}\right)^n = 2.$$

Therefore A is sectorial as a result of Proposition 1.3.1. □

Example 1.3.2. If X, Y are Banach spaces and A is sectorial in X, B is sectorial in Y, then the product operator $(A, B) : D(A) \times D(B) \to X \times Y$, where $(A, B)(x, y) = (Ax, By)$, is sectorial in $X \times Y$.

Proof. Let $\mathcal{S}_{a,\phi_A}$, $\mathcal{S}_{b,\phi_B}$ be the sectors chosen respectively for A and B based on Definition 1.3.1. Define $a' = \min\{a, b\}$, $\mathcal{J} = (A, B)$ and $\mathcal{J}_{-a'} = (A - a'I, B - a'I)$. Then $\{\lambda \in C : Re\,\lambda < 0\}$ is a subset of $\rho(\mathcal{J}_{-a'})$ and

$$\|(\lambda I - \mathcal{J}_{-a'})^{-1}\|_{\mathcal{L}(X\times Y, X\times Y)} \leq \frac{const.}{|\lambda|} \quad \text{for} \quad Re\,\lambda < 0.$$

Hence (A, B) is sectorial in $X \times Y$ as a result of Proposition 1.3.1. □

In many special problems one needs to deal with the perturbations of a generator A of an analytic semigroup. We shall prove below that whenever a perturbation B fulfills an appropriate subordination condition (1.3.19), the *perturbed operator* $A + B$ will still be a generator of an analytic semigroup (see [PAZ 2, Section 3.2], [HE 1, Theorem 1.3.2]).

Proposition 1.3.2. *(Perturbation result) Let $A : D(A) \to X$ be a sectorial operator in a Banach space X and consider a closed, linear operator $B : D(B) \to X$ such that $D(A) \subset D(B) \subset X$ and B is subordinated to A according to the condition*

$$\|Bv\|_X \leq c\|Av\|_X + c'\|v\|_X, \; v \in X. \tag{1.3.19}$$

If the condition (1.3.19) *holds with $c \leq M_0$ (where $M_0 > 0$ is defined in* (1.3.25)*), then the perturbed operator $A + B$ with $D(A + B) = D(A)$ is sectorial in X.*

Proof. Based on Proposition 1.3.1 take $k < 0$, $\omega \in R$ such that, for $A_\omega = A + \omega I$,

$$\{\lambda \in C : Re\,\lambda \leq k\} \subset \rho(A_\omega) \tag{1.3.20}$$

and

$$\|\lambda(\lambda I - A_\omega)^{-1}\|_{\mathcal{L}(X,X)} \leq M \quad \text{for} \quad Re\,\lambda \leq k. \tag{1.3.21}$$

Condition (1.3.19) then reads

$$\|Bv\|_X \leq c\|A_\omega v\|_X + c''\|v\|_X, \; v \in X, \tag{1.3.22}$$

where $c'' = c' + c|\omega|$.

The crucial step of the proof is to show that

$$\exists_{k_0 \le k} \forall_{Re\,\lambda \le k_0}\ 1 \in \rho\Big(B(\lambda I - A_\omega)^{-1}\Big). \tag{1.3.23}$$

For this we shall use (1.3.22) and (1.3.21) estimating as follows:

$$\begin{aligned}
\|B(\lambda I - A_\omega)^{-1}v\|_X &\le c\|A_\omega(\lambda I - A_\omega)^{-1}v\|_X + c''\|(\lambda I - A_\omega)^{-1}v\|_X \\
&\le c\|(\lambda I - A_\omega)(\lambda I - A_\omega)^{-1}v\|_X + c\|\lambda(\lambda I - A_\omega)^{-1}v\|_X \\
&\quad + c''\|(\lambda I - A_\omega)^{-1}v\|_X \\
&\le \left(c(1+M) + \frac{c''M}{|\lambda|}\right)\|v\|_X,\ v \in X.
\end{aligned} \tag{1.3.24}$$

Under the restrictions

$$c \le \frac{1}{2(1+M)} =: M_0,\ 4c''M \le |\lambda| \tag{1.3.25}$$

we have further from (1.3.24) the inequality

$$\|B(\lambda I - A_\omega)^{-1}\|_{\mathcal{L}(X,X)} \le \frac{3}{4}. \tag{1.3.26}$$

Since the spectral radius of a bounded linear operator on X does not exceed its $\mathcal{L}(X,X)$ norm (see [YO 1, Chapter VIII, Section 2]), condition (1.3.26) shows that the number 1 is in the resolvent set of $B(\lambda I - A_\omega)^{-1}$, i.e. (1.3.23) is proved with

$$k_0 := \min\{k, -4c''M\}. \tag{1.3.27}$$

For $Re\,\lambda \le k_0$ we then have

$$\begin{aligned}
[\lambda I - (A_\omega + B)]^{-1} &= \left[\left(I - B(\lambda I - A_\omega)^{-1}\right)(\lambda I - A_\omega)\right]^{-1} \\
&= (\lambda I - A_\omega)^{-1}\left(I - B(\lambda I - A_\omega)^{-1}\right)^{-1},
\end{aligned} \tag{1.3.28}$$

which, in the presence of (1.3.21), (1.3.26) leads to the estimate

$$\begin{aligned}
&\|\,[\lambda I - (A_\omega + B)]^{-1}\,\|_{\mathcal{L}(X,X)} \\
&\le \|(\lambda I - A_\omega)^{-1}\|_{\mathcal{L}(X,X)}\|\left(I - B(\lambda I - A_\omega)^{-1}\right)^{-1}\|_{\mathcal{L}(X,X)} \le \frac{4M}{|\lambda|}.
\end{aligned} \tag{1.3.29}$$

Since $A_\omega + B = (A+B)_\omega$, the operator $A+B$ fulfills the requirements of Proposition 1.3.1(c), which completes the proof of Proposition 1.3.2. □

Example 1.3.3. If A is sectorial in a Banach space X and B is bounded and linear on X, then the perturbed operator $A+B$ is sectorial.

Proof. If B is bounded and linear in X, then (1.3.19) holds with $c = 0$ and $c' = \|B\|_{\mathcal{L}(X,X)}$. The operator $A + B$ is thus sectorial as a consequence of Proposition 1.3.2. □

The following result supplies a large number of examples of sectorial operators.

Proposition 1.3.3. *Let $A : H \supset D(A) \to H$ be a densely defined, linear, self-adjoint operator in a Hilbert space H. If, in addition, A is bounded below in H, that is*

$$\exists_{m\in R}\ \forall_{x\in D(A)}\ \langle Ax, x\rangle_H \geq m\|x\|_H^2, \tag{1.3.30}$$

then A is a sectorial operator in H.

Proof. We shall prove this proposition independently of Proposition 1.3.1 showing directly the validity of Definition 1.3.1. Let us recall that the spectrum of a selfadjoint operator is contained in the real axis. Moreover, since A is bounded below, $\sigma(A)$ must be contained in the interval $[m, \infty)$. This implies in particular that the sector

$$\mathcal{S}_{m,\frac{\pi}{4}} := \left\{\lambda \in C : \frac{\pi}{4} \leq |arg(\lambda - m)| \leq \pi,\ \lambda \neq m\right\} \tag{1.3.31}$$

is contained in the complement of $\sigma(A)$. Therefore, $\mathcal{S}_{m,\frac{\pi}{4}}$ is a subset of $\rho(A)$ and (1.3.2) is satisfied. We shall now prove, for $\lambda \in \mathcal{S}_{m,\frac{\pi}{4}}$, the validity of the estimate (1.3.3).

Let $\lambda \in \mathcal{S}_{m,\frac{\pi}{4}}$ and take $\lambda' := \lambda - m$. Only the following two cases are possible.
Case 1: $\lambda = \lambda' + m$, where $Re(\lambda') < 0$. In this case, since $A - mI$ must be symmetric and nonnegative whereas $-2Re(\lambda') > 0$, we obtain that

$$\begin{aligned}\|(\lambda I - A)x\|_H^2 &= \|\big(\lambda' I - (A - mI)\big)x\|_H^2\\ &= |\lambda'|^2\|x\|_H^2 - 2Re(\lambda')\langle(A - mI)x, x\rangle_H + \|(A - mI)x\|_H^2\\ &\geq |\lambda'|^2\|x\|_H^2.\end{aligned} \tag{1.3.32}$$

Case 2: $\lambda = \lambda' + m$, where $|Im(\lambda')| \geq |Re(\lambda')|$. We then have

$$\begin{aligned}\|(\lambda I - A)x\|_H^2 &= \left\|\big(\lambda' I - (A - mI)\big)x\right\|_H^2\\ &= |Im(\lambda')|^2\|x\|_H^2 + \left\|\big(Re(\lambda')I - (A - mI)\big)x\right\|_H^2\\ &\geq |Im(\lambda')|^2\|x\|_H^2 \geq \frac{|\lambda'|^2}{2}\|x\|_H^2.\end{aligned} \tag{1.3.33}$$

As a result of (1.3.32) and (1.3.33) it is seen that

$$\|(\lambda I - A)x\|_H \geq \frac{|\lambda - m|}{\sqrt{2}}\|x\|_H, \quad \text{for each } \lambda \in \mathcal{S}_{m,\frac{\pi}{4}},\ x \in D(A), \tag{1.3.34}$$

which is the counterpart of (1.3.3). The proof is complete. □

Remark 1.3.4. Let us recall that the linear operator A in a Hilbert space H is *self-adjoint* whenever it is symmetric and its range $R(A)$ coincides with H. Note also that, although boundedness of A from below suffices to prove that A satisfying the assumptions of Proposition 1.3.3 is sectorial, the most important in applications will be the case when A is *positive definite*, i.e. the constant m in condition (1.3.30) is positive. This latter requirement is connected with the need of considerations of fractional powers of A which are not defined, unless $Re\ \sigma(A) > 0$.

Example 1.3.4. An unbounded operator $I - \Delta$ considered in $L^2(R^n)$ with the domain $D(I-\Delta) = H^2(R^n)$ is positive definite and sectorial.

Proof. For functions $\phi, \psi \in C_0^\infty(R^n)$ take an open ball $B_{R^n}(r)$ containing their supports and use twice the integration by parts formula to obtain

(1.3.35)
$$\begin{aligned}\int_{R^n}(I-\Delta)\phi\,\psi dx &= \int_{R^n}\phi\psi dx - \int_{B_{R^n}(r)}\Delta\phi\,\psi dx\\ &= \int_{R^n}\phi\psi dx - \int_{B_{R^n}(r)}\phi\,\Delta\psi dx = \int_{R^n}\phi(I-\Delta)\psi dx.\end{aligned}$$

Furthermore, boundedness of $(I-\Delta)$ from below is a consequence of the following estimate:

(1.3.36)
$$\begin{aligned}\int_{R^n}(I-\Delta)\phi\,\phi dx &= \int_{R^n}\phi^2 dx - \int_{B_{R^n}(r)}\Delta\phi\,\phi dx\\ &= \int_{R^n}\phi^2 dx + \int_{B_{R^n}(r)}|\nabla\phi|^2 dx \geq \int_{R^n}\phi^2 dx,\ \phi \in C_0^\infty(R^n).\end{aligned}$$

Consider next the Fourier transform F and recall that F is an isomorphism on the set $\mathcal{S}(R^n) \subset C^\infty(R^n)$ of rapidly decreasing functions. Denote by F^{-1} its inverse on $\mathcal{S}(R^n)$, take $\phi \in C_0^\infty(R^n)$ and define the element

$$h := \left(F^{-1}\frac{1}{1+|\xi|^2}F\right)\phi. \tag{1.3.37}$$

It is seen that $h \in \mathcal{S}$ and

(1.3.38)
$$\begin{aligned}F[(I-\Delta)h] &= Fh - F[\Delta h]\\ &= \left(\frac{1}{1+|\xi|^2}\right)F\phi + |\xi|^2\left(\frac{1}{1+|\xi|^2}\right)F\phi = F\phi.\end{aligned}$$

Also, by application of F^{-1} to both sides of (1.3.38),

$$(I-\Delta)h = \phi. \tag{1.3.39}$$

Since $C_0^\infty(R^n)$ is dense in $H^2(R^n)$ the conditions (1.3.35), (1.3.36) justify that $I-\Delta$ is a symmetric and positive definite operator on $L^2(R^n)$. Formulas (1.3.38), (1.3.39) show further that the range $R(I-\Delta)$ contains $C_0^\infty(R^n)$ and, therefore, this range is dense in $L^2(R^n)$. This proves directly that $1 \in \rho(\Delta)$. Since the Laplacian considered in $L^2(R^n)$ on the domain $H^2(R^n)$ is closed, the resolvent operator $(I-\Delta)^{-1}$ is bounded and closed, that is its domain $D((I-\Delta)^{-1})$ is closed in $L^2(R^n)$. This shows that

$$R(I-\Delta) = D((I-\Delta)^{-1}) = L^2(R^n) \tag{1.3.40}$$

and, as a consequence of Proposition 1.3.3 (see Remark 1.3.4), $I-\Delta : H^2(R^n) \to L^2(R^n)$ is sectorial in $L^2(R^n)$. □

Example 1.3.5. As shown directly in [HE 1, Section 1.6], the negative Laplacian considered in $L^p(R^n)$, $p \in (1,+\infty)$, with the domain $D(-\Delta) = W^{2,p}(R^n)$ is sectorial and its spectrum coincides with the interval $[0,+\infty)$. Therefore, $(I-\Delta)$ is sectorial and positive in $L^p(R^n)$ (i.e. the elements of its spectrum have positive real parts).

Example 1.3.6. $-\Delta : H^2(\Omega) \cap H_0^1(\Omega) \to L^2(\Omega)$, where $\partial\Omega \in C^2$, is sectorial and positive definite.

Proof. Using elliptic theory (see (1.2.62)) choose $\lambda_0 > 0$ so that $-\Delta + \lambda_0 I$ be onto $L^2(\Omega)$. Integration by parts shows that $-\Delta + \lambda_0 I$ is symmetric and positive definite and hence sectorial as a result of Proposition 1.3.3. Therefore, $-\Delta$ is itself a sectorial operator as an immediate consequence of Definition 1.3.1. From the Poincaré inequality it is seen that $-\Delta$ is positive definite in $L^2(\Omega)$. □

Example 1.3.7. $\Delta^2 : H^4(\Omega) \cap H_0^2(\Omega) \to L^2(\Omega)$, where $\partial\Omega \in C^4$, is sectorial and positive definite.

Proof. The proof is the same as in Example 1.3.6, although positive definiteness of Δ^2 follows from the Smoller inequality (see [SM, Theorem 11.11]):

$$\|\Delta\phi\|^2_{L^2(\Omega)} \geq \mu_1^N \|\nabla u\|^2_{L^2(\Omega)} \geq \mu_1^N \mu_1^D \|\phi\|^2_{L^2(\Omega)}, \quad \phi \in H_0^2(\Omega), \tag{1.3.41}$$

where μ_1^D denotes the smallest positive eigenvalue of $-\Delta$ on Ω with the Dirichlet boundary condition and μ_1^N is a similar number in the Neumann case. □

Example 1.3.8. If the triple $(A, \{B_j\}, \Omega)$ forms a regular elliptic boundary value problem as described in Subsection 1.2.4 then, for some $\lambda_0 > 0$, the operator $A + \lambda_0 I$ acting in $L^p(\Omega)$ with the domain $D(A+\lambda_0 I) = W^{2m,p}_{\{B_j\}}(\Omega)$ is sectorial and $Re\,\sigma(A+\lambda_0 I) > 0$.

Proof. From (1.2.65) and (1.2.66) it is seen that the resolvent $(\lambda I - A)^{-1}$ exists for all $\lambda \in C$ with $Re\,\lambda \leq -\Lambda_0$ and

$$\|(\lambda I - A)^{-1}\|_{\mathcal{L}(L^p(\Omega),L^p(\Omega))} \leq \frac{const.}{|\lambda|}$$

which, based on Proposition 1.3.1, is sufficient for validity of Definition 1.3.1. As a further result of Proposition 1.3.1, $A + \omega I$ is sectorial in $L^p(\Omega)$ for each $\omega \in R$. Therefore, $\lambda_0 > 0$ may be chosen, for which $A + \lambda_0 I$ becomes positive in $L^p(\Omega)$, i.e. the condition $Re\,\sigma(A+\lambda_0 I) > 0$ is satisfied (see Remark 1.3.1). □

Example 1.3.9. Since $(\Delta^2, \{\frac{\partial}{\partial N}, \frac{\partial \Delta}{\partial N}\}, \Omega)$ forms a regular elliptic boundary value problem, Δ^2 considered with the domain $D(\Delta^2) = H^4_{\frac{\partial}{\partial N}, \frac{\partial \Delta}{\partial N}}(\Omega)$ ($\partial\Omega$ of class C^4) is sectorial in $L^2(\Omega)$. More precisely, integration by parts shows that $\Delta^2 + \delta I$ ($\delta > 0$) is symmetric and positive definite so that $\Delta^2 + \delta I$ is a selfadjoint operator in $L^2(\Omega)$.

Example 1.3.10. Consider in $[L^p(\Omega)]^d$ a second order vector valued operator of the form (see [CO], [AL 2, §3])

$$Au = \sum_{i,j=1}^{n} (A_{ij}(x)u_{x_i})_{x_j} + \sum_{i=1}^{n} B_i(x)u_{x_i} \tag{1.3.42}$$

acting on $u = (u_1, \ldots, u_d)$, where $\Omega \subset R^n$ is a bounded smooth domain, and $A_{ij} = [a_{ij}^{\mu\nu}]$, $B_i = [b_i^{\mu\nu}]$ are symmetric $N \times N$ matrices. Assume that

(i) for some $c_0 > 0$ and any real nN-tuple $q = (q_i^\mu)$ with $\mu = 1, \ldots, N$, $i = 1, \ldots, n$,

$$\sum_{i,j=1}^{n} \sum_{\mu,\nu=1}^{N} a_{ij}^{\mu\nu}(x) q_i^\mu q_j^\nu \geq c_0 |q|^2, \; x \in \overline{\Omega},$$

(ii) $A_{ij} = A_{ji}$, for all $i, j = 1, \ldots, n$,
(iii) components $a_{ij}^{\mu\nu}$ of A_{ij} matrices belong to $C^1(\overline{\Omega})$ and B_i components $b_i^{\mu\nu}$ are the elements of $C^0(\overline{\Omega})$.

Then $-A$ with the domain $W^{2,p}(\Omega) \cap W_0^{1,p}(\Omega)$ is sectorial in $L^p(\Omega)$ (see [CO, p. 608]).

Remark 1.3.5. A wide class of triples $(A, \{B_j\}, \Omega)$ forming regular elliptic boundary value problems has been described in [AM 1, Theorem 6.6]. As was pointed out later in [AM 4] considerations concerning item (ii) of this theorem contained too weak assumptions.

1.3.3. Fractional powers of sectorial operators.

Assume that X is a Banach space and

$$A \text{ is a sectorial operator in } X \text{ with } Re\, \sigma(A) > 0. \tag{1.3.43}$$

A satisfying (1.3.3) will be called a *sectorial, positive operator* (see Remark 1.3.3). In particular, a self-adjoint, positive definite operator in a Hilbert space satisfies (1.3.3).

Theorem 1.3.2 allows us to define, for $\alpha \in (0, +\infty)$, the operators $A^{-\alpha} : X \to X$ by the integral formula (see [HE 1], [FR 1])

$$A^{-\alpha} v = \frac{1}{\Gamma(\alpha)} \int_0^{+\infty} t^{\alpha-1} e^{-At} v dt. \tag{1.3.44}$$

Proposition 1.3.4. *Suppose that* (1.3.43) *holds.* $A^{-\alpha}$, $\alpha \in (0,+\infty)$, *are well defined linear bounded operators on* X *giving a one-to-one correspondence between* X *and the range* $R(A^{-\alpha})$. *Also,* $A^{-1}: D(A) \to X$ *coincides with the inverse of* A *and*

$$A^{-\alpha}A^{-\beta} = A^{-(\alpha+\beta)}, \quad \text{for } \alpha,\beta > 0.$$

As seen above, each $A^{-\alpha}$, $\alpha > 0$, is invertible. We further denote these inverse operators by A^{α} and use the symbol X^{α} for the domain of definition of A^{α}, that is $X^{\alpha} := R(A^{-\alpha})$. We also extend the notion of the power operator to the case when $\alpha = 0$, taking $A^0 := I$ (identity map) on $X^0 := X$.

The following results are well known (see [FR 1, Part 2, Section 14], [HE 1, Section 1.4]).

Proposition 1.3.5. *Let* (1.3.43) *be satisfied. Then* X^{α} $(\alpha \in [0,+\infty))$ *with the norm* $\|v\|_{X^{\alpha}} := \|A^{\alpha}v\|_X$ *are Banach spaces whereas* $A^{\alpha}: X^{\alpha} \to X$ *are linear closed and densely defined operators in* X *satisfying*

$$A^{\alpha}A^{\beta} = A^{\beta}A^{\alpha} = A^{\alpha+\beta}, \quad \alpha,\beta \geq 0. \tag{1.3.45}$$

Furthermore, X^{α} *is a dense subset of* X^{β} *for* $\alpha \geq \beta \geq 0$, *the inclusions*

$$X^{\alpha} \subset X^{\beta}, \quad \alpha > \beta \geq 0, \tag{1.3.46}$$

are dense and continuous and, additionally, they are compact provided that A *has compact resolvent.*

Proposition 1.3.6. *Assume that* (1.3.43) *holds. For* $\alpha \geq 0$, $t \geq 0$, $A^{\alpha}e^{-At}$ *is a bounded linear operator on* X, *such that*

$$\|A^{\alpha}e^{-At}\|_{\mathcal{L}(X,X)} \leq c_{\alpha}\frac{e^{-at}}{t^{\alpha}}, \quad t > 0, \tag{1.3.47}$$

$$A^{\alpha}e^{-At} = e^{-At}A^{\alpha} \quad \text{on } X^{\alpha},\ t \geq 0;$$

here $a > 0$ *is such that* $Re\,\sigma(A) > a$.

For a sectorial operator A with $Re\,\sigma(A) > 0$ the following rule of raising a power A^{β} to a power α holds (see [KO 1, Theorem 10.6], [WAT]).

Proposition 1.3.7. *Let* A *satisfy the conditions of Definition* 1.3.1 *with some* $a \in R$ *and* $\phi \in (0,\frac{\pi}{2})$, *let also* $Re(\sigma(A)) > 0$. *Then we have*

$$(A^{\beta})^{\alpha} = A^{\alpha\beta}, \tag{1.3.48}$$

for arbitrary $\beta \in (0,\frac{\pi}{\phi})$ *and* $\alpha > 0$.

We remark that (1.3.48) may not be true in general. For example, $(A^2)^{\frac{1}{2}}$ may not be equal to A for A^2 and A generating C^0 semigroups (see [YO 2] for details). However, if A satisfies (1.3.43), then half of the angle opening ϕ of its sector $\mathcal{S}_{a,\phi}$ fulfills $0 < \phi < \frac{\pi}{2}$. Therefore, Proposition 1.3.7 yields in particular the formula

$$(A^2)^\alpha = A^{2\alpha}, \quad \text{for each } \alpha > 0. \tag{1.3.49}$$

Below we show that *sectoriality* of the operator A remains valid for A restricted to any of the fractional order spaces X^β, $\beta \geq 0$, provided it holds for A considered on the basic phase space X. This property will be useful in the studies of the smooth solutions to the abstract Cauchy problem (2.1.1).

Proposition 1.3.8. *Let X^1 be a domain of a sectorial operator $A_{|X} : X^1 \subset X \to X$ acting in a Banach space X and satisfying $Re\,\sigma(A_{|X}) > 0$. Let $\beta > 0$ and $A_{|X^\beta} : X^{1+\beta} \subset X^\beta \to X^\beta$, equipped with the domain $X^{1+\beta} \subset X^1$, denote the restriction of $A_{|X}$ to the fractional power space $X^\beta \subset X$ so that*

$$A_{|X}x = A_{|X^\beta}x \quad for \quad x \in X^{1+\beta}. \tag{1.3.50}$$

Then $A_{|X^\beta}$ is a sectorial operator in X^β and $Re\,\sigma(A_{|X^\beta}) > 0$.

Proof. From Proposition 1.3.5, the inclusion $X^{1+\beta} \subset X^\beta$ is dense for each $\beta > 0$. $A_{|X^\beta}$ is thus a linear and densely defined operator in X^β, $\beta > 0$. Furthermore, $A_{|X^\beta}$ is closed in X^β, which is the direct consequence of the closedness of $A_{|X}$ in X.

Let us now estimate the resolvent of $A_{|X^\beta}$ in X^β. Since $A_{|X}$ is sectorial in X, there are $a \in R$ and $\phi \in (0, \frac{\pi}{2})$ (a, ϕ fixed from now on) such that the sector $\mathcal{S}_{a,\phi}$ is contained in the resolvent set $\rho(A_{|X})$ of $A_{|X}$ and

$$\|(\lambda I - A_{|X})^{-1}x\|_X \leq \frac{M}{|\lambda - a|}\|x\|_X, \ \lambda \in \mathcal{S}_{a,\phi}, \ x \in X. \tag{1.3.51}$$

We shall show for the restriction $A_{|X^\beta}$ of $A_{|X}$ that

$$\mathcal{S}_{a,\phi} \subset \rho(A_{|X^\beta}),$$

$$\|(\lambda I - A_{|X^\beta})^{-1}y\|_{X^\beta} \leq \frac{M}{|\lambda - a|}\|y\|_{X^\beta}, \ \lambda \in \mathcal{S}_{a,\phi}, \ y \in X^\beta.$$

If $\lambda \in \mathcal{S}_{a,\phi}$, then to each $y \in X^\beta$ corresponds a unique $x \in X^1$ satisfying $(\lambda I - A_{|X})x = A^\beta_{|X}y$. Applying $A^{-\beta}_{|X}$ to both sides of the latter equality and noting that

$$A^{-\beta}_{|X}(\lambda I - A_{|X})x = (\lambda I - A_{|X})A^{-\beta}_{|X}x, \ x \in X^1, \tag{1.3.52}$$

we conclude, in the presence of (1.3.50), that the equation $(\lambda I - A_{|X^\beta})\tilde{x} = y$ has a unique solution $\tilde{x} \in X^{1+\beta}$ for each $y \in X^\beta$. Consequently, the inverse $(\lambda I - A_{|X^\beta})^{-1}$ is defined on X^β and, by (1.3.50),

$$(\lambda I - A_{|X^\beta})^{-1}y = (\lambda I - A_{|X})^{-1}y, \ y \in X^\beta. \tag{1.3.53}$$

Based on (1.3.53), (1.3.52) and the estimate (1.3.51), we verify finally that

$$
\begin{aligned}
\|(\lambda I - A_{|X^\beta})^{-1}y\|_{X^\beta} &= \|(\lambda I - A_{|X})^{-1}y\|_{X^\beta} \\
&= \|A^\beta_{|X}(\lambda I - A_{|X})^{-1}y\|_X = \|(\lambda I - A_{|X})^{-1}A^\beta_{|X}y\|_X \\
&\leq \frac{M}{|\lambda - a|}\|A^\beta_{|X}y\|_X = \frac{M}{|\lambda - a|}\|y\|_{X^\beta},\ \lambda \in S_{a,\phi},\ y \in X^\beta.
\end{aligned}
$$

Similar calculations show that $Re\,\sigma(A_{|X^\beta}) > 0$ if $Re\,\sigma(A_{|X}) > 0$. The proof is complete. □

Description of X^α spaces via interpolation arguments. Fix $\alpha \in (0,1)$ and $v \in D(A)$. Recalling (1.3.44), we find that

(1.3.54)

$$
\begin{aligned}
A^\alpha v = A^{-(1-\alpha)}A^{1-\alpha}A^\alpha v &= \frac{1}{\Gamma(1-\alpha)}\int_0^{+\infty} t^{-\alpha}e^{-At}A^{1-\alpha}A^\alpha v dt \\
&= \frac{1}{\Gamma(1-\alpha)}\int_0^{+\infty} t^{-\alpha}e^{-At}Av dt.
\end{aligned}
$$

Since A is a negative generator of an analytic semigroup such that $Re\,\sigma(A) > 0$, the resolvent of $-A$ may be written in the integral form

$$
(sI + A)^{-1}w = \int_0^{+\infty} e^{-st}e^{-At}w dt,\ w \in X,\ s \geq 0. \tag{1.3.55}
$$

Using (1.3.55) and the equality

$$
t^{-\alpha} = \frac{1}{\Gamma(\alpha)}\int_0^{+\infty} s^{\alpha-1}e^{-ts}ds
$$

we obtain further from (1.3.54)

(1.3.56)

$$
\begin{aligned}
A^\alpha v &= \frac{1}{\Gamma(1-\alpha)\Gamma(\alpha)}\int_0^{+\infty}\left(\int_0^{+\infty} s^{\alpha-1}e^{-ts}ds\right)e^{-At}Av dt \\
&= \frac{1}{\Gamma(1-\alpha)\Gamma(\alpha)}\int_0^{+\infty} s^{\alpha-1}\left(\int_0^{+\infty} e^{-ts}e^{-At}Av dt\right)ds \\
&= \frac{1}{\Gamma(1-\alpha)\Gamma(\alpha)}\int_0^{+\infty} s^{\alpha-1}(sI + A)^{-1}Av ds \\
&= \frac{1}{\Gamma(1-\alpha)\Gamma(\alpha)}\int_0^{+\infty} s^{\alpha-1}A(sI + A)^{-1}v ds.
\end{aligned}
$$

Formula (1.3.56) coincides for real α satisfying inequality $0 < \alpha < 1$ with the definition of fractional powers A^α of positive operators stated in [TR, Remark 1.15.1/2]. Following further the approach of [TR, 1.15.3], i.e. extending (1.3.56) to the formula (see [TR, 1.15.1(1)]) valid for α from the whole complex plane, we obtain the description of X^α as the complex *interpolation spaces* under the additional assumption that the purely imaginary powers A^{it} of A (with real t) are known to be uniformly bounded. The result [TR, Theorem 1.15.3]) then reads:

Proposition 1.3.9. *The following interpolation formula holds -*

$$[X^\alpha, X^\beta]_\theta = X^{(1-\theta)\alpha+\theta\beta}, \quad for \ \alpha,\beta \geq 0, \ \theta \in (0,1) \tag{1.3.57}$$

- together with the corresponding moments inequality

$$\|v\|_{X^{(1-\theta)\alpha+\theta\beta}} \leq c_\theta \|v\|^\theta_{X^\beta} \|v\|^{1-\theta}_{X^\alpha}, \ v \in X^\alpha \cap X^\beta \ \alpha,\beta \geq 0, \ \theta \in (0,1), \tag{1.3.58}$$

provided A is a sectorial operator in a Banach space X such that $Re\,\sigma(A) > 0$ and $\|A^{it}\|_{\mathcal{L}(X,X)} \leq const.(\varepsilon)$ for all $t \in [-\varepsilon,\varepsilon]$. Here $[\cdot,\cdot]_\theta$ denotes a complex interpolation space (see [TR, 1.9.3]*).*

Since $X = X^0$ and $D(A) = X^1$, choosing $\alpha = 0$, $\beta = 1$ in Proposition 1.3.9, we conclude that:

Corollary 1.3.4. *Under the assumptions of Proposition* 1.3.9, X^θ *$(\theta \in (0,1))$ are intermediate spaces between X and $D(A)$, that is,*

$$X^\theta = D(A^\theta) = [X, D(A)]_\theta \tag{1.3.59}$$

and

$$\|v\|_{X^\theta} \leq c_\theta \|Av\|^\theta_X \|v\|^{1-\theta}_X, \ v \in D(A). \tag{1.3.60}$$

Some applications of interpolation theory. The latter result is important in applications. In Subsection 1.3.2 we have introduced a number of sectorial positive operators. To see the action of Corollary 1.3.4 let us reconsider some previously studied examples.

Take in the Hilbert space $L^2(R^n)$ a positive definite, selfadjoint operator $A = I - \Delta$ with the domain $D(A) = H^2(R^n)$ (see Example 1.3.4). In this Hilbert case A^{-it} are unitary operators and, in particular, Proposition 1.3.9 holds (see [TR, 1.18.10]). Hence, condition (1.3.59) implies that

$$X^\alpha = [L^2(R^n), H^2(R^n)]_\alpha = H^{2\alpha}_2(R^n), \ \alpha \in (0,1),$$

where the second equality is written based on the characterization given in [TR, 2.4.2 (11)]. In the general case of $A = I - \Delta$ on $L^p(R^n)$ we can follow the calculations of [TR, 2.5.3] that lead to the formula

$$(I-\Delta)^\alpha \phi = F^{-1}(1+|\xi|^2)^\alpha F\phi, \ \phi \in S(R^n), \ \alpha > 0, \tag{1.3.61}$$

(F being the Fourier transform and $S(R^n)$ denoting the set of rapidly decreasing functions). Since $S(R^n)$ is dense in $H^{2\alpha}_p(R^n) = \{\phi \in L^p(R^n) : \|\phi\|_{H^{2\alpha}_p(R^n)} = \|F^{-1}(1+|\xi|^\alpha)F\phi\|_{L^p(R^n)} < \infty\}$ and $(I-\Delta)^\alpha$ is closed, we must have the inclusion $H^{2\alpha}_p(R^n) \subset D\big((I-\Delta)^\alpha\big)$. Based on the *lift property* (see [TR, Theorem 2.3.4]) we observe further that (1.3.61) holds for $\phi \in H^{2\alpha}_p(R^n)$ and, moreover,

$$\begin{aligned} X^\alpha &= (I-\Delta)^{-\alpha}\big(L^p(R^n)\big) = (I-\Delta)^{-\alpha}\left(F^{-1}(1+|\xi|^2)^\alpha F\big(H^{2\alpha}_p(R^n)\big)\right) \\ &= H^{2\alpha}_p(R^n), \ \alpha > 0. \end{aligned} \tag{1.3.62}$$

The above result can be generalized to elliptic operators on the whole of R^n whose top order coefficients are merely assumed to be bounded and uniformly continuous functions (see [P-S], [A-H-S]).

Unfortunately, the characterizations of the domains X^α, $\alpha \in (0,1)$, of fractional powers of elliptic operators on bounded domains Ω which are known in the literature are not yet fully satisfactory. Therefore, we introduce an extra assumption and then present the conditions under which this characterization holds.

α-Condition. A regular elliptic boundary value problem $(A, \{B_j\}, \Omega)$ is said to satisfy the *α-condition*, whenever

$$D(A^\alpha) = [L^p(\Omega), W^{2m,p}_{\{B_j\}}(\Omega)]_\alpha \subset W^{2m\alpha,p}(\Omega), \quad \alpha \in (0,1), \tag{1.3.63}$$

where $[\cdot,\cdot]_\alpha$ denotes a *complex interpolation space.*

Note that the α-condition holds if the imaginary powers of A are bounded. The first equality in (1.3.63) is then a consequence of Proposition 1.3.9 (see [TR, Section 1.15.3]). It is also known (see [TR, Section 4.3.1]) that, for bounded $\Omega \subset R^n$ satisfying the cone condition,

$$[L^p(\Omega), W^{2m,p}(\Omega)]_\alpha = H^{2m\alpha}_p(\Omega), \alpha \in (0,1).$$

Directly from the definition of the complex interpolation space (see [TR, Section 1.9.1]), we thus have the embedding

$$[L^p(\Omega), W^{2m,p}_{\{B_j\}}(\Omega)]_\alpha \subset [L^p(\Omega), W^{2m,p}(\Omega)]_\alpha, \quad \alpha \in (0,1),$$

leading to the inclusion in (1.3.63).

In the case of the higher $2m$-th order elliptic operators ($m > 1$), known results guaranteeing validity of the α-condition require that both the coefficients of the operators A, B_j, $j = 1, 2, \ldots, m$, and the boundary $\partial\Omega$ are of class C^∞ (see [SEE], [TR]). It seems that in many situations such restrictive assumptions are rather technical and may be relaxed. One may see [A-H-S], [P-S], [S-T], [GU 1], for recent developments within this field. We also remark that satisfactory (from the point of view of regularity assumptions) results are still not known in general and the problem attracts the attention of many mathematicians.

Below we describe three known sufficient conditions for validity of the α-condition. These cases cover all the examples reported later in Chapter 6.

CASE I

We start with the results of [SEE] (see also [TR]) concerning the regular elliptic boundary value problems $(A, \{B_j\}, \Omega)$ of arbitrary order $2m$ under the additional assumption, which will be valid throughout the whole of Case I, that the coefficients a_σ of A are in $C^\infty(\overline{\Omega})$, the coefficients b^j_σ of B_j belong to $C^\infty(\partial\Omega)$, and the boundary $\partial\Omega$ is of class C^∞.

Certainly, A is sectorial in $L^p(\Omega)$ and we may assume here without loss of generality that A is positive. For such problems, from [TR, Theorem 4.9.1], it follows that

$$\|A^{it}\|_{\mathcal{L}(L^p(\Omega),L^p(\Omega))} \le const.e^{const._1|t|}, \ t \in R.$$

Therefore, using (1.3.59) we obtain in this case that

$$X^\alpha = [L^p(\Omega), W^{2m,p}_{\{B_j\}}(\Omega)]_\alpha, \ \alpha \in (0,1) \tag{1.3.64}$$

(see [TR, Sec. 4.9.2], also the source paper [SEE]). However, the expected equality

$$[L^p(\Omega), W^{2m,p}_{\{B_j\}}(\Omega)]_\alpha = H^{2m\alpha}_{p,\{B_j\}}(\Omega), \ \alpha \in (0,1), \tag{1.3.65}$$

is not always true (see [TR, 4.3.3 (7)]) but holds under the additional requirement that

$$2m\alpha - \frac{1}{p} \ne m_j \ \text{ for all } \ j = 1, \dots, m. \tag{1.3.66}$$

Here m_j denotes the order of B_j, whereas $H^{2m\alpha}_{p,\{B_j\}}(\Omega)$ is defined as in (1.2.67) by the formula

$$H^{2m\alpha}_{p,\{B_j\}}(\Omega) := \{\phi \in H^{2m\alpha}_p(\Omega) : \forall_{i \in \{j:\ m_j < 2m\alpha - \frac{1}{p}\}} \ B_i\phi_{|\partial\Omega} = 0\}. \tag{1.3.67}$$

If (1.3.66) is violated, we have merely the proper inclusion

$$[L^p(\Omega), W^{2m,p}_{\{B_j\}}(\Omega)]_\alpha \subset H^{2m\alpha}_{p,\{B_j\}}(\Omega), \ \alpha \in (0,1); \tag{1.3.68}$$

and the structure of $[L^p(\Omega), W^{2m,p}_{\{B_j\}}(\Omega)]_\alpha$ in this case is more complicated (see [SEE], [TR, 4.3.3 (9)]).

Let A be a sectorial, positive operator on $L^q(\Omega)$ given by a regular $2m$-th order elliptic boundary value problem $(A, \{B_j\}, \Omega)$ with C^∞ smooth data. The domain of A then satisfies the inclusions

$$W^{2m,q}_0(\Omega) \subset D(A) = W^{2m,q}_{\{B_j\}}(\Omega) \subset W^{2m,q}(\Omega). \tag{1.3.69}$$

Using the characterization of the space X^α given in (1.3.64), (1.3.68) we find for all $\alpha \in [0,1]$ that

$$X^\alpha \subset W^{2m\alpha,q}(\Omega), \ \text{ if } \alpha \in [0,1] \text{ and either } q \ge 2 \text{ or } 2m\alpha \in N \tag{1.3.70}$$

(for the restriction on parameters in (1.3.70) recall that $H^s_q(\Omega) \subset B^s_{q,q}(\Omega)$ if and only if $q \ge 2$; see [TR, Remark 2.3.3/4, Theorem 4.6.1(b)]).

We can extend the embedding (1.3.70) to all values of $\alpha \ge 0$ using the regularity theory of elliptic operators. Based on the isomorphism property of A (see [TR, Theorem 5.5.1(b)] with $\lambda = 0$) we have the following condition:

(1.3.71)

$$\begin{aligned} &\text{if } s \ge 0,\ u \in D(A) \text{ and } Au \in W^{s,q}(\Omega), \text{ then } u \in W^{2m+s,q}(\Omega) \\ &\text{and } \|u\|_{W^{2m+s,q}(\Omega)} \le const.\|Au\|_{W^{s,q}(\Omega)}. \end{aligned}$$

The condition (1.3.71) allows us to justify inductively for $k = 1, 2, \dots$, that

$$\text{if } \alpha \in [k-1, k) \text{ and either } q \geq 2 \text{ or } 2m\alpha \in N, \quad \text{then } X^\alpha \subset W^{2m\alpha, q}(\Omega), \tag{1.3.72}$$

which ensures the validity of (1.3.70) for all $\alpha > 0$. Indeed, the case $k = 1$ is obvious from (1.3.70). For the induction assume that (1.3.72) holds for k and take $u \in X^\alpha$ where $\alpha \in [k, k+1)$. Then $\alpha = k + \theta_\alpha$ with $\theta_\alpha \in [0, 1)$, $u \in X^\alpha \subset D(A)$ and $Au \in X^{k-1+\theta_\alpha} \subset W^{2m(k-1+\theta_\alpha), q}(\Omega)$. Therefore, from (1.3.71), $u \in W^{2m(k+\theta_\alpha), q}(\Omega) = W^{2m\alpha, q}(\Omega)$, which completes the induction.

As a consequence of the above considerations we obtain

Proposition 1.3.10. *Let A be a sectorial operator on $L^q(\Omega)$, $1 < q < +\infty$, given by a regular elliptic boundary value problem $(A, \{B_j\}, \Omega)$ satisfying the α-condition. Then the following inclusions hold:*

$$X^\alpha \subset \begin{cases} W^{t,p}(\Omega) & \text{if } 2m\alpha - \frac{n}{q} \geq t - \frac{n}{p},\ 2 \leq q \leq p < +\infty, \\ C^{k+\mu}(\overline{\Omega}) & \text{if } 2m\alpha - \frac{n}{q} \geq k + \mu,\ k \in N,\ \mu \in (0,1), \end{cases} \tag{1.3.73}$$

for any $\alpha \in [0, 1]$. In the first embedding the condition $q \leq p$ will be neglected when $2m\alpha > t$. Furthermore, the assumption $q \geq 2$ may be changed to $q > 1$ whenever $2m\alpha \in N$. The second embedding holds also with $\mu = 0$, provided that the strict inequality $2m\alpha - \frac{n}{q} > k$ is satisfied. Also, (1.3.73) holds with all $\alpha > 0$ provided that Ω is a bounded C^∞ domain and coefficients of A and B_j $(j = 1, \dots, m)$ are infinitely many times continuously differentiable in $\overline{\Omega}$ and $\partial\Omega$ respectively.

Proof. The proof is a direct consequence of (1.3.70), Proposition 1.2.1 and the inclusion (1.2.43). For the requirement $q \geq 2$ note that for fractional $2m\alpha$ and $q \neq 2$ we have, by (1.2.38), $W^{2m\alpha, q}(\Omega) = B^{2m\alpha}_{q,q}(\Omega)$ and hence (1.3.70) fails for $q < 2$ (see [TR, Remark 2.3.3/4, Theorem 4.6.1(b)]). Finally, to include all $\alpha \geq 0$ more regularity of the data, needed for the validity of (1.3.71), is required. □

Remark 1.3.6. Note that, for $\alpha = k + \frac{r}{2m}$ $(k \in N, r = 0, 1, \dots, 2m-1)$, given (1.3.64), (1.3.65) we have

$$\|u\|_{X^\alpha} = \|A^{\frac{r}{2m}} A^k u\|_{L^q(\Omega)} \leq c\|A^k u\|_{H^r_q(\Omega)} = c\|A^k u\|_{W^{r,q}(\Omega)}. \tag{1.3.74}$$

Furthermore

$$\|A^k u\|_{W^{r,q}(\Omega)} \leq c'\|u\|_{W^{2mk+r,q}(\Omega)}, \tag{1.3.75}$$

where (1.3.75) follows inductively from [TR, Theorem 5.3.4]. Connecting (1.3.74), (1.3.75) and (1.3.73) we thus observe that $\|\cdot\|_{W^{2m\alpha,q}(\Omega)}$ is for such an α an equivalent norm in the space X^α.

Remark 1.3.7. It should also be emphasized that the structure of X^α $(\alpha \geq 0)$ strongly involves boundary conditions. In the above considerations, related to a regular $2m$-th order elliptic boundary value problem $(A, \{B_j\}, \Omega)$ with C^∞ data, we have $D(A) = W^{2m,q}_{\{B_j\}}(\Omega)$, $X^\theta \subset W^{2m\theta,q}_{\{B_j\}}(\Omega)$ ($\theta \in [0,1]$) so that we obtain the equivalence

$$\phi \in X^{k+\theta} \ (k \in N,\ \theta \in [0,1)) \ \text{ if and only if}$$
$$\phi, A\phi, \dots, A^{k-1}\phi \in W^{2m,q}_{\{B_j\}}(\Omega) \ \text{ and } \ A^k\phi \in H^{2m\theta}_{q,\{B_j\}}(\Omega).$$

As an illustration may serve here the second order Dirichlet problem $(-\Delta, I, \Omega)$ with $\partial\Omega \in C^\infty$. In this simple example the fractional power spaces $X^{k+\theta}$ ($k \in N$, $\theta \in (\frac{1}{2mq}, 1)$) may be characterized as follows:

$$X^{k+\theta} = \{\phi \in W^{2m(k+\theta),q}(\Omega) : \phi = \Delta\phi = \dots = \Delta^k\phi = 0 \ \text{ on } \ \partial\Omega\}.$$

Remark 1.3.8. Often it is inconvenient to assume the C^∞ smoothness of the data. Under the more natural smoothness assumption that $(A, \{B_j\}, \Omega)$ is a regular elliptic boundary value problem (Definition 1.2.1) as a consequence of the subordination inequality of [FR 1, Lemma 17.1, p. 177] (see also [HE 1, §6.4, Exercise 11]) a version of Proposition 1.3.10 will be proved (see [HE 1, Theorem 1.6.1]). However, under this assumption the inequalities describing the range of parameters in (1.3.73) should be sharp; that is $2m\alpha - \frac{n}{q} > t - \frac{n}{p}$ or $2m\alpha - \frac{n}{q} > k + \mu$ respectively. We lose also the nice characterization of X^α as the complex interpolation space intermediate between X and $D(A)$.

CASE II

Also when X is a Hilbert space, description of fractional order spaces X^α is quite complete. If A is a selfadjoint positive operator in a Hilbert space X, then for $\alpha, \beta \geq 0$ (see [TR, 1.18.10]),

$$[D(A^\alpha), D(A^\beta)]_\theta = D(A^{\alpha(1-\theta)+\beta\theta}),\ \theta \in (0,1),$$

i.e. Proposition 1.3.9 holds. We remark that in this case A^{it} are *unitary operators.*

Following the results of [GU 1], if $\Omega \subset R^n$ is a bounded domain with $C^{2m+\varepsilon}$ boundary ($\varepsilon > 0$), the coefficients a_σ of A are ε-(Hölder-continuous) in $\overline{\Omega}$, and coefficients b^j_σ of the boundary operators B_j belong to $C^{2m-m_j+\varepsilon}(\overline{\Omega})$, then for any $\theta \in (0,1)$ such that $2\theta - \frac{1}{2}$ is not equal m_j for any $j = 1, \dots, m$, we have the equality (see [GU 1, Sections 4, 3], also [LU 1, p. 114])

$$(L^2(\Omega), D(A))_{\theta,2} = H^{2m\theta}_{2,\{B_j\}}(\Omega),$$

where $(\cdot,\cdot)_{\theta,2}$ denotes the real interpolation space. Moreover, since (see [TR, Remark 1.19.3])

$$(L^2(\Omega), D(A))_{\theta,2} = [L^2(\Omega), D(A)]_\theta,$$

we have finally the characterization

$$D(A^\theta) = [L^2(\Omega), D(A)]_\theta = H^{2m\theta}_{2,\{B_j\}}(\Omega),$$

where $\theta \in (0,1)$ and $2m\theta - \frac{1}{2}$ is not equal to any m_j (m_j, $j = 1, \dots, m$, denoting, as before, the order of B_j). Again, if $2m\theta - \frac{1}{2} = m_j$ for some $j = 1, \dots, m$, then we have only the proper inclusion $D(A^\theta) \subset H^{2m\theta}_{2,\{B_j\}}(\Omega)$ (see [GU 1, p. 449] for the structure of $(L^2(\Omega), B^{2m}_{2,2,\{B_j\}})_{\theta,2}$ in that case).

Setting in particular $A = \Delta^2 + \rho I$ on $L^2(\Omega)$ ($\rho > 0$, $\partial\Omega \in C^{4+\varepsilon}$) and choosing $D(A) = \{\phi \in H^4(\Omega) : \frac{\partial \phi}{\partial N} = \frac{\partial(\Delta\phi)}{\partial N} = 0 \text{ on } \partial\Omega\}$ we obtain

$$D(A^\theta) = (L^2(\Omega), D(A))_{\theta,2} = H^{4\theta}_{2,\{\frac{\partial}{\partial N}, \frac{\partial(\Delta)}{\partial N}\}}(\Omega), \tag{1.3.76}$$

except the values $\theta \in \{\frac{1}{4}, \frac{3}{4}\}$.

A variant of the characterization of the interpolation space $[L^2(\Omega), D(A)]_\theta$ belongs to [GR, Theorem 8.1]. It is assumed there that the boundary $\partial\Omega$ belongs to C^{2m} while the coefficients of B_j are in $C^{2m}(\overline{\Omega})$ (see [GR] for the precise assumptions).

Case III

We now discuss the result of [P-S] concerning the second order elliptic boundary value problem. Consider a sectorial operator on $L^p(\Omega)$, $1 < p < +\infty$, corresponding to

$$Au = -\sum_{i,j=1}^{n} a_{ij}(x)\frac{\partial^2 u}{\partial x_i \partial x_j} + \sum_{i=1}^{n} b_i(x)\frac{\partial u}{\partial x_i} + c(x)u, \tag{1.3.77}$$

where either $\Omega \subset R^n$ is a bounded domain with $\partial\Omega \in C^2$ or $\Omega = R^n$. We set $D(A) = W^{2,p}(\Omega) \cap W^{1,p}_0(\Omega)$ and assume that:

(i) the matrix $[a_{ij}(x)]_{n\times n}$ is real, symmetric and uniformly elliptic:

$$\exists_{a_0>0}\ \forall_{x\in\overline{\Omega}}\ \forall_{\xi\in R^n}\ a_0|\xi|^2 \le \sum_{i=1}^{n} a_{ij}(x)\xi_i\xi_j \le a_0^{-1}|\xi|^2.$$

(ii) for some $\alpha \in (0,1)$, $a_{ij} \in C^\alpha(\overline{\Omega})$ and, if $\Omega = R^n$, the limit $a^\infty_{ij} := \lim_{|x|\to+\infty} a_{ij}(x)$ exists, fulfilling

$$|a_{ij}(x) - a^\infty_{ij}| \le C|x|^{-\alpha} \text{ for } x \in R^n, |x| \ge 1.$$

(iii) the coefficients b_i, c admit the representations

$$b_i(x) = \sum_{j=1}^{N} b_i^j(x),\ c(x) = \sum_{j=1}^{N} c^j(x),\ i = 1,\dots,n,$$

with $b_i^j \in L^{r_j}(\Omega)$, $c^j \in L^{s_j}(\Omega)$, $p \le r_j \le +\infty$, $p \le s_j \le +\infty$, $r_j > n$, $s_j > \frac{n}{2}$, and

$$b_i \in L^{r'}(\Omega),\ c \in L^{s'}(\Omega),\ i = 1,\dots,n,\ 1 \le r' < n,\ 1 \le s' < \frac{n}{2}.$$

Note that for bounded Ω the condition (iii) holds if, in particular, b_i and c are in $L^\infty(\Omega)$.

Under the assumptions (i)-(iii), for sufficiently large $\delta > 0$, the operator $A + \delta I$ considered on $W^{2,p}(\Omega) \cap W^{1,p}_0(\Omega)$, $1 < p < +\infty$, fulfills the estimate (see [P-S])

$$\|A^{it}\|_{\mathcal{L}(L^p(\Omega),L^p(\Omega))} \le Ce^{C't},\ t \in R, \tag{1.3.78}$$

so that the α-condition holds. The extension of the above result can be found in [A-H-S]. Also, the recent paper [S-T] should be mentioned where the case of the Neumann boundary condition is considered restricted, however, to the space of functions with zero average.

Finally, imaginary powers of the *Stokes operator* A_q were studied in [GI-SO], [G-G-S] (A_q is defined in (8.3.4)). In each of the four cases - $\Omega = R^n$ or Ω a halfspace or $\Omega \subset R^n$ a bounded domain with $\partial\Omega \in C^{2+\mu}$, $0 < \mu < 1$, or Ω an exterior domain with $\partial\Omega \in C^{2+\mu}$ - and for any $1 < q < +\infty$, we have an estimate

$$\|A_q^{is}\|_{\mathcal{L}(L^q(\Omega),L^q(\Omega))} \leq Ce^{\varepsilon|s|},$$

with $\varepsilon > 0$ and $C = C(\Omega, q, \varepsilon)$ independent of s. Therefore, for any of the above choices of Ω the α-condition is satisfied for the Stokes operator.

Domains of fractional powers of the perturbed operator. Let us finally come back to Proposition 1.3.2 in which a perturbation result for sectorial operators was stated. Having shown that the perturbed operator is sectorial it is important to observe how the perturbation influences the domains of its fractional powers. For perturbations described in Proposition 1.3.2 it is easy to conclude that the corresponding fractional power spaces remain unchanged.

Corollary 1.3.5. *Under the assumptions of Proposition* 1.3.2 *and additional requirements that both A and $A+B$ are positive operators with bounded imaginary powers, the following equality holds:*

$$D\big((A+B)^\alpha\big) = D(A^\alpha),\ \alpha \in (0,1). \tag{1.3.79}$$

Proof. Since $D(A) \subset D(B)$, we have $D(A+B) = D(A)$. Using then Corollary 1.3.4 we obtain

$$D(A^\alpha) = [X, D(A)]_\alpha = [X, D(A+B)]_\alpha = D\big((A+B)^\alpha\big),\ \alpha \in (0,1).$$

The proof is complete. □

For more complete discussion concerning positivity and boundedness of imaginary powers of the perturbed operator one may consult [AM 5].

Bibliographical Notes.

- **Section 1.1.** Preliminary concepts of Section 1.1 are based on the monograph [HA 2]. We also refer the reader partially to the nice description given in [LA 2]. A slightly different approach can be found in [TE 1]. A number of examples and counterexamples relative to dissipative systems are contained in [CH-HA].
- **Subsection 1.2.1.** An interesting review of the elementary inequalities of Subsection 1.2.1 is given in [GI-TR]. The uniform Gronwall inequality is taken from [TE 1] whereas Volterra type inequalities may be found in [HE 1].

- **Subsections 1.2.2 and 1.2.3.** Results reported in Subsections 1.2.2, 1.2.3 are treated mostly in [TR]. However, mentioned here also should be an elegant paper [HE 2] in which a concept of the *net smoothness* of the elements from Sobolev spaces is given.
- **Section 1.3.** This section is based on the monographs [FR 1], [HE 1] and [LU 1]. We remark that the most complete description concerning the powers of positive operators may be found in [TR], [AM 5] and in the source papers [KO 1], [KO 2].
- **Acknowledgments.** Mention should be made of the latest monograph [TA 2], which contains the detailed proof of the Nirenberg-Gagliardo type inequalities as well as the description of the properties of elliptic operators like e.g. the estimate (1.2.66) (see [TA 2, Theorem 5.5]). In this context we remark that, following the explanation given us by Professor Hiroki Tanabe, the assumption on λ in [TA 2, Theorem 5.5] should read $Re\,\lambda \leq 0$, $|\lambda| > C_p$. Let us express also our gratitude to Professor Yoshikazu Giga for pointing us to the papers [GI-SO], [G-G-S] in which the estimates of imaginary powers of the Stokes operator were studied under natural smoothness assumptions on $\partial\Omega$. We are finally grateful to Mister R. Czaja for helping us to avoid a number of misprints in Chapter 1.

CHAPTER 2

The abstract Cauchy problem

A number of initial boundary value problems originating in the applied sciences will be considered further in this book in the form of an abstract evolutionary equation with a sectorial operator in the main part. A particular advantage of studying problems of this type is that the estimates, necessary to ensure the existence of the global solutions and, next, to provide the control over their asymptotics, become much easier to obtain. For instance, the semigroup is known (in advance) to be compact if only the main part operator has a compact resolvent. Furthermore, the knowledge of any estimate of solutions asymptotically independent of initial conditions is then sufficient for the existence of the global attractor.

This chapter contains the basic facts concerning the abstract Cauchy problem for a sectorial equation. We introduce the Cauchy integral formula equivalent to this problem, prove existence of the local solutions and study their regularity properties.

2.1. Evolutionary equation with sectorial operator

Consider the Cauchy problem

$$\begin{cases} \dot{u} + Au = F(u), \ t > 0, \\ u(0) = u_0, \end{cases} \tag{2.1.1}$$

where A is sectorial in a Banach space X. Without lack of generality we may assume that A is a sectorial, positive operator (i.e. $Re\,\sigma(A) > 0$). Since the resolvent set of a closed linear operator is known to be open on the complex plane, we have

$$Re(\sigma(A)) > a, \quad \text{for some positive } a,$$

so that the power spaces X^α with $\alpha > 0$ are defined in the usual way as the ranges of appropriate power operators $A^{-\alpha}$ (see Section 1.3.3). In the situation described we shall always take it that for fixed $\alpha \in [0,1)$ the nonlinear term F in (2.1.1) is Lipschitz continuous on bounded subsets of X^α. The latter means equivalently that there exists a nondecreasing function $L : [0,+\infty) \to [0,+\infty)$, such that the estimate

$$\|F(v) - F(w)\|_X \le L(r)\|v - w\|_{X^\alpha}$$

holds for each $v, w \in B_{X^\alpha}(r)$, where $B_{X^\alpha}(r)$ denotes an open ball in X^α centered at zero with radius r.

For simplification of further references we introduce the following assumption.

Assumption 2.1.1. *Let X be a Banach space, $A : D(A) \to X$ a sectorial and positive operator in X and let $F : X^\alpha \to X$ be Lipschitz continuous on bounded subsets of X^α for some $\alpha \in [0,1)$.*

Then the notion of a *local X^α solution* of (2.1.1) is formulated.

Definition 2.1.1. *Let X be a Banach space, $\alpha \in [0,1)$ and u_0 be an element of X^α. If, for some real $\tau > 0$, a function $u \in C([0,\tau), X^\alpha)$ satisfies*

$u(0) = u_0$,
$u \in C^1((0,\tau), X)$,
$u(t)$ belongs to $D(A)$ for each $t \in (0,\tau)$,
the first equation in (2.1.1) holds in X for all $t \in (0,\tau)$,

*then u is called a **local X^α solution** of (2.1.1).*

In Corollary 2.3.1 below will be shown that the local X^α solution has additional regularity properties:

$$u \in C((0,\tau), X^1) \text{ and } \dot{u} \in C((0,\tau), X^\gamma),\ \gamma \in [0,1).$$

Our main concern in this section will be the proof of the following local existence result (see [HA 2, Section 4.2], [HE 1, Chapter 3]).

Theorem 2.1.1. *Under Assumption 2.1.1, for each $u_0 \in X^\alpha$, there exists a unique X^α solution $u = u(t,u_0)$ of (2.1.1) defined on its maximal interval of existence $[0,\tau_{u_0})$ which means that either $\tau_{u_0} = +\infty$, or*

$$\text{(2.1.2)} \qquad \text{if } \tau_{u_0} < +\infty \text{ then } \limsup_{t \to \tau_{u_0}^-} \|u(t,u_0)\|_{X^\alpha} = +\infty.$$

As a consequence of the embeddings between fractional power spaces (see (1.3.46)), if the function F is Lipschitz continuous on bounded subsets of X^α ($\alpha \in [0,1)$), then it possesses this property as a map from X^β into X for each $\beta \in [\alpha,1)$. This allows for the extension of many results concerning the X^α solutions onto X^β spaces (i.e. X^β solutions) with arbitrary $\beta \in [\alpha,1)$. In particular the conclusion below is valid.

Corollary 2.1.2. *Under Assumption 2.1.1, for each $\beta \in [\alpha,1)$ $u_0 \in X^\beta$, there exists a unique X^β solution $u = u(t,u_0)$ of (2.1.1) defined on its maximal interval of existence $[0,\tau_{u_0})$.*

Remarks on integration in Banach spaces. When h is a function acting from a measure space $(Y, \mathcal{M}, m)$ into a Banach space X, h is known to be m-integrable on $\mathcal{D} \in \mathcal{M}$ if and only if there is a sequence of finitely valued functions $h_n : \mathcal{D} \to X$, such that

$$(\alpha) \ \lim_{n\to\infty} \|h_n(y) - h(y)\|_X = 0 \ \text{ for a.e. } y \in \mathcal{D},$$

$$(\beta) \ \lim_{n\to\infty} \int_{\mathcal{D}} \|h_n(y) - h(y)\|_X dy = 0.$$

We recall that $h_n : \mathcal{D} \to X$ is a *finitely valued function* if there are elements $x_1, \ldots, x_{k_n} \in X$ and disjoint sets $\mathcal{D}_1, \ldots, \mathcal{D}_{k_n} \in \mathcal{M}$ such that

- $\bigcup_{i=1}^{k_n} \mathcal{D}_i = \mathcal{D}$,
- $h_n = \sum_{i=1}^{k_n} x_i \chi_{\mathcal{D}_i}$, $\chi_{\mathcal{D}_i}$ being the characteristic function of $\mathcal{D}_i$,
- $m(\{y \in \mathcal{D} : h_n(y) \neq 0\}) < +\infty$.

Consequently, the integral $\int_{\mathcal{D}} h(y)dy$ is an element of X defined by

$$\int_{\mathcal{D}} h(y)dy := \lim_{n\to+\infty} \int_{\mathcal{D}} h_n(y)dy, \tag{2.1.3}$$

where

$$\int_{\mathcal{D}} h_n(y)dy = \sum_{i=1}^{k_n} c_i, \qquad c_i = \begin{cases} x_i m(\mathcal{D}_i), & x_i \neq 0,\ i = 1, \ldots, k_n, \\ 0, & x_i = 0. \end{cases}$$

Note that the limit (2.1.3) is independent of the choice of a sequence of simple functions h_n. The condition (α) means that h is a strongly measurable function. The latter is then crucial in order to show that f is Bochner integrable if and only if $\|f\|_X$ is m-integrable (see [YO 1, Chapter V, Section 5]), which is important in calculations.

There is no need to consider the above notion of the *Bochner integral* in its whole generality since we are going to deal mostly with continuous functions on real line segments. Therefore, let us focus here on a less general situation when m is the Lebesgue measure and $\mathcal{D}$ denotes a Lebesgue measurable subset of $Y = R^n$. In such a case, whenever $h : \mathcal{D} \to X$ is continuous, the functions h and $\|h\|_X$ are known to be simultaneously integrable or not. Furthermore, if h is also bounded, the notion of the integral of h over the closed interval in R^n may be introduced with the use of the classical Riemann approximations. The primary notion of the Bochner integral may thus be seen to coincide in this case with the usual Riemann type integral.

In the case $Y = R$, particularly important in further calculations will be the following result of elementary analysis. It concerns in particular the *improper integrals* which will appear in considerations of the next section.

Proposition 2.1.1. *Let $h : (t_1, t_2) \to X$ be a continuous function. Then the integral $\int_{t_1}^{t_2} h(y)dy$ exists if and only if $\int_{t_1}^{t_2} \|h(y)\|_X dy < \infty$. Moreover if h is integrable on (t_1, t_2), then*

$$\int_{t_1}^{t_2} h(y)dy = \lim_{(t,\tau)\to(t_1,t_2)} \int_t^{\tau} h(y)dy \tag{2.1.4}$$

and, if in addition $h \in C([t_1,t_2],X) \cap C^1((t_1,t_2),X)$, *then*

$$h(t_2) - h(t_1) = \int_{t_1}^{t_2} h'(y)dy. \tag{2.1.5}$$

Remark 2.1.1. Note that the integral in the equality (2.1.5) may have only the improper sense, i.e. $\int_{t_1}^{t_2} h'(y)dy = \lim_{(t,\tau)\to(t_1,t_2)} \int_t^\tau h'(y)dy$. Also $\int_{t_1}^{t_2} h(y)dy$ in (2.1.4) has only the improper sense unless (as in Proposition 2.1.1) it is assumed that $\int_{t_1}^{t_2} h(y)dy$ exists. Nevertheless (2.1.4) points out that if $\int_{t_1}^{t_2} h(y)dy$ exists, then it may be considered as an improper integral. Therefore there is no need to introduce any special notation for improper integrals and we shall denote it hereafter in a standard way.

Proposition 2.1.2. *Let* $B : D(B) \to X$ *be a closed linear operator in a Banach space* X *and* $h : [0,t) \to X$ *and* $B \circ h : [0,t) \to X$ *be continuous functions. Assume further that* $h(y) \in D(B)$ *for each* $y \in [0,t)$ *and there exist improper integrals* $\int_0^t h(y)dy$ *and* $\int_0^t Bh(y)dy$. *Then* $\int_0^t h(y)dy \in D(B)$ *and*

$$B\int_0^t h(y)dy = \int_0^t Bh(y)dy.$$

These few remarks will be useful in further considerations of the integral form of the problem (2.1.1). For more details see [H-P], [YO 1], [FR 1].

2.2. Variation of constants formula

Lemma 2.2.1. *(Integral Cauchy formula) Let Assumption* 2.1.1 *hold and* $u \in C([0,\tau),X^\alpha)$. *Then* u *is a local* X^α *solution of* (2.1.1) *if and only if* u *satisfies in* X *the integral equation*

$$u(t) = e^{-At}u_0 + \int_0^t e^{-A(t-s)}F(u(s))ds, \quad \text{for } t \in [0,\tau). \tag{2.2.1}$$

Proof. Let us proceed to the proof that $u \in C([0,\tau),X^\alpha)$ satisfying (2.2.1) is a local X^α solution of (2.1.1).

Step 1. First we shall derive the following supportive estimate:

$$\|(e^{-At} - I)v\|_X \le \frac{c_{1-\delta}}{\delta} t^\delta \|v\|_{X^\delta}, \ \delta \in (0,1], \ v \in X^\delta. \tag{2.2.2}$$

This is a consequence of the equality

$$\begin{aligned}(e^{-At} - I)v &= \int_0^t \frac{d}{ds}(e^{-As}v)ds \\ &= -\int_0^t Ae^{-As}vds = -\int_0^t A^{1-\delta}e^{-As}A^\delta vds,\end{aligned}$$

from which we find

$$\begin{aligned}\|(e^{-At} - I)v\|_X &\le \int_0^t \|A^{1-\delta}e^{-As}\|_{\mathcal{L}(X,X)}\|A^\delta v\|_X ds \\ &\le c_{1-\delta}\|v\|_{X^\delta}\int_0^t \frac{e^{-as}}{s^{1-\delta}}ds \le \frac{c_{1-\delta}}{\delta}t^\delta\|v\|_{X^\delta}.\end{aligned}$$

Step 2. With the use of (2.2.2) we shall show the following local Hölder condition:

$$\forall_{0<t_0<t_1<\tau}\ \forall_{\delta\in(0,1)}\ \exists_{const.>0}\ \forall_{t,\bar{t}\in(t_0,t_1)}\quad \|u(t)-u(\bar{t})\|_{X^\alpha}\le const.|t-\bar{t}|^\delta, \tag{2.2.3}$$

where $const.$ depends on α,δ,t_0,t_1.

Let $0<h<t_1<\tau$ be fixed. From (2.2.1) we obtain

$$\begin{aligned} u(t+h)-u(t) &= (e^{-Ah}-I)e^{-At}u_0 \\ &+\left(\int_0^h+\int_h^{t+h}\right)e^{-A(t+h-s)}F(u(s))ds-\int_0^t e^{-A(t-s)}F(u(s))ds \\ &= (e^{-Ah}-I)e^{-At}u_0+\int_0^h e^{-A(t+h-s)}F(u(s))ds \\ &+\int_0^t e^{-A(t-s)}\big(F(u(s+h))-F(u(s))\big)ds,\ t\in(0,t_1-h). \end{aligned}$$

Next using the estimate (2.2.2) with $0<\delta<1$ and Lipschitz continuity of F we get

$$\begin{aligned} \|u(t+h)-u(t)\|_{X^\alpha} &\le \|(e^{-Ah}-I)e^{-At}A^\alpha u_0\|_X \\ &+\int_0^h c_\alpha\frac{e^{-a(t+h-s)}}{(t+h-s)^\alpha}\|F(u(s))\|_X ds \\ &+\int_0^t c_\alpha\frac{e^{-a(t-s)}}{(t-s)^\alpha}\|F(u(s+h))-F(u(s))\|_X ds \\ &\le h^\delta\frac{c_{1-\delta}c_\delta}{\delta t^\delta}\|u_0\|_{X^\alpha}+\frac{c_\alpha\sup_{s\in[0,t_1]}\|F(u(s))\|_X}{t^\alpha}h \\ &+c_\alpha L_{t_1}\int_0^t\frac{1}{(t-s)^\alpha}\|u(s+h)-u(s)\|_{X^\alpha}ds \\ &\le\frac{h^\delta}{t^{\delta_{max}}}\left(\frac{c_{1-\delta}c_\delta}{\delta}t_1^{\delta_{max}-\delta}\|u_0\|_{X^\alpha}+c_\alpha t_1^{1-\delta}t_1^{\delta_{max}-\alpha}\sup_{s\in[0,t_1]}\|F(u(s))\|_X\right) \\ &+c_\alpha L_{t_1}\int_0^t\frac{1}{(t-s)^\alpha}\|u(s+h)-u(s)\|_{X^\alpha}ds,\ t\in(0,t_1-h), \end{aligned}$$

where $\delta_{max}=\max\{\alpha,\delta\}$.

Lemma 1.2.9 now implies

$$t^{\delta_{max}}\|u(t+h)-u(t)\|_{X^\alpha}\le const._{t_1,\delta,\alpha}h^\delta,\ t\in(0,t_1-h)$$

(where $const._{t_1,\delta,\alpha}$ is independent of h), and finally,

$$\|u(t+h)-u(t)\|_{X^\alpha}\le\frac{const._{t_1,\delta,\alpha}}{t_0^{\delta_{max}}}h^\delta,\ t\in(t_0,t_1-h), \tag{2.2.4}$$

which shows the validity of (2.2.3).

Step 3. Condition (2.2.3) allows us to justify that

$$\int_0^t e^{-A(t-s)}F(u(s))ds \in D(A),\ t\in(0,\tau), \tag{2.2.5}$$

$$u(t)\in D(A) \text{ for } t\in(0,\tau) \tag{2.2.6}$$

and also

$$\int_0^t Ae^{-A(t-s)}F(u(s))ds = A\int_0^t e^{-A(t-s)}F(u(s))ds,\ t\in(0,\tau), \tag{2.2.7}$$

where the integral on the left side of (2.2.7) is considered as an improper integral.

Indeed, $e^{-A(t-s)}F(u(s))\in D(A)$ for $s\in[0,t)$ and from (2.2.3) and local Lipschitz continuity of F we have

$$\begin{aligned}
&\int_{t_0}^t \|Ae^{-A(t-s)}\left[F(u(t))-F(u(s))\right]\|_X ds \\
&\le \int_{t_0}^t c_1\frac{e^{-a(t-s)}}{t-s}\|F(u(t))-F(u(s))\|_X ds \\
&\le c_1\int_{t_0}^t \frac{1}{t-s}L\|u(t)-u(s)\|_{X^\alpha} ds \\
&\le c_1L\int_{t_0}^t \frac{1}{t-s}const.(t-s)^\delta ds \\
&= \frac{c_1 const. L}{\delta}(t-t_0)^\delta;
\end{aligned} \tag{2.2.8}$$

whereas

$$\begin{aligned}
\int_{t_0}^t Ae^{-A(t-s)}F(u(t))ds &= \int_{t_0}^t \frac{d}{ds}\left(e^{-A(t-s)}F(u(t))\right)ds \\
&= (I-e^{-A(t-t_0)})F(u(t)).
\end{aligned} \tag{2.2.9}$$

This shows the existence of the improper integral $\int_0^t Ae^{-A(t-s)}F(u(s))ds$. Certainly, there exists $\int_0^t e^{-A(t-s)}F(u(s))ds$ as the integral of a continuous function, so that (2.2.5) and (2.2.7) follow directly from Proposition 2.1.2 (note continuity of the function $[0,t)\ni s\to Ae^{-A(t-s)}F(u(s))\in X$). Therefore (2.2.6) follows from (2.2.5) and equation (2.2.1).

Step 4. For

$$u_1(t) = \int_0^t e^{-A(t-s)}\Big(F(u(s))-F(u(t))\Big)ds, \tag{2.2.10}$$

we are now able to prove that

$$Au_1\in C((0,\tau),X^\gamma) \text{ for each } \gamma\in(0,1) \tag{2.2.11}$$

(note that $u_1(t)\in D(A)$ for $t\in[0,\tau)$ as a consequence of (2.2.8) and Proposition 2.1.2). More precisely, we shall justify the stronger condition

$$\begin{aligned}
&\forall_{0<\gamma<\delta<1}\ \forall_{0<\varepsilon<\bar\varepsilon<\tau}\ \exists_{const.>0}\ \forall_{t_1,t_2\in(\varepsilon,\bar\varepsilon)} \\
&\|Au_1(t_2)-Au_1(t_1)\|_{X^\gamma}\le const.|t_2-t_1|^{\delta-\gamma},
\end{aligned} \tag{2.2.12}$$

where *const.* depends on $\varepsilon, \overline{\varepsilon}, \gamma, \delta$.

Take $0 < \gamma < \delta < 1,\ 0 < \varepsilon \le t_1 < t_2 < \overline{\varepsilon} < \tau$ and consider the equality

$$\begin{aligned}
&Au_1(t_2) - Au_1(t_1)\\
&= \int_0^{t_1} A\left(e^{-A(t_2-s)} - e^{-A(t_1-s)}\right)\left(F(u(s)) - F(u(t_1))\right)ds\\
&\quad + \left(e^{-A(t_2-t_1)} - e^{-At_2}\right)\left(F(u(t_1)) - F(u(t_2))\right)\\
&\quad + \int_{t_1}^{t_2} Ae^{-A(t_2-s)}\left(F(u(s)) - F(u(t_2))\right)ds\\
&=: \mathcal{J}_1 + \mathcal{J}_2 + \mathcal{J}_3.
\end{aligned}$$

Roughly speaking, it suffices to show that $\|\mathcal{J}_k\|_{X^\gamma}$, $k = 1, 2, 3$, are proportional to $(t_2 - t_1)^{\delta-\gamma}$. In the calculations below, conditions (2.2.2), (2.2.3), Lipschitz continuity of F and the usual estimates for analytic semigroups will be used several times. We have the following estimates:

(2.2.13)

$$\begin{aligned}
\|\mathcal{J}_1\|_{X^\gamma} &\le \int_0^{\frac{\varepsilon}{2}} \|A^\gamma A(e^{-A(t_2-t_1)} - I)e^{-A(t_1-s)}\left(F(u(s)) - F(u(t_1))\right)\|_X ds\\
&\quad + \int_{\frac{\varepsilon}{2}}^{t_1} \|A^\gamma A(e^{-A(t_2-t_1)} - I)e^{-A(t_1-s)}\left(F(u(s)) - F(u(t_1))\right)\|_X ds\\
&=: \mathcal{J}_{11} + \mathcal{J}_{12},
\end{aligned}$$

where further

$$\begin{aligned}
\mathcal{J}_{11} &\le \int_0^{\frac{\varepsilon}{2}} \|(e^{-A(t_2-t_1)} - I)A^\gamma Ae^{-A(t_1-s)}\left(F(u(s)) - F(u(t_1))\right)\|_X ds\\
&\le \int_0^{\frac{\varepsilon}{2}} \frac{c_{1-(\delta-\gamma)}}{\delta-\gamma}(t_2-t_1)^{\delta-\gamma}\|A^{\delta-\gamma}A^\gamma Ae^{-A(t_1-s)}\left(F(u(s)) - F(u(t_1))\right)\|_X ds\\
&\le \left(\int_0^{\frac{\varepsilon}{2}} \frac{c_{1-\delta+\gamma}}{\delta-\gamma}\frac{c_{1+\delta}}{(t_1-s)^{1+\delta}} 2\sup_{s\in[0,\overline{\varepsilon}]}\|F(u(s))\|_X ds\right)(t_2-t_1)^{\delta-\gamma}\\
&\le 2\sup_{s\in[0,\overline{\varepsilon}]}\|F(u(s))\|_X \frac{c_{1-\delta+\gamma}}{\delta-\gamma}\frac{c_{1+\delta}}{\delta}\left(\frac{2}{\varepsilon}\right)^\delta (t_2-t_1)^{\delta-\gamma},
\end{aligned}$$

and

$$\mathcal{J}_{12} \le \left(\int_{\frac{\varepsilon}{2}}^{t_1} \frac{c_{1-\delta+\gamma}}{\delta-\gamma}\frac{c_{1+\delta}}{(t_1-s)^{1+\delta}} L_{\overline{\varepsilon}} const._{\overline{\varepsilon},\varepsilon,\delta_1}(t_1-s)^{\delta_1} ds\right)(t_2-t_1)^{\delta-\gamma}$$

(note that δ_1 appearing above may be chosen between δ and 1 based on (2.2.3), so the integral in the brackets above is bounded). Similarly we estimate

(2.2.14)

$$\begin{aligned}
\|\mathcal{J}_2\|_{X^\gamma} &\le \|A^\gamma e^{-A(t_2-t_1)}(e^{-At_1} - I)\left(F(u(t_1)) - F(u(t_2))\right)\|_X\\
&\le (c_0+1)\frac{c_\gamma}{(t_2-t_1)^\gamma} L_{\overline{\varepsilon}} const._{\overline{\varepsilon},\varepsilon,\delta}(t_2-t_1)^\delta
\end{aligned}$$

and

(2.2.15)

$$\|\mathcal{J}_3\|_{X^\gamma} \leq \int_{t_1}^{t_2} \|A^\gamma A e^{-A(t_2-s)}\Big(F(u(s)) - F(u(t_2))\Big)\|_X ds$$
$$\leq \int_{t_1}^{t_2} \frac{c_{1+\gamma}}{(t_2-s)^{1+\gamma}} L_{\overline{\varepsilon}} const._{\overline{\varepsilon},\varepsilon,\delta}(t_2-s)^\delta ds$$
$$= c_{1+\gamma} L_{\overline{\varepsilon}} const._{\overline{\varepsilon},\varepsilon,\delta} \frac{(t_2-t_1)^{\delta-\gamma}}{\delta-\gamma}.$$

From (2.2.13)-(2.2.15) each $\|\mathcal{J}_k\|_{X^\gamma}$ $(k = 1,2,3)$ is seen to be proportional to $(t_2-t_1)^{\delta-\gamma}$. Condition (2.2.12) is thus proved.

Step 5. In the presence of (2.2.5) it is easy to see that the right side derivative $\dot{u}_+(t)$ satisfies for $t \in (0,\tau)$ the equality

$$\dot{u}_+(t) + A\left(e^{-At}u_0 + \int_0^t e^{-A(t-s)}F(u(s))ds\right) = F(u(t)). \tag{2.2.16}$$

Indeed, passing to the limit with $h \to 0+$ in the formula

$$\frac{1}{h}(u(t+h) - u(t)) = \frac{1}{h}(e^{-Ah} - I)e^{-At}u_0$$
$$+ \frac{1}{h}(e^{-Ah} - I)\int_0^t e^{-A(t-s)}F(u(s))ds + \frac{1}{h}\int_t^{t+h} e^{-A(t+h-s)}F(u(s))ds$$

we come directly to (2.2.16).

Step 6. Note that

(2.2.17)

$$Au_1(t) = A\int_0^t e^{-A(t-s)}F(u(s))ds - \int_0^t Ae^{-A(t-s)}F(u(t))ds$$
$$= A\int_0^t e^{-A(t-s)}F(u(s))ds - F(u(t)) + e^{-At}F(u(t)).$$

Connecting (2.2.16), (2.2.17) we observe that $\dot{u}_+$ may be expressed by the equality

$$\dot{u}_+ = -Ae^{-At}u_0 + e^{-At}F(u(t)) - Au_1(t). \tag{2.2.18}$$

The functions $(0,\tau) \ni t \to Ae^{-At}u_0 \in X^\gamma$ and $(0,\tau) \ni t \to e^{-At}F(u(t)) \in X^\gamma$ are continuous, whereas from (2.2.11) we have $Au_1 \in C((0,\tau), X^\gamma)$. Therefore, the right side derivative $\dot{u}_+ : (0,\tau) \to X$ is continuous. However, u must then be continuously differentiable on $(0,\tau)$; that is

$$\dot{u}(t) = \dot{u}_+(t) \ \text{ and } \ u \in C^1((0,\tau), X). \tag{2.2.19}$$

To justify (2.2.19) it suffices to consider an auxiliary function $\omega : [t_0,\tau) \to X$ ($t_0 \in (0,\tau)$ is arbitrarily chosen):

$$\omega(t) = \int_{t_0}^t \dot{u}_+(s)ds + u(t_0), \ t \in [t_0,\tau).$$

Certainly, $\omega \in C^1([t_0,\tau),X)$ and $\dot{\omega}_+ \equiv \dot{u}_+$, so that for any bounded linear functional $x^* \in X^*$, the right side derivative $[x^*(\omega(t)) - x^*(u(t))]'_+$ is identically equal to zero on $[t_0,\tau)$. As a consequence of the equality $\omega(t_0) = u(t_0)$ we obtain further

$$x^*(\omega(t) - u(t)) = 0, \ t \in [t_0,\tau), \ x^* \in X^*.$$

Therefore, by standard Hahn-Banach theorem arguments, ω is seen to coincide with u.

In Steps 3, 5, 6 all requirements of Definition 2.1.1 were verified. We shall prove further that $u \in C([0,\tau),X^\alpha)$ being a local X^α solution of (2.1.1) satisfies the Cauchy formula (2.2.1).

For $0 < s < t < \tau$, we obtain from (2.1.1) that

$$e^{-A(t-s)}\dot{u}(s) + e^{-A(t-s)}Au(s) = e^{-A(t-s)}F(u(s)). \tag{2.2.20}$$

Note that the function $R^+ \times R^+ \ni (t,s) \to e^{-At}u(s) \in X$ has continuous first order partial derivatives and consider the composite function $(0,t) \ni s \to e^{-A(t-s)}u(s) \in X$. Differentiating, we get

$$\begin{aligned} \frac{d}{ds}\left(e^{-A(t-s)}u(s)\right) &= Ae^{-A(t-s)}u(s) + e^{-A(t-s)}\dot{u}(s) \\ &= e^{-A(t-s)}\dot{u}(s) + e^{-A(t-s)}Au(s). \end{aligned} \tag{2.2.21}$$

As a consequence of (2.2.20) and the formula (2.2.21) we have

$$\frac{d}{ds}\left(e^{-A(t-s)}u(s)\right) = e^{-A(t-s)}F(u(s)), \ \text{ for } \ 0 < s < t < \tau, \tag{2.2.22}$$

so that (2.2.1) follows by integration of (2.2.22) over $(0,t)$ (see Proposition 2.1.1).

The proof of Lemma 2.2.1 is complete. □

2.3. Local X^α solutions

This section is mostly devoted to the proof of Theorem 2.1.1. We shall use here the standard approach, based on considerations of the integral equation (2.2.1) in a Banach space $C([0,\delta],X^\alpha)$ equipped with the supremum norm:

$$\|v\|_{C([0,\delta],X^\alpha)} := \sup_{t\in[0,\delta]} \|v(t)\|_{X^\alpha}.$$

The proof of existence is based on the *Banach fixed point theorem* applied to a contraction Ψ acting on the appropriately chosen closed ball $B_{C([0,\delta],X^\alpha)}(u_0,r)$ in the space $C([0,\delta],X^\alpha)$. Additionally, in the final part of this section the solutions of (2.1.1) will also be shown to depend continuously on the initial data.

Proof. Step 1. We start with the proof of uniqueness. Consider $u_1,u_2 \in C([0,\delta],X^\alpha)$ with $u_1(0) = u_2(0) = u_0$ such that

$$u_i(t) = e^{-At}u_0 + \int_0^t e^{-A(t-s)}F(u_i(s))ds \ t \in [0,\delta], \ i = 1,2.$$

Then, since F is Lipschitz continuous on bounded sets, we obtain

$$\|u_1(t)-u_2(t)\|_{X^\alpha} \le \int_0^t \|A^\alpha e^{-A(t-s)}\|_{\mathcal{L}(X,X)}\|F(u_1(s))-F(u_2(s))\|_X ds$$
$$\le L_\delta c_\alpha \int_0^t \frac{1}{(t-s)^\alpha}\|u_1(s)-u_2(s)\|_{X^\alpha} ds.$$

Therefore, from Lemma 1.2.9, u_1, u_2 are equal on $[0,\delta]$, which proves the uniqueness result.

Step 2. It is clear that each $u_0 \in X^\alpha$ may be identified with a constant (in time) function $u_0 \in C([0,\delta],X^\alpha)$. Therefore fix $u_0 \in X^\alpha$, $r>0$ and consider the following integral transformation $\Psi : B_{C([0,\delta],X^\alpha)}(u_0,r) \to C([0,\delta],X^\alpha)$:

$$B_{C([0,\delta],X^\alpha)}(u_0,r) \ni v \xrightarrow{\Psi} \Psi(v)(t) := e^{-At}u_0 + \int_0^t e^{-A(t-s)}F(v(s))ds \ \text{ for } t\in[0,\delta],$$

where $B_{C([0,\delta],X^\alpha)}(u_0,r)$ denotes a closed ball in a Banach space $C([0,\delta],X^\alpha)$ centered on u_0 with radius $r>0$ (the value of the parameter $\delta>0$ will be further restricted by (2.3.6)). Adapting considerations of Lemma 2.2.1 connected with the proof of (2.2.3), we see that $\Psi(v)(t)$ is a continuous function of variable $t>0$ with values in X^α. For the proof of right side continuity at $t=0$ it suffices to note that since F is bounded on bounded sets of X^α and $v \in B_{C([0,\delta],X^\alpha)}(u_0,r)$ we have

(2.3.1)
$$\|\Psi(v)(t)-u_0\|_{X^\alpha} \le \|A^\alpha(e^{-At}-I)u_0\|_X + \int_0^t \|A^\alpha e^{-A(t-s)}F(v(s))\|_X ds$$
$$\le \|(e^{-At}-I)A^\alpha u_0\|_X + \sup_{\|v\|_{X^\alpha}\le r+\|u_0\|_{X^\alpha}} \|F(v)\|_X c_\alpha \int_0^t \frac{e^{-a(t-s)}}{(t-s)^\alpha}ds$$
$$\le \|(e^{-At}-I)A^\alpha u_0\|_X + \sup_{\|v\|_{X^\alpha}\le r+\|u_0\|_{X^\alpha}} \|F(v)\|_X c_\alpha \frac{t^{1-\alpha}}{1-\alpha},\ t\in[0,\delta].$$

The function Ψ is thus properly defined.

Step 3. Now, we shall choose $\delta>0$ for which the range of Ψ will be contained in $B_{C([0,\delta],X^\alpha)}(u_0,r)$; that is Ψ will be considered as a mapping

(2.3.2)
$$\Psi : B_{C([0,\delta],X^\alpha)}(u_0,r) \to B_{C([0,\delta],X^\alpha)}(u_0,r).$$

Recalling (2.3.1) we obtain

(2.3.3)
$$\|\Psi(v)(t)-u_0\|_{X^\alpha} \le \varepsilon\|u_0\|_{X^\alpha} + \sup_{\|v\|_{X^\alpha}\le r+\|u_0\|_{X^\alpha}} \|F(v)\|_X c_\alpha \frac{\delta^{1-\alpha}}{1-\alpha}$$
$$\le \frac{r}{2}+\frac{r}{2} \ \text{ for } \ t\in[0,\delta],$$

provided that $\delta \leq \delta_1$. The number δ_1 is chosen relatively to $\varepsilon := \frac{r}{2\|u_0\|_{X^\alpha}}$ according to continuity of $e^{-At} : X \to X$ at $t = 0$ and, additionally, it satisfies the inequality

$$\text{(2.3.4)} \qquad \sup_{\|v\|_{X^\alpha} \leq r + \|u_0\|_{X^\alpha}} \|F(v)\|_X c_\alpha \frac{\delta_1^{1-\alpha}}{1-\alpha} \leq \frac{r}{2}.$$

Step 4. We shall introduce below the second restriction for δ, requiring of Ψ to fulfill the contraction property:

$$\exists_{\omega \in [0,1)} \forall_{v_1, v_2 \in B_{C([0,\delta], X^\alpha)}(u_0, r)} \|\Psi(v_1) - \Psi(v_2)\|_{C([0,\delta],X^\alpha)} \leq \omega \|v_1 - v_2\|_{C([0,\delta],X^\alpha)}.$$

Take $v_1, v_2 \in B_{C([0,\delta],X^\alpha)}(u_0, r)$ and fix any $\omega \in (0,1)$. As a consequence of Lipschitz continuity of F on bounded sets we have

$$\|\Psi(v_1)(t) - \Psi(v_2)(t)\|_{X^\alpha} \leq \int_0^t \|A^\alpha e^{-A(t-s)}\|_{\mathcal{L}(X,X)} \|F(v_1(s)) - F(v_2(s))\|_X ds$$

$$\leq L_{2r} \sup_{s \in [0,t]} \|v_1(s) - v_2(s)\|_{X^\alpha} c_\alpha \int_0^t \frac{e^{-a(t-s)}}{(t-s)^\alpha} ds$$

$$\leq L_{2r} \sup_{s \in [0,t]} \|v_1(s) - v_2(s)\|_{X^\alpha} c_\alpha \frac{\delta^{1-\alpha}}{1-\alpha} \leq \omega \sup_{s \in [0,\delta]} \|v_1(s) - v_2(s)\|_{X^\alpha},$$

provided that δ does not exceed δ_2 where

$$\text{(2.3.5)} \qquad L_{2r} c_\alpha \frac{\delta_2^{1-\alpha}}{1-\alpha} = \omega.$$

Step 5. It is thus seen that for

$$\text{(2.3.6)} \qquad \delta_0 := \min\{\delta_1, \delta_2\},$$

where δ_1, δ_2 were defined in (2.3.4) and (2.3.5), Ψ is a contraction on a complete metric space $B_{C([0,\delta_0],X^\alpha)}(u_0, r)$. Therefore, from the Banach fixed point theorem there exists a unique element $u \in B_{C([0,\delta_0],X^\alpha)}(u_0, r)$ for which $\Psi(u) = u$, that is such that u satisfies the integral equation

$$\text{(2.3.7)} \qquad u(t) = e^{-At} u_0 + \int_0^t e^{-A(t-s)} F(u(s)) ds \quad \text{for} \quad t \in [0, \delta_0].$$

Step 6. We shall now characterize the maximal interval of existence of the solution to (2.3.7). Consider the set $\mathcal{I}_{u_0}$:

$$\mathcal{I}_{u_0} := \left\{ \delta > 0 : \exists_{u^\delta \in C([0,\delta],X^\alpha)} \forall_{t \in [0,\delta]} \, u^\delta(t) = e^{-At} u_0 + \int_0^t e^{-A(t-s)} F(u^\delta(s)) ds \right\}.$$

Considerations of the previous step show that $\mathcal{I}_{u_0}$ is nonempty. Furthermore, if $\delta_1, \delta_2 \in \mathcal{I}_{u_0}$ and $\delta_1 < \delta_2$ then from the uniqueness property (Step 1)

$$\text{(2.3.8)} \qquad u^{\delta_1} = u^{\delta_2} \quad \text{on} \quad [0, \delta_1].$$

Define further

$$\text{(2.3.9)} \qquad \tau_{u_0} := \sup \mathcal{I}_{u_0}$$

and observe that

$$\tau_{u_0} \notin \mathcal{I}_{u_0}. \tag{2.3.10}$$

Indeed, assume that $\tau_{u_0} \in \mathcal{I}_{u_0}$. Then there is $u^{\tau_{u_0}} \in C([0,\tau_{u_0}],X^\alpha)$ satisfying

$$u^{\tau_{u_0}}(t) = e^{-At}u_0 + \int_0^t e^{-A(t-s)}F(u^{\tau_{u_0}}(s))ds, \ t \in [0,\tau_{u_0}]. \tag{2.3.11}$$

On the other side, the Banach fixed point theorem ensures (see Steps 2-5) the existence of some $w \in C([0,\varepsilon],X^\alpha)$ which fulfills

$$w(t) = e^{-At}u^{\tau_{u_0}}(\tau_{u_0}) + \int_0^t e^{-A(t-s)}F(w(s))ds, \ t \in [0,\varepsilon]. \tag{2.3.12}$$

Define

$$u(t) = \begin{cases} u^{\tau_{u_0}}(t), & t \in [0,\tau_{u_0}], \\ w(t-\tau_{u_0}), & t \in [\tau_{u_0},\tau_{u_0}+\varepsilon]. \end{cases}$$

Certainly $u \in C([0,\tau_{u_0}+\varepsilon],X^\alpha)$ and $\tau_{u_0}+\varepsilon \in \mathcal{I}_{u_0}$, which contradicts (2.3.9).
Step 7. The results of the previous step ensure that $\mathcal{I}_{u_0} = (0,\tau_{u_0})$ and the function $u : [0,\tau_{u_0}) \to X^\alpha$ given by the condition

$$u(t,u_0) := u^\delta(t) \ \text{ whenever } \delta \in \mathcal{I}_{u_0} \text{ and } t \in [0,\delta]$$

is well defined for any $u_0 \in X^\alpha$, $u(\cdot,u_0) \in C([0,\tau_{u_0}),X^\alpha)$ and $u(\cdot,u_0)$ satisfies the integral equation

$$u(t,u_0) = e^{-At}u_0 + \int_0^t e^{-A(t-s)}F(u(s,u_0))ds, \ \text{ for each } \ t \in [0,\tau_{u_0}).$$

It then follows from Lemma 2.2.1 that $u(\cdot,u_0)$ is a local X^α solution of (2.1.1) on $[0,\tau_{u_0})$ which, as a result of Step 1, is unique in the smoothness class considered of the solution. It remains to show that $[0,\tau_{u_0})$ is a *maximal interval of existence* in the sense that

$$\text{if } \tau_{u_0} < +\infty \ \text{ then } \ \limsup_{t\to\tau_{u_0}^-} \|u(t,u_0)\|_{X^\alpha} = +\infty.$$

Assume to the contrary that

$$\sup_{t\in[0,\tau_{u_0})} \|u(t,u_0)\|_{X^\alpha} \le const. \ \text{ for some } \tau_{u_0} < +\infty$$

and define

$$M := \sup_{t\in[0,\tau_{u_0})} \|F(u(t,u_0))\|_X.$$

Then calculations of Lemma 2.2.1 (Step 2.) may be adapted and (2.2.3) (used with $\delta = \alpha$ and $t_0 = \frac{\tau_{u_0}}{2}$, $t_1 = \tau_{u_0}$) gives

(2.3.13)
$$\|u(t,u_0) - u(\bar{t},u_0)\|_{X^\alpha}$$
$$\leq const.(\alpha, \tau_0, \|u_0\|_{X^\alpha}, M)(t-\bar{t})^\alpha, \quad \frac{\tau_{u_0}}{2} < \bar{t} < t < \tau_{u_0}.$$

It is seen from (2.3.13), with the use of Cauchy sequences, that there exists $\lim_{t\to\tau_{u_0}^-} u(t,u_0)$. Therefore, $u : [0,\tau_{u_0}) \to X^\alpha$ may be extended in a natural way to a function $\overline{u} \in C([0,\tau_{u_0}], X^\alpha)$ and $\overline{u}$ satisfies (2.3.11). The latter implies however that $\tau_{u_0} \in \mathcal{I}_{u_0}$, which contradicts (2.3.10). The proof of Theorem 2.1.1 is complete. □

Additional properties of X^α solutions. From considerations of Lemma 2.2.1 the additional smoothness of the time derivative of u may be derived.

Corollary 2.3.1. *Let Assumption* 2.1.1 *hold and* $u \in C([0,\tau), X^\alpha)$. *If* u *satisfies in* X *the integral equation* (2.2.1), *then*

$$u \in C((0,\tau), X^1), \tag{2.3.14}$$

$$\dot{u} \in C((0,\tau), X^\gamma), \ \gamma \in [0,1). \tag{2.3.15}$$

Proof. Since $u \in C((0,\tau), X^\alpha) \cap C^1((0,\tau), X)$ and $F : X^\alpha \to X$ is continuous, $Au = F(u) - \dot{u}$ belongs to $C((0,\tau), X)$ as stated in (2.3.14).

Recall now the definition of u_1 in (2.2.10):

$$u_1(t) = \int_0^t e^{-A(t-s)}\Big(F(u(s)) - F(u(t))\Big)ds.$$

Then (2.2.18) reads

$$\dot{u} = -Ae^{-At}u_0 + e^{-At}F(u(t)) - Au_1(t)$$

and (2.3.15) follows from continuity of the functions $(0,\tau) \ni t \to Ae^{-At}u_0 \in X^\gamma$, $(0,\tau) \ni t \to e^{-At}F(u(t)) \in X^\gamma$ ($\gamma \in [0,1)$) and from condition (2.2.11). □

The proposition below shows that the solutions of (2.1.1) originating in a bounded set B exist at least until some positive time $\tau_B > 0$.

Proposition 2.3.1. *Under Assumption* 2.1.1, *for any bounded set* $B \subset X^\alpha$ *there is a time* $\tau_B > 0$ *such that solutions* $u(t,u_0)$ *of* (2.1.1) *with* $u_0 \in B$ *exist and are bounded in* X^α *uniformly for* $t \in [0,\tau_B]$ *and* $u_0 \in B$.

Proof. Write $\rho = \sup_{\phi\in B}\|\phi\|_{X^\alpha}$, let $B_1 = B_{X^\alpha}(0, c_0\rho + 1)$ be a ball in X^α, where c_0 is a constant appearing in the estimate

$$\|e^{-At}\|_{\mathcal{L}(X,X)} \leq c_0 e^{-at}, \ t \geq 0,$$

and define $b := \sup_{v\in B_1}\|F(v)\|_X$.

Based on the integral equation corresponding to (2.1.1), for $u_0 \in B$ and as long as $u(t,u_0) \in B_1$, we find

(2.3.16)
$$\begin{aligned}\|u(t,u_0)\|_{X^\alpha} &\le c_0 e^{-at}\|u_0\|_{X^\alpha} \\ &\quad + \int_0^t \|A^\alpha e^{-A(t-s)}\|_{\mathcal{L}(X,X)}\|F(u(s,u_0))\|_X ds \\ &\le c_0\rho + b\int_0^t c_\alpha \frac{e^{-a(t-s)}}{(t-s)^\alpha} ds \le c_0\rho + bc_\alpha \int_0^t \frac{dy}{y^\alpha} \\ &= c_0\rho + \frac{bc_\alpha}{1-\alpha} t^{1-\alpha}.\end{aligned}$$

Hence the solutions starting from B will not leave B_1 at least until the second term in the estimate (2.3.16) reaches the value 1 (this term is increasing, and equal to zero for $t=0$). Such a time τ_B is given by the condition

$$\frac{bc_\alpha}{1-\alpha}\tau_B^{1-\alpha} = 1. \tag{2.3.17}$$

For $t \in [0,\tau_B]$ we thus have an estimate

$$\sup_{u_0 \in B} \|u(t,u_0)\|_{X^\alpha} \le c_0\rho + 1, \tag{2.3.18}$$

completing the proof. □

The above result will be useful in the proof that the solutions of (2.1.1) depend continuously (and uniformly for t varying in compact intervals of time) on the initial conditions.

Proposition 2.3.2. *Let Assumption* 2.1.1 *be satisfied,* $\{u_n\} \subset X^\alpha$ *and* $u_n \to u_0$ *in* X^α. *Then* $T_0 := \inf\{\tau_{u_n} : n = 1,2,\dots\}$ *is positive,* $\tau_{u_0} \ge T_0$ *(*τ_{u_n} *being the lifetime of* $u(t,u_n)$*) and*

$$\forall_{T \in (0,T_0)} \sup_{t\in[0,T]} \|u(t,u_n) - u(t,u_0)\|_{X^\alpha} \to 0 \quad when \quad n \to \infty. \tag{2.3.19}$$

Proof. Since $\{u_n\} \subset X^\alpha$ is convergent to u_0 in X^α there is $\rho > 0$ such that

$$\|u_n - u_0\|_{X^\alpha} < \rho \quad \text{for} \quad n \in N \tag{2.3.20}$$

and hence, as a result of Proposition 2.3.1, $T_0 := \inf\{\tau_{u_n} : n = 1,2,\dots\}$ is a positive number. For any $T \in (0,T_0)$ and $n = 1,2,\dots$, we shall next show the validity of the following implication:

(2.3.21)
$$\begin{aligned}&\text{if } T' \in [0,T] \text{ and } \sup_{t\in[0,T']} \|u(t,u_n) - u(t,u_0)\|_{X^\alpha} \le \rho \text{ then} \\ &\sup_{t\in[0,T']} \|u(t,u_n) - u(t,u_0)\|_{X^\alpha} \le \|u_n - u_0\|_{X^\alpha} const.(\alpha,\rho,T).\end{aligned}$$

To justify this condition we estimate the difference of solutions $\big(u(t,u_n)-u(t,u_0)\big)$ based on the integral equation (2.2.1). The Lipschitz condition for F ensures that

$$\|F(u(t,u_n))-F(u(t,u_0))\|_X \le L_\rho \|u(t,u_n)-u(t,u_0)\|_{X^\alpha},\ t\in[0,T'],$$

and we obtain

(2.3.22)
$$\begin{aligned}
&\|u(t,u_n)-u(t,u_0)\|_{X^\alpha}\\
&\le \|e^{-At}(u_n-u_0)+\int_0^t e^{-A(t-s)}[F(u(s,u_n))-F(u(s,u_0))]ds\|_{X^\alpha}\\
&\le c_0\|u_n-u_0\|_{X^\alpha}+c_\alpha L_\rho\int_0^t \frac{e^{-a(t-s)}}{(t-s)^\alpha}\|u(s,u_n)-u(s,u_0)\|_{X^\alpha}ds,\ t\in[0,T'].
\end{aligned}$$

From (2.3.22) it follows, given Lemma 1.2.9, that

$$\begin{aligned}
&\sup_{t\in[0,T']}\|u(t,u_n)-u(t,u_0)\|_{X^\alpha}\\
&\le c_0\|u_n-u_0\|_{X^\alpha} const.(c_\alpha L_\rho,\alpha,T').
\end{aligned}\tag{2.3.23}$$

Since, from Lemma 1.2.9, *const.* appearing at the right side in (2.3.23) is increasing with respect to its last argument we can increase it to the value $const.(c_\alpha L_\rho,\alpha,T)$, which proves the required implication.

Now fix any $T\in(0,T_0)$, $n=1,2,\dots$, and consider the solution $u(t,u_n)$. From (2.3.20) and continuity of X^α solutions at $t=0$ there exists $t_n>0$ such that

$$\|u(t,u_n)-u(t,u_0)\|_{X^\alpha}<\rho \ \text{ for }\ t\in[0,t_n].$$

Let T_n' denote the (hypothetical) first positive time at which

$$\|u(T_n',u_n)-u(T_n',u_0)\|_{X^\alpha}=\rho.$$

If $T_n'\le T$ then, using the implication (2.3.21), we find

$$\sup_{t\in[0,T_n']}\|u(t,u_n)-u(t,u_0)\|_{X^\alpha}\le\|u_n-u_0\|_{X^\alpha} const.(\alpha,\rho,T).$$

However, since

$$\|u_n-u_0\|_{X^\alpha} const.(\alpha,\rho,T)\le\frac{\rho}{2}\ \text{ for }\ n\ge n_T,$$

we arrive, for each $n\ge n_T$, at a contradiction with the hypothesis that $T_n'\le T$.

Therefore the following condition must be valid:

$$\forall_{T\in(0,T_0)}\ \exists_{n_T\in N}\ \forall_{n\ge n_T}\ \sup_{t\in[0,T]}\|u(t,u_n)-u(t,u_0)\|_{X^\alpha}<\rho.\tag{2.3.24}$$

As a consequence of (2.3.24), (2.3.21) we have further

$$\forall_{T\in(0,T_0)}\ \exists_{n_T\in N}\ \forall_{n\ge n_T}\ \sup_{t\in[0,T]}\|u(t,u_n)-u(t,u_0)\|_{X^\alpha}\le\|u_n-u_0\|_{X^\alpha} const.(\alpha,\rho,T)$$

and consequently

$$\lim_{n\to\infty} \sup_{t\in[0,T]} \|u(t,u_n) - u(t,u_0)\|_{X^\alpha} = 0 \quad \text{for each} \quad T \in (0,T_0).$$

The above considerations ensure that $\tau_{u_0} \geq T_0$. The proof is complete. □

Bibliographical Notes.

- **Section 2.1.** Formulation of the Cauchy problem was given in [HE 1] with the improvement of [MIK]. It was also noted in [HA 2].
- **Sections 2.2 and 2.3.** These sections are based on [HE 1] and [LU 1]. We should recall additionally the monograph [PAZ 2] and the manuscript [RB 2].

CHAPTER 3

Semigroups of global solutions

We continue the study of the abstract Cauchy problem (2.1.1),

$$\begin{cases} \dot{u} + Au = F(u), \ t > 0, \\ u(0) = u_0, \end{cases}$$

under the basic Assumption 2.1.1. Our concerns here are global solutions of (2.1.1) and the conditions guaranteeing the existence of a semigroup of global X^α solutions corresponding to (2.1.1).

Definition 3.0.1. *A function* $u = u(t)$ *is called a* ***global*** X^α ***solution*** *of* (2.1.1) *if it fulfills the requirements of Definition* 2.1.1 *with* $\tau = +\infty$.

According to the local existence result of Theorem 2.1.1 for the proof of global X^α solvability of (2.1.1) it suffices to show that the norm $\|u(t,u_0)\|_{X^\alpha}$ does not *blow up* in a finite time T. That is,

$$\forall_{T>0} \limsup_{t \to T^-} \|u(t)\|_{X^\alpha} < +\infty.$$

Various criteria for the existence of global solutions are connected in general with the appropriate limitation of the growth of the nonlinear term F.

Sublinear growth restriction. Global solvability of (2.1.1) follows provided that F fulfills the sublinear growth restriction:

$$\|F(v)\|_X \le const.(1 + \|v\|_{X^\alpha}), \quad \text{for} \quad v \in X^\alpha. \tag{3.0.1}$$

Proof. Indeed, take $u_0 \in X^\alpha$ and denote by $u = u(t,u_0)$ the local X^α solution of (2.1.1) obtained in Theorem 2.1.1. This solution is known to exist on some interval $[0, \tau_{u_0})$ which is maximal in the sense of (2.1.2). Assume now that τ_{u_0} is finite. Certainly, u satisfies the integral equation

$$u(t,u_0) = e^{-At}u_0 + \int_0^t e^{-A(t-s)} F(u(s,u_0))ds \quad \text{for} \quad t \in [0, \tau_{u_0}).$$

Applying (3.0.1) and the usual estimates for analytic semigroups we then get

$$\|u(t,u_0)\|_{X^\alpha} \le \|e^{-At}\|_{\mathcal{L}(X,X)}\|u_0\|_{X^\alpha} + \int_0^t \|A^\alpha e^{-A(t-s)}\|_{\mathcal{L}(X,X)}\|F(u(s,u_0))\|_X ds$$

$$\le c_0\|u_0\|_{X^\alpha} + c_\alpha\, const. \int_0^t \frac{e^{-a(t-s)}}{(t-s)^\alpha}(1+\|u(s,u_0)\|_{X^\alpha})ds, \ t \in [0,\tau_{u_0}).$$

Consider further $y(t) := \|u(t,u_0)\|_{X^\alpha}$. Regularity of u ensures that $y : [0,\tau_{u_0}) \to [0,+\infty)$ is a continuous function. From the above calculations it is also seen that y satisfies the Volterra type integral inequality

$$y(t) \le a + b\int_0^t \frac{1}{(t-s)^\beta} y(s)ds, \quad \text{for } t \in [0,\tau_{u_0}),$$

with $a = c_0\|u_0\|_{X^\alpha} + c_\alpha\, const. \int_0^{\tau_{u_0}} \frac{e^{-a(t-s)}}{(t-s)^\alpha}$ and $b = c_\alpha const.$ Therefore, as a consequence of Lemma 1.2.9, we obtain that

$$\sup_{t\in[0,\tau_{u_0})} y(t) \le a\ const.(b,\beta,\tau_{u_0}),$$

which shows that (2.1.2) fails. However, the latter contradicts that τ_{u_0} is finite, so that the existence of global solutions of (2.1.1) is proved. □

Unfortunately condition (3.0.1), as very restrictive, is rarely satisfied. It, nevertheless, cannot be significantly weakened unless some additional knowledge of the solutions of (2.1.1) is provided. For this purpose we shall introduce in further considerations an additional Banach space Y and assume that solutions of (2.1.1) are known to be estimated in the norm of this space. This allows us to transform (3.0.1) into a more general condition (3.1.4) similar in spirit, which is applicable to a large number of equations from the applied sciences.

3.1. Generation of nonlinear semigroups

If to any initial data $u_0 \in X^\alpha$ corresponds a global X^α solution $u(t,u_0)$ of (2.1.1) then a C^0 *semigroup* $\{T(t)\}$ *corresponding to* (2.1.1) may be defined on X^α via the relation

(3.1.1) $$T(t)u_0 = u(t,u_0),\ t \ge 0.$$

Remark 3.1.1. Recall Definition 1.1.1 and consider briefly the properties of the family of mappings $\{T(t)\}$ defined in (3.1.1). From the definition of the X^α solution to (2.1.1) it follows immediately that $T(0)u_0 = u(0,u_0) = u_0$. Since (2.1.1) is autonomous, the property

(3.1.2) $$T(t_1)T(t_2)u_0 = T(t_1+t_2)u_0,\ t_1,t_2 \ge 0,$$

is a consequence of the *uniqueness* of solutions. Finally continuity of the map

$$[0,+\infty)\times X^\alpha \ni (t,u_0) \to T(t)u_0 = u(t,u_0) \in X^\alpha$$

follows from Proposition 2.3.2 and the fact that $u(\cdot,u_0) \in C([0,+\infty),X^\alpha)$ (note that Proposition 2.3.2 reads that $T(t)u_0$ is continuous as a function of u_0 uniformly with respect to t varying in compact subsets of $[0,+\infty)$).

We need now to formulate effective conditions guaranteeing global X^α solvability of (2.1.1). But since the construction of the attractor for (2.1.1) will be our main concern here, it is reasonable to find for (2.1.1) such a set of requirements which would guarantee, besides the existence of $\{T(t)\}$, an additional property that orbits of bounded sets would be bounded. Note that, by Proposition 2.3.1 and Lemma 1.1.2, a compact C^0 semigroup corresponding to (2.1.1) which possesses a global attractor must have bounded orbits of bounded sets.

3.1.1. Semigroups on X^α. Consider the two conditions:

(A_1) Relation (3.1.1) *defines on X^α, corresponding to* (2.1.1), *a C^0 semigroup $\{T(t)\}$ of global X^α solutions having orbits of bounded sets bounded.*

(A_2) It is possible to choose

- *a Banach space Y, with $D(A) \subset Y$,*
- *a locally bounded function $c : [0, +\infty) \longrightarrow [0, +\infty)$,*
- *a nondecreasing function $g : [0, +\infty) \longrightarrow [0, +\infty)$,*
- *a certain number $\theta \in [0, 1)$,*

such that, for each $u_0 \in X^\alpha$, both conditions

$$\|u(t, u_0)\|_Y \leq c(\|u_0\|_{X^\alpha}),\ t \in (0, \tau_{u_0}), \tag{3.1.3}$$

and

$$\|F(u(t, u_0))\|_X \leq g(\|u(t, u_0)\|_Y)(1 + \|u(t, u_0)\|^\theta_{X^\alpha}),\ t \in (0, \tau_{u_0}), \tag{3.1.4}$$

hold.

Remark 3.1.2. The subordination condition (3.1.4) may also be formulated in another form, as in [C-D 1]:

$$\|F(u(t, u_0))\|_X \leq g(\|u(t, u_0)\|_Y)(1 + \|u(t, u_0)\|_{X^{\alpha'}}),\ t \in (0, \tau_{u_0}), \tag{3.1.5}$$

with some $\alpha' \in [0, 1)$, provided that $Y \subset X$. Indeed, such equivalence follows through Proposition 1.3.9 and the estimate

$$\|u(t, u_0)\|_{X^{\alpha'}} \leq c_{\frac{\alpha'}{\alpha}} \|u(t, u_0)\|_X^{1-\frac{\alpha'}{\alpha}} \|u(t, u_0)\|_{X^\alpha}^{\frac{\alpha'}{\alpha}}, \tag{3.1.6}$$

where $\alpha \in (\alpha', 1)$ is arbitrary, since the first right factor in (3.1.6) may be attached to g in (3.1.5) without destroying its properties. The implication from (3.1.4) to (3.1.5) follows from Young's inequality.

Remark 3.1.3. The subordination condition in [LU 1, p. 272] is, at first glance, less restrictive:

$$\|F(u(t, u_0))\|_X \leq c(\|u(t, u_0)\|_X)(1 + \|u(t, u_0)\|^\gamma_{X^\alpha}),\ t \in (0, \tau_{u_0}), \tag{3.1.7}$$

since the exponent γ, belonging to the interval $(1, \frac{1}{\alpha})$, is strictly greater than 1. However, thanks to (3.1.6), for $Y \subset X$ the condition (3.1.7) may be extended to the form (3.1.5)

$$\|F(u(t, u_0))\|_X \leq g(\|u(t, u_0)\|_Y)(1 + \|u(t, u_0)\|_{X^{\alpha''}}),\ t \in (0, \tau_{u_0}),$$

with $\alpha'' = \alpha\gamma$. Since $\gamma \in (1, \frac{1}{\alpha})$, such α'' is evidently between α and 1.

We shall prove the following global existence result.

Theorem 3.1.1. *Under Assumption* 2.1.1, *conditions* (A_1) *and* (A_2) *are equivalent.*

Proof. Let us take $u_0 \in X^\alpha$, and $t > 0$. From the integral equation

$$u(t, u_0) = e^{-At}u_0 + \int_0^t e^{-A(t-s)} F\Big(u(s, u_0)\Big) ds, \tag{3.1.8}$$

we obtain

(3.1.9)

$$\begin{aligned}
&\|u(t, u_0)\|_{X^\alpha} \\
&\leq \|A^\alpha e^{-At}u_0\|_X + \int_0^t \|A^\alpha e^{-A(t-s)}\|_{\mathcal{L}(X,X)} \|F\Big(u(s, u_0)\Big)\|_X ds \\
&\leq c_0 e^{-at}\|u_0\|_{X^\alpha} + \int_0^t c_\alpha \frac{e^{-a(t-s)}}{(t-s)^\alpha} g(\|u(s, u_0)\|_Y)(1 + \|u(s, u_0)\|^\theta_{X^\alpha}) ds.
\end{aligned}$$

Using further the estimate (3.1.3) and recalling that $\int_0^\infty \frac{e^{-ay}}{y^\alpha} dy = \frac{\Gamma(1-\alpha)}{a^{1-\alpha}}$ we get from (3.1.9)

(3.1.10)

$$\begin{aligned}
\|u(t, u_0)\|_{X^\alpha} \leq & c_0\|u_0\|_{X^\alpha} \\
& + c_\alpha g(c(\|u_0\|_{X^\alpha})) \frac{\Gamma(1-\alpha)}{a^{1-\alpha}} \left(1 + \sup_{\tau \in [0,t]} \|u(\tau, u_0)\|^\theta_{X^\alpha}\right).
\end{aligned}$$

Since the right side of the above inequality is nondecreasing in t, (3.1.10) may be strengthened to the condition

$$\begin{aligned}
\sup_{\tau \in [0,t]} \|u(\tau, u_0)\|_{X^\alpha} \leq & c_0\|u_0\|_{X^\alpha} \\
& + g(c(\|u_0\|_{X^\alpha})) c_\alpha \frac{\Gamma(1-\alpha)}{a^{1-\alpha}} \left(1 + \sup_{\tau \in [0,t]} \|u(\tau, u_0)\|^\theta_{X^\alpha}\right).
\end{aligned}$$

As a consequence, $z := \sup_{\tau \in [0,t]} \|u(\tau, u_0)\|_{X^\alpha}$ fulfills the relation

$$0 \leq z \leq b(u_0)(1 + z^\theta), \tag{3.1.11}$$

where $b(u_0) := \max\{c_0\|u_0\|_{X^\alpha}, g(c(\|u_0\|_{X^\alpha})) c_\alpha \frac{\Gamma(1-\alpha)}{a^{1-\alpha}}\}$. However, any number z satisfying (3.1.11) is bounded by a positive root $z_1(u_0)$ of an algebraic equation

$$b(u_0)(1 + z^\theta) - z = 0,$$

so that $\sup_{\tau \in [0,t]} \|u(\tau, u_0)\|_{X^\alpha}$ remains bounded by $z_1(u_0)$ for all $t > 0$. Hence, by Theorem 2.1.1, for each $u_0 \in X^\alpha$, the problem (2.1.1) has a global solution $u(t, u_0)$ and, moreover,

$$\sup_{\tau \in [0,\infty)} \|u(\tau, u_0)\|_{X^\alpha} \leq z_1(u_0). \tag{3.1.12}$$

A strongly continuous semigroup of operators $T(t) : X^\alpha \longrightarrow X^\alpha$, $t \geq 0$, defined in (3.1.1), is then generated by (2.1.1). Furthermore, as follows from (3.1.10)-(3.1.12), the orbit $\gamma^+(\mathcal{B}_{X^\alpha}(R))$ is bounded in X^α whenever the function $c(\cdot)$ is locally bounded on $[0,\infty)$. This is because $b(u_0)$ depends only on the radius R of a ball $B_{X^\alpha}(R)$ and is independent of the choice of a particular $u_0 \in B_{X^\alpha}(R)$. This justifies that (A_2) implies (A_1).

The proof of the converse is formal. If (3.1.1) defines, corresponding to (2.1.1), a C^0 semigroup $\{T(t)\}$ of global X^α solutions having orbits of bounded sets bounded, then for each $r > 0$ there exists $m_r > 0$ such that

$$\sup_{v \in B_{X^\alpha}(r)} \sup_{t \geq 0} \|u(t,v)\|_{X^\alpha} = m_r,$$

whereas $F : X^\alpha \to X$ is bounded on bounded sets as a consequence of Assumption 2.1.1. Then (A_2) holds with $Y = X^\alpha$, $c(s) = m_s$, $g(s) = \sup_{\|v\|_{X^\alpha} \leq s} \|F(v)\|_X$ and $\theta = 0$. The proof is complete. □

Remark 3.1.4. An implicit assumption of the above theorem is the inclusion $X^\alpha \subset Y$ (see the estimate (3.1.3)). Nevertheless, we skipped expressing it directly since, as seen further in Theorem 3.2.2, condition (3.1.3) may also be formulated in a weaker form (3.2.5), in which validity of this inclusion is not needed.

Remark 3.1.5. It should also be pointed again (see Corollary 2.1.2) that, under Assumption 2.1.1, the nonlinear term F is Lipschitz continuous on bounded subsets of X^β whenever $\beta \in [\alpha, 1)$. Therefore, if condition (A_2) holds then, simultaneously, (2.1.1) generates a C^0 semigroup $\{T_\beta(t)\}$ having orbits of bounded sets bounded on each of the spaces X^β, $\beta \in [\alpha, 1)$. We often use this property in further considerations (especially in examples), choosing the parameter β from the admissible scale $[\alpha, 1)$ as it is needed in a particular situation. Nevertheless, since the semigroup $\{T_\beta(t)\}$ is defined by solutions $u(t, u_0)$, $u_0 \in X^\beta$, which, generically, are not sensitive to the choice of β we shall usually omit the symbol β in the subscript of $\{T(t)\}$. The above concepts will be extended in Subsection 4.3 where the solutions to (2.1.1) are studied in the scales of fractional power spaces.

3.1.2. Metric phase spaces. Let us focus further on the situation when the semigroup of global solutions generated by (2.1.1) needs to be considered on a proper subset $V \subset X^\alpha$. One of the possible reasons is that the semigroup may not be dissipative on the whole of X^α, as it is in the case when the set of equilibria of $\{T(t)\}$ is unbounded (the Cahn-Hilliard equation may serve here as an example). Nevertheless also possible is the situation when some initial conditions produce solutions which blow up in finite or infinite time although, for the others, trajectories are globally bounded and, moreover, enter asymptotically a common subset of the phase space on which (2.1.1) is considered. Such situations, when the global attractor may occur only on a metric subspace of the starting phase space, seem to be much more natural for parabolic problems and may frequently be observed in applications. This justifies the further restriction of $\{T(t)\}$ to an appropriate complete metric subspace V of the original phase space X^α.

Based on considerations of Theorem 3.1.1 we shall obtain below a similar result concerning the existence of a C^0 semigroup corresponding to (2.1.1) on the subset $V \subset X^\alpha$. Before we formulate such a result in Corollary 3.1.2 we shall make some observations which, step by step, allow us to introduce this corollary.

Observation 3.1.1. *Let Assumption* 2.1.1 *be satisfied and* $u_0 \in X^\alpha$. *Then* $u(t, u_0)$ *is defined and uniformly bounded for* $t \geq 0$ *if, and only if, there are a Banach space* $Y \supset D(A)$ *and a nondecreasing function* $g : [0, \infty) \longrightarrow [0, \infty)$ *such that*

(3.1.13)
$$\exists_{c>0}\ \forall_{t\in(0,\tau_{max}(u_0))}\ \|u(t, u_0)\|_Y \leq c$$

and

(3.1.14)
$$\exists_{\theta\in[0,1)}\ \forall_{t\in(0,\tau_{max}(u_0))} \|F(u(t, u_0))\|_X \leq g(\|u(t, u_0)\|_Y)(1 + \|u(t, u_0)\|^\theta_{X^\alpha}).$$

Observation 3.1.2. *Let Assumption* 2.1.1 *be satisfied and* $V \subset X^\alpha$. *Then* $u(t, u_0)$ *exists globally in time for each* $u_0 \in V$ *if, and only if, there are a Banach space* $Y \supset D(A)$ *and a nondecreasing function* $g : [0, \infty) \longrightarrow [0, \infty)$ *such that*

(3.1.15)
$$\forall_{u_0\in V}\ \exists_{c(u_0)>0}\ \forall_{t\in(0,\tau_{max}(u_0))}\ \|u(t, u_0)\|_Y \leq c(u_0)$$

and

(3.1.16)
$$\exists_{\theta\in[0,1)}\ \forall_{u_0\in V}\ \forall_{t\in(0,\tau_{max}(u_0))}$$
$$\|F(u(t, u_0))\|_X \leq g(\|u(t, u_0)\|_Y)(1 + \|u(t, u_0)\|^\theta_{X^\alpha}).$$

Observation 3.1.2 contains criteria for boundedness of trajectories of points. In Observation 3.1.3 below, a similar result concerning orbits of bounded sets is given.

Observation 3.1.3. *Let Assumption* 2.1.1 *be satisfied and* $V \subset X^\alpha$. *Then the following conditions are equivalent:*

(3.1.17)
$u(t, u_0)$ does not blow up for $u_0 \in V$ and for each bounded subset B of X^α the set $\{u(t, u_0) : u_0 \in B \cap V, t \geq 0\}$ is bounded in X^α,

(3.1.18)
there are a Banach space $Y \supset D(A)$ and a nondecreasing function $g : [0, \infty) \to [0, \infty)$ such that (3.1.15), (3.1.16) *hold and the function $c : V \longrightarrow [0, \infty)$ in the condition* (3.1.15) *takes bounded subsets of V into bounded subsets of* $[0, \infty)$.

The above observations allow us to formulate the following conclusion.

Corollary 3.1.2. *Let Assumption* 2.1.1 *be satisfied, and V be any subset of* X^α. *If the requirements of (A_2) hold for $u_0 \in V$, that is* *it is possible to choose a Banach space Y, with $D(A) \subset Y$, a locally bounded function $c : [0,+\infty) \to [0,+\infty)$, a nondecreasing function $g : [0,+\infty) \longrightarrow [0,+\infty)$, and a certain number $\theta \in [0,1)$, such that for $u_0 \in V$ both* (3.1.3) *and* (3.1.4) *are satisfied,*

then local solutions corresponding to $u_0 \in V$ exist for all $t \geq 0$. If, in addition,

$$u_0 \in V \text{ implies that } u(t,u_0) \in V \text{ as long as it exists}$$

then relation (3.1.1) *defines on V, corresponding to* (2.1.1), *a C^0 semigroup $\{T(t)\}$ of global X^α solutions having orbits of bounded sets bounded.*

3.2. Smoothing properties of the semigroup

We shall proceed to the study of the smoothing action of the semigroup generated by (2.1.1) on X^α.

Smoothing effects. If we take Assumption 2.1.1 and denote by $T(t)u_0$ a local X^α solution of (2.1.1) originating at $u_0 \in X^\alpha$ then, thanks to Proposition 2.3.1, for any $t \in (0,\tau_B)$ the image $T(t)B$ of a bounded set $B \subset X^\alpha$ will be properly defined.

Below, we shall prove that for each $t \in (0,\tau_B)$ the image $T(t)B$ (which, from Definition 2.1.1, is contained in $D(A) = X^1$) remains bounded in $X^{\alpha+\varepsilon}$ whenever $\alpha+\varepsilon < 1$.

Proposition 3.2.1. *Under Assumption* 2.1.1, *let $T(t)u_0$ denote a local X^α solution of* (2.1.1) *originating at $u_0 \in X^\alpha$. Then*

$$\forall_{\substack{B\subset X^\alpha \\ B \text{ bounded}}} \forall_{t\in(0,\tau_B)}\ T(t)B \text{ is bounded in } X^{\alpha+\varepsilon} \text{ for each } \alpha+\varepsilon<1.$$

Proof. For bounded $B \subset X^\alpha$, $\rho > 0$ such that $B \subset B_{X^\alpha}(\rho)$, any $u_0 \in B$ and $t \in (0,\tau_B)$ (where τ_B is chosen as in Proposition 2.3.1) we have

$$\begin{aligned}
&\|A^\varepsilon u(t,u_0)\|_{X^\alpha} \\
&\leq \|A^\varepsilon e^{-At}\|_{\mathcal{L}(X,X)}\|u_0\|_{X^\alpha} + \int_0^t \|A^{\alpha+\varepsilon}e^{-A(t-s)}\|_{\mathcal{L}(X,X)}\|F(u(s,u_0))\|_X ds \\
&\leq c_\varepsilon \frac{e^{-at}}{t^\varepsilon}\rho + b\int_0^t c_{\alpha+\varepsilon}\frac{e^{-ay}}{y^{\alpha+\varepsilon}}dy,
\end{aligned}$$

where $b := \sup_{v\in B_{X^\alpha}(0,c\rho+1)} \|F(v)\|_X$. Therefore, $T(t)B$ is bounded in $X^{\alpha+\varepsilon}$, which completes the proof of Proposition 3.2.1. □

Actually, a stronger estimate of the solutions in X^1 can be shown. However in that case the estimate is much more sophisticated.

Lemma 3.2.1. *Under the assumptions of Proposition* 3.2.1 *we have*

$$\forall_{\substack{B\subset X^\alpha \\ B \text{ bounded}}} \forall_{t\in(0,\tau_B)}\ T(t)B \text{ is bounded in } X^1.$$

Proof. As shown in Step 3 of the proof of Lemma 2.2.1,

$$A\int_0^t e^{-A(t-s)}F(u(s))ds = \int_0^t Ae^{-A(t-s)}F(u(s))ds, \ t \in (0, \tau_{u_0}),$$

where, for fixed $0 < t_0 < t < t_1 < \tau_{u_0}$, we have

$$\left\|\int_{t_0}^t Ae^{-A(t-s)}F(u(s))ds\right\|_X$$
$$\leq \|I - e^{-A(t-t_0)}\|_{\mathcal{L}(X,X)}\|F(u(t))\|_X + c_1 const._{\alpha,t_0,t_1}(t-t_0)^\alpha,$$

and

$$\left\|\int_0^{t_0} Ae^{-A(t-s)}F(u(s))ds\right\|_X$$
$$\leq \int_0^{t_0} c_1 \frac{e^{-a(t-s)}}{t-s}\|F(u(s))\|_X ds \leq c_1 \sup_{s\in[0,t_0]} \|F(u(s))\|_X \ln\frac{t}{t-t_0}.$$

Since $F : X^\alpha \to X$ is Lipschitz continuous on bounded sets, the above formulas ensure that, for fixed $t \in (0, \tau_{u_0})$,

$$\left\|A\int_0^t e^{-A(t-s)}F(u(s))ds\right\|_X \leq const.\left(\alpha, t, \sup_{s\in[0,t]}\|u(s)\|_{X^\alpha}, F\right).$$

From Proposition 2.3.1, if $B \subset X^\alpha$ is bounded, then

$$\exists_{\tau_B>0}\ \exists_{M_B>0}\ \forall_{u_0\in B}\ \forall_{t\in[0,\tau_B)}\ \sup_{s\in[0,t]}\|u(s,u_0)\|_{X^\alpha} < M_B,$$

so that

$$\forall_{t\in(0,\tau_B)}\ \forall_{u_0\in B}\ \left\|A\int_0^t e^{-A(t-s)}F(u(s,u_0))ds\right\|_X \leq const.(\alpha, t, M_B, F).$$

Based on (2.2.1) we thus have

$$\forall_{\substack{B\subset X^\alpha \\ B \text{ bounded}}}\ \exists_{\tau_B>0}\ \forall_{t\in(0,\tau_B)}\ \forall_{u_0\in B}$$
$$\|u(t,u_0)\|_{X^1} \leq \frac{c_{1-\alpha}e^{-at}}{t^{1-\alpha}}\|u_0\|_{X^\alpha} + const.(\alpha, t, M_B, F),$$

which completes the proof. □

Below we are going to prove a smoothing result different in spirit, which will be particularly useful in consideration of problems where the resolvent of A is not compact.

Theorem 3.2.1. *Let Assumption* 2.1.1 *be satisfied and* (3.1.1) *define a C^0 semigroup $T(t) : X^\alpha \to X^\alpha$, $t > 0$, of global X^α solutions of* (2.1.1). *Then, for each $t > 0$, $T(t)$ takes sets that are bounded in X^α and precompact in X into precompact subsets of X^α.*

Proof. We should show that if B is a bounded subset of X^α which, in addition, is precompact in X then for each $t > 0$ the image $T(t)B$ is precompact in X^α.

Let us choose $t > 0$ and consider any sequence $\{u_n\} \subset B$. It suffices to show that $\{T(t)u_n\}$ has a subsequence convergent in X^α. Since $\{u_n\}$ as a subset of B is bounded then, from compactness of $cl_X B$ in X, there exists a subsequence $\{u_{n_k}\}$ of $\{u_n\}$ which is convergent in X. In addition, since the sequence $\{u_{n_k}\} \subset B$ and B is bounded in X^α, the set $\{T(s)u_{n_k} : s \in [0, \tau_B], k \in N\}$ (where $\tau_B \in (0, t)$ is chosen as in Proposition 2.3.1) is bounded in X^α. Therefore, via Assumption 2.1.1, there exists a Lipschitz constant $L(\tau_B, B)$ such that

$$(3.2.1)\quad \|F(T(s)u_{n_r}) - F(T(s)u_{n_k})\|_X \leq L(\tau_B, B)\|T(s)u_{n_r} - T(s)u_{n_k}\|_{X^\alpha},$$

for all $s \in [0, \tau_B]$ and $r, k \in N$. Let us write the integral equation for the difference $[T(\tau)u_{n_r} - T(\tau)u_{n_k}]$:

(3.2.2)

$$\begin{aligned} T(\tau)u_{n_r} - T(\tau)u_{n_k} &= e^{-A\tau}[u_{n_r} - u_{n_k}] \\ &+ \int_0^\tau e^{-A(\tau-s)}[F(T(s)u_{n_r}) - F(T(s)u_{n_k})]ds, \ \tau \in (0, \tau_B). \end{aligned}$$

Estimating (3.2.2) in a standard way, and applying condition (3.2.1), we obtain

(3.2.3)

$$\begin{aligned} &\|T(\tau)u_{n_r} - T(\tau)u_{n_k}\|_{X^\alpha} \leq \|A^\alpha e^{-A\tau}\|_{\mathcal{L}(X,X)}\|u_{n_r} - u_{n_k}\|_X \\ &+ \int_0^\tau \|A^\alpha e^{-A(\tau-s)}\|_{\mathcal{L}(X,X)}\|F(T(s)u_{n_r}) - F(T(s)u_{n_k})\|_X ds \\ &\leq c_0 \frac{e^{-a\tau}}{\tau^\alpha}\|u_{n_r} - u_{n_k}\|_X \\ &+ \int_0^\tau c_\alpha \frac{e^{-a(\tau-s)}}{(\tau-s)^\alpha} L(\tau_B, B)\|T(s)u_{n_r} - T(s)u_{n_k}\|_{X^\alpha} ds \\ &\leq \frac{c_0\|u_{n_r} - u_{n_k}\|_X}{\tau^\alpha} \\ &+ c_\alpha L(\tau_B, B)\int_0^\tau \frac{1}{(\tau-s)^\alpha}\|T(s)u_{n_r} - T(s)u_{n_k}\|_{X^\alpha} ds, \ \tau \in (0, \tau_B). \end{aligned}$$

Let us choose any $0 < t^* < \min\{\frac{\tau_B}{2}, t\}$ satisfying

$$\frac{c_\alpha L(\tau_B, B){t^*}^{1-\alpha}}{2^{1-2\alpha}} \frac{2}{1-\alpha} \leq \frac{1}{2}.$$

As a consequence of Lemma 1.2.9, inequality (3.2.3) implies the estimate

$$(3.2.4)\qquad \|[T(t^*)u_{n_r} - T(t^*)u_{n_k}]\|_{X^\alpha} \leq \frac{2c_0\|u_{n_r} - u_{n_k}\|_X}{t^{*\alpha}}.$$

Since $\{u_{n_k}\}$ was a Cauchy sequence in the topology of X, it is seen from (3.2.4) that $\{T(t^*)u_{n_k}\}$ is also a Cauchy sequence in X^α so that $\{T(t^*)u_{n_k}\}$ is convergent in the topology of X^α. Therefore $\{T(t)u_{n_k}\} = \{T(t-t^*)T(t^*)u_{n_k}\}$ is convergent in X^α as a result of the continuity of $T(t)$. The proof is complete. □

Further existence result. As a particular consequence of the smoothing action described above we shall strengthen the existence result of Theorem 3.1.1. For this, let us introduce the following set of requirements (A_2'), being a counterpart of (A_2).

(A_2') *There are given*

- *a Banach space* Y, *with* $D(A) \subset Y$,
- *a function* $c: [0,+\infty) \times (0,+\infty) \longrightarrow R$ *for which* $c(\cdot, s)$ *is locally bounded for each fixed* $s > 0$,
- *a nondecreasing function* $g: [0,+\infty) \longrightarrow [0,+\infty)$,
- *a number* $\theta \in [0,1)$,

such that, for each $u_0 \in X^\alpha$, *both*

$$\|u(t,u_0)\|_Y \leq c(\|u_0\|_{X^\alpha}, s),\ 0 < s \leq t < \tau_{u_0}, \tag{3.2.5}$$

and

$$\|F(u(t,u_0))\|_X \leq g(\|u(t,u_0)\|_Y)(1 + \|u(t,u_0)\|^\theta_{X^\alpha}),\ t \in (0, \tau_{u_0}), \tag{3.2.6}$$

hold.

Such a formulation of the requirements shows that the knowledge of the Y estimate of solutions is important only for t strictly positive. In particular, the Y estimate may be singular at $t = 0$ and this, in fact, does not influence the behavior of the system in the phase space X^α.

We proceed now to the proof of the following.

Theorem 3.2.2. *Under Assumption* 2.1.1, *the conditions* (A_1), (A_2') *are equivalent.*

Proof. Let us take $R > 0$, $\|u_0\|_{X^\alpha} < R$, a quantity $\tau_{B_{X^\alpha}(R)}$ satisfying condition (2.3.17) in the proof of Proposition 2.3.1 and any $t \in [\tau_{B_{X^\alpha}(R)}, \tau_{u_0})$. We write an integral equation equivalent to (2.1.1) in the form

$$u(t,u_0) = e^{-At}u_0 + \left(\int_0^{\tau_{B_{X^\alpha}(R)}} + \int_{\tau_{B_{X^\alpha}(R)}}^t\right) e^{-A(t-s)} F\Big(u(s,u_0)\Big)ds, \tag{3.2.7}$$

and obtain that

(3.2.8)
$$\|u(t,u_0)\|_{X^\alpha} \leq c_0\|u_0\|_{X^\alpha} + 1 + \int_{\tau_{B_{X^\alpha}(R)}}^t c_\alpha \frac{e^{-a(t-s)}}{(t-s)^\alpha} g(\|u(s,u_0)\|_Y)(1 + \|u(s,u_0)\|^\theta_{X^\alpha})ds.$$

Application of (3.2.5) to the right hand side of (3.2.8) leads to the estimate

(3.2.9)
$$\|u(t,u_0)\|_{X^\alpha} \leq c_0 R + 1 + g(c(\|u_0\|_{X^\alpha}, \tau_{B_{X^\alpha}(R)}))c_\alpha \frac{\Gamma(1-\alpha)}{a^{1-\alpha}}\left(1 + \sup_{\tau\in[0,t]} \|u(\tau,u_0)\|^\theta_{X^\alpha}\right).$$

Therefore, $z := \sup_{\tau\in[0,t]} \|u(\tau,u_0)\|_{X^\alpha}$ fulfills the relation (3.1.11) in the proof of Theorem 3.1.1 with $b(u_0)$ defined now as

$$b(u_0) := \max\left\{c_0R+1, g(c(\|u_0\|_{X^\alpha}, \tau_{B_{X^\alpha}(R)}))c_\alpha\frac{\Gamma(1-\alpha)}{a^{1-\alpha}}\right\}$$

We are thus in the same position as in the proof of Theorem 3.1.1, so that the proof can be finished analogously. □

3.3. Compactness results

One of the advantages of the parabolic regular initial boundary value problems is that the corresponding semigroups are usually compact. The latter provides the strong control over the trajectories of a system and, essentially, is related to the compactness of the resolvent of the operator appearing in the main part of the equation.

We start with the necessary and sufficient condition for compactness of a C^0 semigroup $\{S(t)\}$ of bounded linear operators due to A. Pazy [PAZ 1].

Proposition 3.3.1. *A C^0 semigroup $\{S(t)\}$ of bounded linear operators is compact for $t>0$ if and only if it is continuous in the uniform operator topology for $t>0$ and the resolvent of its infinitesimal generator Λ is a compact linear operator for every λ in the resolvent set $\rho(\Lambda)$ of Λ.*

In this vein we continue next the study of the problem (2.1.1).

Theorem 3.3.1. *Let Assumption 2.1.1 hold, the resolvent of A be compact and the relation (3.1.1) define a C^0 semigroup $T(t): X^\alpha \to X^\alpha$, $t>0$, of global X^α solutions of (2.1.1). Then, for each $t>0$, $T(t)$ is a compact map on X^α.*

Proof. Let us fix $t>0$ and bounded subset B of X^α. Then choose τ_B according to Proposition 3.2.1 and take $t_1 \in (0,\min\{t,\tau_B\})$. From Proposition 3.2.1 it may be seen that $T(t_1)B$ is bounded in $X^{\alpha+\varepsilon}$ for $\alpha+\varepsilon<1$. Since the resolvent of A is compact, we know that the inclusion $X^{\alpha+\varepsilon} \subset X^\alpha$ is compact and $T(t_1)B$ is thus a precompact subset of X^α. Continuity of the map $T(t-t_1)$ in X^α ensures then that the image $T(t)B = T(t-t_1)T(t_1)B$ will be precompact in X^α. The proof is complete. □

Remark 3.3.1. Such a theorem was proved by J. K. Hale (see [HA 2, Theorem 4.2.2]). The proof given there was based on the compactness of maps e^{-At} for $t>0$, which property follows from the compactness of embeddings between the domains of fractional powers of A (ensured by the compactness of the resolvent of A). Theorem 3.3.1 may also be derived from Theorem 3.2.1 based on the compactness of the inclusion $X^\alpha \subset X$. However, the property of $T(t)$ described in Theorem 3.2.1 is more general. As shown below, it will be possible to prove compactness of $T(t)$ on a suitably chosen subspace of X^α, despite the generic noncompactness of the resolvent of A. It is also worth noting that it was not assumed in Theorem 3.3.1 that $T(t)$ takes bounded sets into bounded sets. As may be seen from Theorem 3.3.1 this latter property is generic for sectorial equations in the case when the resolvent of A is compact and is a particular consequence of the smoothing action of $T(t)$ (see Corollary 3.3.2 below).

Corollary 3.3.2. *Under the assumptions of Theorem 3.3.1, the semigroup $\{T(t)\}$ generated by (2.1.1) on X^α is completely continuous. More precisely, for each bounded subset B of X^α and each $T \geq 0$, the following conditions hold:*

$$\text{the union } \bigcup_{0\leq t\leq T} T(t)B \text{ is bounded in } X^\alpha, \tag{3.3.1}$$

$$\text{the union } \bigcup_{t_1\leq t\leq T} T(t)B \text{ is a precompact set in } X^\alpha \text{ for each } t_1 \in (0,T). \tag{3.3.2}$$

Proof. Since, by Theorem 3.3.1, $\{T(t)\}$ is a compact semigroup on X^α, (3.3.2) is a consequence of Lemma 1.1.2. Condition (3.3.1) follows then from (3.3.2) and Proposition 2.3.1. □

The compactness result of Theorem 3.3.1 was proved under the essential assumption that the resolvent of A was compact. Nevertheless, as shown below, Theorem 3.2.1 allows us to make $T(t)$ compact on a suitable metric subspace of X^α even in the case when this latter assumption is not satisfied.

Assumption 3.3.1. *Consider a Banach space Z and the product Banach space $X^\alpha \cap Z$ with the usual norm*

$$\|\cdot\|_{X^\alpha\cap Z} := \|\cdot\|_{X^\alpha} + \|\cdot\|_Z.$$

Let $X^\alpha \cap Z$ be compactly embedded in X, and assume that there is a subset V_α of the product $Z \cap X^\alpha$ such that

(i) V_α is a bounded set in Z,
(ii) $T(t)V_\alpha \subset V_\alpha$, for $t \geq 0$ (that is V_α is positively invariant with respect to $T(t)$).

Now the following generalization of Theorem 3.2.1 can be proved.

Theorem 3.3.3. *Let Assumption 2.1.1 be satisfied and (3.1.1) define a C^0 semigroup $T(t): X^\alpha \to X^\alpha$, $t > 0$, of global X^α solutions of (2.1.1). If, in addition, Assumption 3.3.1 is satisfied then, for each $t > 0$, a map $T(t): cl_{X^\alpha}V_\alpha \to cl_{X^\alpha}V_\alpha$ is compact.*

Proof. From (ii) in Assumption 3.3.1, the C^0 semigroup $\{T(t)\}$ corresponding to (2.1.1) can be restricted to a complete metric space $cl_{X^\alpha}V_\alpha$, so that we have a family of mappings

$$T(t): cl_{X^\alpha}V_\alpha \to cl_{X^\alpha}V_\alpha,\ t \geq 0.$$

Let $B \subset cl_{X^\alpha}V_\alpha$ be bounded, nonempty, and $d(B)$ denote the diameter of B. For fixed $v_0 \in B$ the following inclusions hold:

$$B \subset \{v \in cl_{X^\alpha}V_\alpha : \|v - v_0\|_{X^\alpha} < d(B) + 1\} =: K, \quad B \subset cl_{X^\alpha}(K \cap V_\alpha).$$

Since $T(t)$ is continuous in the topology of X^α we obtain further that

$$T(t)B \subset T(t)(cl_{X^\alpha}[K \cap V_\alpha]) \subset cl_{X^\alpha}T(t)(K \cap V_\alpha). \tag{3.3.3}$$

From (i) in Assumption 3.3.1 and the definition of K, the product $K \cap V_\alpha$ is seen to be bounded in both X^α and Z. Therefore, $K \cap V_\alpha$ is precompact in X and, as a consequence of Theorem 3.2.1, the image $T(t)(K \cap V_\alpha)$ is precompact in X^α. From the inclusion (3.3.3) the set $T(t)B$ is precompact in X^α, which completes the proof of Theorem 3.3.3. □

Corollary 3.3.4. *Let $\{T(t)\}$ be a C^0 semigroup in a metric space X and W be a dense subset of X. Then the following two conditions are equivalent:*

(i) for each open ball $B_X(r)$ the set $T(t)[W \cap B_X(r)]$ is precompact in X,
(ii) for each bounded subset B of X, the set $T(t)B$ is precompact in X.

Bibliographical notes.

- **Section 3.1.** The first condition for global solvability of (2.1.1) of the type considered in (A_2) can be found in [FR 1, p. 180]. It was next generalized in [WA 1], [AM 1] to a form similar to (A_2), again in connection with global solvability of semilinear parabolic problems. Then it was observed in [C-D 3] that this condition, slightly modified, is also sufficient for the existence of a global attractor for problems of such type (compare Corollary 4.2.2 of Chapter 4). Moreover (see [C-C-D 1], [C-D 4], [C-D 6]) this condition proved to be applicable in various examples taken from the applied sciences (see Chapter 6 for detailes).
- **Section 3.3.** Usually compactness of the semigroup is proved under the assumption that the resolvent operator is compact (see [HA 2, Theorem 4.2.2]). As shown in Theorem 3.3.3, in some cases, compactness of the semigroup may be obtained without this assumption.

CHAPTER 4

Construction of the global attractor

In Chapter 3 we have formulated necessary and sufficient conditions for the problem (2.1.1),

$$\begin{cases} \dot{u} + Au = F(u), \ t > 0, \\ u(0) = u_0, \end{cases}$$

to generate a C^0 semigroup $\{T(t)\}$ of global X^α solutions with orbits of bounded sets bounded. Some common situations when this semigroup is compact have also been discussed. In the following considerations we are going to establish *point dissipativeness* of $\{T(t)\}$ which is now a crucial step to show the existence of a global attractor.

4.1. Dissipativeness of $\{T(t)\}$

In condition (A_2) we required that a sufficiently smooth introductory estimate of solutions is known and, simultaneously, the nonlinear term F is subordinated to a θ-th power ($\theta \in [0,1)$) of the sectorial operator A. In some examples these properties may hold only for solutions originating in a subset V of the phase space X^α on which (2.1.1) is studied. Therefore in Subection 3.1.2, the requirements of (A_2) were considered further for solutions starting from $u_0 \in V$ (instead of $u_0 \in X^\alpha$). It was next concluded in Corollary 3.1.2 that if (A_2) is fulfilled for $u_0 \in V$, then the problem (2.1.1) generates on V a C^0 semigroup $\{T(t)\}$ of global X^α solutions with orbits of bounded sets bounded.

The goal of the present section is to show that the semigroup will be point dissipative provided that the introductory estimate (3.1.3) of (A_2) remains *asymptotically independent of initial conditions.*

Theorem 4.1.1. *Let Assumption* 2.1.1 *be satisfied and V be a subset of X^α such that all requirements of (A_2) hold for $u_0 \in V$. If the condition* (3.1.3) *of (A_2) has the additional property that*

(4.1.1) $$\exists_{const.>0} \ \forall_{u_0 \in V} \ \limsup_{t\to+\infty} \|u(t,u_0)\|_Y < const.,$$

then the solutions of (2.1.1) *satisfy the inequality*

$$\exists_{const.'>0}\ \forall_{u_0\in V}\ \limsup_{t\to+\infty}\|u(t,u_0)\|_{X^\alpha}\le const.'. \tag{4.1.2}$$

Proof. As a result of (4.1.1), for each $u_0\in V$ it is possible to choose a positive time t_{u_0} such that, for any $t>\tau>t_{u_0}$,

$$\sup_{s\in[\tau,t)}\|u(s,u_0)\|_Y\le const., \tag{4.1.3}$$

where *const.* is independent of $u_0\in V$. Let us write corresponding to (2.1.1) an integral equation in the form

$$u(t,u_0)=e^{-At}u_0+\left(\int_0^\tau+\int_\tau^t\right)e^{-A(t-s)}F\big(u(s,u_0)\big)ds. \tag{4.1.4}$$

As a result of (3.1.4), for each $u_0\in V$, we obtain

(4.1.5)

$$\begin{aligned}\|u(t,u_0)\|_{X^\alpha}&\le\|A^\alpha e^{-At}u_0\|_X+\int_0^\tau\|A^\alpha e^{-A(t-s)}\|_{\mathcal{L}(X,X)}\|F\big(u(s,u_0)\big)\|_X ds\\&+\int_\tau^t\|A^\alpha e^{-A(t-s)}\|_{\mathcal{L}(X,X)}g(\|u(s,u_0)\|_Y)(1+\|u(s,u_0)\|^\theta_{X^\alpha})ds\\&\le c_0e^{-at}\|u_0\|_{X^\alpha}+\sup_{\|v\|_{X^\alpha}\le z_1(u_0)}\|F(v)\|_X\int_{t-\tau}^t c_\alpha\frac{e^{-ay}}{y^\alpha}dy\\&+g(const.)\left(1+\sup_{s\in[\tau,t]}\|u(s,u_0)\|^\theta_{X^\alpha}\right)\int_0^{t-\tau}c_\alpha\frac{e^{-ay}}{y^\alpha}dy,\qquad t>\tau>t_{u_0},\end{aligned}$$

where the quantity $z_1(u_0)$ was introduced above the formula (3.1.12).

For any $\varepsilon>0$ choose $\tau:=\tau_0(\varepsilon)>t_{u_0}$, for which

$$\sup_{s\in[\tau_0(\varepsilon),+\infty)}\|u(s,u_0)\|_{X^\alpha}\le\limsup_{t\to+\infty}\|u(t,u_0)\|_{X^\alpha}+\varepsilon. \tag{4.1.6}$$

Note further that the first two terms of the sum on the right side of (4.1.5) tend to zero when $t\longrightarrow+\infty$. Therefore, passing in (4.1.5) to the upper limit when $t\to+\infty$, we get

(4.1.7)

$$\limsup_{t\to+\infty}\|u(t,u_0)\|_{X^\alpha}\le g(const.)c_\alpha\frac{\Gamma(1-\alpha)}{a^{1-\alpha}}\left(1+\sup_{s\in[\tau_0(\varepsilon),+\infty)}\|u(s,u_0)\|^\theta_{X^\alpha}\right).$$

Making use of the condition (4.1.6) and writing

$$C:=g(const.)c_\alpha\frac{\Gamma(1-\alpha)}{a^{1-\alpha}} \tag{4.1.8}$$

we obtain from (4.1.7) that

$$\limsup_{t\to+\infty}\|u(t,u_0)\|_{X^\alpha}\le C\left(1+\left(\varepsilon+\limsup_{t\to+\infty}\|u(t,u_0)\|_{X^\alpha}\right)^\theta\right)$$

and, since $\varepsilon > 0$ could be arbitrarily small, also

$$\limsup_{t\to+\infty} \|u(t,u_0)\|_{X^\alpha} \le C\left(1+\left(\limsup_{t\to+\infty} \|u(t,u_0)\|_{X^\alpha}\right)^\theta\right). \tag{4.1.9}$$

As seen from (4.1.9), the quantity $z := \limsup_{t\to+\infty} \|u(t,u_0)\|_{X^\alpha}$ fulfills the relation

$$z \le C(1+z^\theta),$$

which implies that

$$\limsup_{t\to+\infty} \|u(t,u_0)\|_{X^\alpha} \le z_0, \tag{4.1.10}$$

with z_0 being a positive root of the equation $C(1+z^\theta) - z = 0$.

Looking back at the definition of C in (4.1.8), it is seen that z_0 remains independent of $u_0 \in V$. Therefore, condition (4.1.10) guarantees that for any $\delta > 0$ and $u_0 \in V$ there exists a corresponding positive time $t_0(u_0, \delta)$ in which a solution $u(t, u_0)$ enters a ball

$$B_{X^\alpha}(z_0+\delta) = \{v \in X^\alpha : \|v\|_{X^\alpha} \le z_0 + \delta\}$$

and remains in it for all $t > t_0(u_0)$. In particular, a ball $B_{X^\alpha}(z_0)$ attracts points of V, which completes the proof of Theorem 4.1.1. □

Corollary 4.1.2. *If the function g which appears in condition* (3.1.4) *is continuous and satisfies the equality $g(0) = 0$, then for each $u_0 \in X^\alpha$*

$$\limsup_{t\to+\infty} \|u(t,u_0)\|_Y = 0 \;\; \textit{implies that} \;\; \limsup_{t\to+\infty} \|u(t,u_0)\|_{X^\alpha} = 0.$$

Corollary 4.1.3. *If instead of* (4.1.1) *the stronger condition*

$$\exists_{const.>0}\, \forall_{r>0} \limsup_{t\to+\infty} \sup_{u_0\in B_{X^\alpha}(r)} \|u(t,u_0)\|_Y \le const. \tag{4.1.11}$$

holds, then the semigroup $\{T(t)\}$ is bounded dissipative on X^α.

Corollary 4.1.4. *Under Assumption* 2.1.1 *the following two conditions are equivalent.*

(i) *The relation* (3.1.1) *defines on X^α a C^0 semigroup $\{T(t)\}$ of global X^α solutions of* (2.1.1) *which has bounded orbits of bounded sets and is point dissipative (or bounded dissipative) in X^α.*

(ii) *The requirements of* (A_2) *are satisfied with the estimate* (3.1.3) *satisfying additionally condition* (4.1.1) *(or* (4.1.11)*) with $V = X^\alpha$.*

Proof. The implication $(ii) \Longrightarrow (i)$ follows immediately from Theorems 3.1.1, 4.1.1. To observe $(i) \Longrightarrow (ii)$ recall (see proof of Theorem 3.1.1) that, since orbits of bounded sets are bounded, the requirements of (A_2) hold with $Y = X^\alpha$, $c(s) = \sup_{v\in B_{X^\alpha}(s)} \sup_{t\ge0} \|u(t,v)\|_{X^\alpha}$, $g(s) = \sup_{\|v\|_{X^\alpha}\le s} \|F(v)\|_X$ and $\theta = 0$. In particular we thus have (3.1.3) with $Y = X^\alpha$. As a consequence of dissipativeness in X^α, this estimate is asymptotically independent of $u_0 \in X^\alpha$. □

Remark 4.1.1. Without any significant changes in the proof Theorem 4.1.1 and the corollaries above remain true if (A_2') is assumed instead of (A_2). Also we point out that if, instead of absorption of points in the topology of Y (4.1.1), the absorption of bounded sets (4.1.11) were assumed, then similar considerations would allow us to show the absorption of bounded sets in the topology of X^α.

4.2. Existence of a global attractor - abstract setting

We are now able to formulate the results concerning the existence of the global attractor for the semigroup corresponding to (2.1.1) in fractional power spaces.

The case of compact resolvent. In this case we know in advance (see Theorem 3.3.1) that the semigroup is compact and, recalling Corollary 1.1.6, we only need yet to know that $\{T(t)\}$ is point dissipative.

Theorem 4.2.1. *Let Assumption* 2.1.1 *be satisfied, the resolvent of A be compact, and V be any subset of X^α such that*

(i) V is closed in the topology of X^α,
(ii) all the requirements of (A_2) (or, respectively, of (A_2')) hold for $u_0 \in V$,
(iii) $T(t)V \subset V$ for each $t \geq 0$,
(iv) the estimate (3.1.3) *in (A_2) (or, respectively* (3.2.5), *of (A_2')) is asymptotically independent of $u_0 \in V$.*

Then the semigroup $\{T(t)\}$ restricted to a complete metric space V has a global attractor.

Proof. For the proof it suffices to refer to Corollary 1.1.6; although we should note that, as a consequence of Corollary 3.1.2, the semigroup corresponding to (2.1.1) on V has bounded orbits of bounded sets.

From assumption (iv) and Theorem 4.1.1, $\{T(t)\}$ is point dissipative on V; whereas as a result of Theorem 3.3.1 $T(t) : V \to V$ is a compact map for each $t > 0$. Therefore, $\{T(t)\}$ is a point dissipative and compact C^0 semigroup on V so that the existence of a global attractor for $\{T(t)\}$ follows from Corollary 1.1.6. The proof is complete. □

Corollary 4.2.2. *Let Assumption* 2.1.1 *be satisfied and A have compact resolvent. Then the following two conditions are equivalent.*

(i) The relation (3.1.1) *defines on X^α a C^0 semigroup $\{T(t)\}$ of global X^α solutions of the problem* (2.1.1) *which has a global attractor in X^α.*
(ii) Condition (A_2) (or respectively (A_2')) holds with the estimate (3.1.3) *asymptotically independent of $u_0 \in X^\alpha$.*

Proof. The implication $(ii) \Longrightarrow (i)$ follows immediately from Theorem 4.2.1. To prove that $(i) \Longrightarrow (ii)$ note first that since the resolvent of A is compact if the global attractor exists the orbits of bounded sets need to be bounded (see Corollary 3.3.2). Also, this semigroup must be point dissipative. Therefore (ii) follows as in Corollary 4.1.4. □

Let us now consider a more special case, when (2.1.1) provides the existence of the Lyapunov function for $\{T(t)\}$ (see Definition 1.1.7). In this case, based on Corollary 3.1.2 and Corollary 1.1.5, the existence of the global attractor may then be established as follows.

Theorem 4.2.3. *Let Assumption* 2.1.1 *be satisfied and the resolvent of A be compact. Let further V be any subset of X^α such that*

(i) all the requirements of (A_2) (or of (A_2')) hold for $u_0 \in V$,
(ii) V is closed in the topology of X^α,
(iii) $T(t)V \subset V$ for each $t \geq 0$,
(iv) the set $\{v \in V : T(t)v = v \text{ for all } t \geq 0\}$ is bounded in V.

If there is a Lyapunov function then $\{T(t)\}$ restricted to a complete metric space V has a global attractor.

Problems with noncompact resolvent. Unlike the case previously studied problems with noncompact resolvent are more difficult to handle. The lack of compactness of the Sobolev type embeddings between fractional power spaces, and, hence, generic noncompactness of the trajectories, are usually not easy to overcome. Nevertheless the existence of a global attractor will still follow within the technique introduced above, provided that one more estimate of solutions is additionally known. Therefore, we consider in this case an auxiliary Banach space Z and we assume that the solutions of (2.1.1) remain bounded in the norm of this space. Since the semigroup will become compact on some metric subspace $V \subset X^\alpha$, the existence of the global attractor for the problem (2.1.1) will follow analogously to the case described previously.

Theorem 4.2.4. *Let Assumption* 2.1.1 *be satisfied and consider a Banach space Z such that for the product space $X^\alpha \cap Z \subset X$ equipped with the norm*

$$\|\cdot\|_{X^\alpha \cap Z} = \|\cdot\|_{X^\alpha} + \|\cdot\|_Z$$

the inclusion

$$X^\alpha \cap Z \subset X$$

is compact. Let further V be a subset of $X^\alpha \cap Z$ such that

(a) all the requirements of (A_2) (or, respectively, of (A_2')) hold for $u_0 \in cl_{X^\alpha} V$,
(b) V is bounded in Z,
(c) $T(t)V \subset V$ for each $t \geq 0$,
(d) the estimate (3.1.3) *in (A_2) (or, respectively,* (3.2.5) *of (A_2')) is asymptotically independent of $u_0 \in cl_{X^\alpha} V$.*

Then the semigroup $\{T(t)\}$ restricted to a complete metric space $cl_{X^\alpha} V$ has a global attractor.

Proof. Similarly to the proof of Theorem 4.2.1, all conditions required in Corollary 1.1.6 are satisfied. The only difference in this case, in comparison with Theorem 4.2.1, is that we need to use Theorem 3.3.3 to prove compactness of $T(t)$ in $cl_{X^\alpha}V$. □

4.3. Global solvability and attractors in X^α scales

The approach used in this book allows us to study the solutions of the problem (2.1.1) not in a single space X^α, but in a family of spaces X^β where the parameter β varies in an interval or half line. As already observed in Corollary 2.1.2, if there exists $\alpha \in [0,1)$ such that the nonlinear term $F: X^\alpha \to X$ is Lipschitz continuous on bounded sets, then the inclusion $X^\beta \subset X^\alpha$, $\beta \in [\alpha,1)$, immediately ensures the analogous property for $F: X^\beta \to X$, $\beta \in [\alpha,1)$ and, consequently, local X^β solvability of (2.1.1) is established for any $\beta \in [\alpha,1)$. Furthermore, if the condition (A_2) holds for some $\alpha \in [0,1)$, then the same inclusions $X^\beta \subset X^\alpha$, $\beta \in [\alpha,1)$, allow us to get immediately an extension of (3.1.4) to X^β spaces,

$$\|F(u(t,u_0))\|_X \leq g(\|u(t,u_0)\|_Y)(1+c\|u(t,u_0)\|^\theta_{X^\beta}),$$

which, through Theorem 3.1.1 (or Theorem 3.2.2), leads to the global X^β solvability of (2.1.1) for any $\beta \in [\alpha,1)$. This proves

Corollary 4.3.1. *Under Assumption* 2.1.1, *the validity of* (A_2) *is necessary and sufficient for the problem* (2.1.1) *to generate, for each* $\beta \in [\alpha,1)$, *a* C^0 *semigroup* $T_\beta(t): X^\beta \to X^\beta$, $t \geq 0$, *of global* X^β *solutions with bounded orbits of bounded sets.*

As mentioned in Remark 3.1.5, if $\beta' \geq \beta$, then

$$T_\beta(t)u_0 = T_{\beta'}(t)u_0 \ \text{ for all } \ u_0 \in X^{\beta'}, t \geq 0. \tag{4.3.1}$$

We thus usually omit the symbol β in the subscript of $\{T(t)\}$.

It may be next observed that if Theorem 4.1.1 holds for some $\alpha \in [0,1)$ and $V = X^\alpha$, then formula (4.1.2) may be strengthened to the condition

$$\forall_{\beta\in[\alpha,1)}\ \exists_{const.'>0}\ \forall_{u_0\in X^\beta}\ \limsup_{t\to+\infty} \|u(t,u_0)\|_{X^\beta} \leq const.' \tag{4.3.2}$$

Therefore, under the basic Assumption 2.1.1, we obtain the following generalization of Corollary 4.1.4.

Corollary 4.3.2. *The problem* (2.1.1) *generates on each* X^β, $\beta \in [\alpha,1)$, *a point dissipative* C^0 *semigroup of global* X^β *solutions to* (2.1.1) *which has bounded orbits of bounded sets, if and only if the requirements of* (A_2) *hold with the estimate* (3.1.3) *satisfying the inequality* (4.1.1) *with* $V = X^\alpha$.

Then Corollary 4.2.2 reads that:

Corollary 4.3.3. *If Assumption* 2.1.1 *holds and A has compact resolvent, then corresponding to* (2.1.1) *a C^0 semigroup of the global X^β solutions exists on each X^β, $\beta \in [\alpha, 1)$, and possesses a global attractor $\mathcal{A}_\beta$ in X^β, $\beta \in [\alpha, 1)$, if and only if condition (A_2) holds with the estimate* (3.1.3) *asymptotically independent of $u_0 \in X^\alpha$.*

We shall next prove

Observation 4.3.1. *The global attractors constructed above in various X^β topologies coincide, that is,*

$$\mathcal{A}_\beta = \mathcal{A}_\alpha, \quad \textit{for all} \quad \beta \in [\alpha, 1).$$

Proof. Since $\mathcal{A}_\beta$ is a compact and invariant subset of X^α which is attracted by $\mathcal{A}_\alpha$, we obtain the inclusion $\mathcal{A}_\beta \subset \mathcal{A}_\alpha$, $\beta \in [\alpha, 1)$. The opposite inclusion follows from Proposition 3.2.1 which shows that the image $T(1)\mathcal{A}_\alpha$ is a bounded subset of X^β, $\alpha \leq \beta < 1$. Therefore $\mathcal{A}_\alpha$, as a bounded and invariant subset of X^β which is attracted by $\mathcal{A}_\beta$, must be contained in $\mathcal{A}_\beta$. □

The above arguments show that, once having local X^α solvability, global X^α solvability, or a global attractor in X^α, we have the same result in X^β for any $\beta \in [\alpha, 1)$ provided that the usual assumptions on A and F are satisfied. Similar reasoning also holds for the solutions to (2.1.1) in metric phase spaces in which case one should consider a corresponding family of metric phase spaces $V_\beta = V \cap X^\beta$, $\beta \in [\alpha, 1)$. Since such considerations are contained further in Sections 8.4, 8.5 in connection with the concept of *dissipativenes in two spaces* (see [HA 2]), this is a matter we will not pursue here.

Bibliographical notes.

- **Sections 4.1 and 4.2.** Throughout these sections we follow in general the monograph [HA 2] and the results of [C-D 3], [C-D 4]. J. K. Hale approach [HA 2] to the existence of a global attractor requires only *point dissipativeness* of the semigroup, which property is easier to verify in special examples than the *absorbing property* required in [TE 1]. For other theorems dealing with existence of global attractors one may refer e.g. to [B-V 2].
- **Section 4.3.** Considerations of this section correspond to those of [C-C-D 1, Section 2.0]. However, we explicitly formulate here necessary and sufficient conditions for the global solvability and existence of a global attractor to (2.1.1). Some extension of the latter results into metric phase spaces may be found in [C-D 5] and further in Sections 8.4, 8.5.

CHAPTER 5

Application of abstract results to parabolic equations

In the present chapter we want to explain the connections between the abstract problem (2.1.1) considered earlier and its realization in the form of an initial boundary value problem for a $2m$-th order uniformly parabolic semilinear equation.

5.1. Formulation of the problem

Consider the equation

$$u_t = -Au + f(x, d^{m_0}u), \ (t, x) \in R^+ \times \Omega, \tag{5.1.1}$$

where

$$A = \sum_{|\xi|, |\zeta| \leq m} (-1)^{|\zeta|} D^{\zeta}(a_{\xi,\zeta}(x) D^{\xi}) \tag{5.1.2}$$

denotes a $2m$-th order uniformly strongly elliptic operator in a bounded domain $\Omega \subset R^n$ with $\partial\Omega \in C^{2m}$. The function $f : \overline{\Omega} \times R^{d_0} \longrightarrow R$ (where $d_0 = \frac{(n+m_0)!}{n!(m_0)!}$ is the number of all multi-indices β with $|\beta| \leq m_0$) is continuous and *locally Lipschitz continuous* with respect to each of its functional arguments separately uniformly for $x \in \Omega$ and $d^{m_0}u$, $m_0 \leq 2m - 1$, stands for the vector

$$\left(u, \frac{\partial u}{\partial x_1}, \dots, \frac{\partial u}{\partial x_n}, \frac{\partial^2 u}{\partial x_1^2}, \dots, \frac{\partial^{m_0} u}{\partial x_n^{m_0}} \right) \tag{5.1.3}$$

of the spatial partial derivatives of u of order not exceeding m_0.

The notion of local Lipschitz continuity appearing above should be understood in the sense of the following definition.

Definition 5.1.1. *A real function $h = h(y_1, \dots, y_s)$ defined on a subset of R^s is said to satisfy a local Lipschitz condition with respect to each of the variables $y_1, \dots, y_s$ separately, if for all $i = 1, \dots, s$ and each compact subset $\overline{B}$ of the domain of definition of this function there exists a positive constant $L = L(i, \overline{B})$ such that the inequality*

$$|h(y_1, \dots, y_i, \dots, y_s) - h(y_1, \dots, \overline{y}_i, \dots, y_s)| \leq L|y_i - \overline{y}_i|$$

holds for all $(y_1, \dots, y_i, \dots, y_s)$, $(y_1, \dots, \overline{y}_i, \dots, y_s)$ belonging to $\overline{B}$.

A function $h = h(x, y_1, \dots, y_s)$ defined on $\Omega \times R^s$ ($\Omega \subset R^n$) is said to be locally Lipschitz continuous with respect to $y_1, \dots, y_s$ uniformly for $x \in \Omega$, if for each $i = 1, \dots, s$ and each compact subset $\overline{B} \subset R^s$ there exists $L = L(i, \overline{B}) > 0$ such that, for all $(x, y_1, \dots, y_i, \dots, y_s), (x, y_1, \dots, \overline{y}_i, \dots, y_s) \in \Omega \times \overline{B}$,

$$|h(x, y_1, \dots, y_i, \dots, y_s) - h(x, y_1, \dots, \overline{y}_i, \dots, y_s)| \le L|y_i - \overline{y}_i|.$$

Remark 5.1.1. The simple example of the function $h = h(x, y_1, \dots, y_s)$ locally Lipschitz continuous with respect to $y_1, \dots, y_s$ uniformly for $x \in \Omega$ is a map $h = h_1(x)h_2(y_1, \dots, y_s)$, where h_1 is bounded in Ω and h_2 satisfies the local Lipschitz condition with respect to each of the variables $y_1, \dots, y_s$ separately.

Together with (5.1.1) the following initial-boundary conditions are considered:

$$\begin{cases} u(0, x) = u_0(x) \text{ in } \Omega, \\ B_1 u = \dots = B_m u = 0 \text{ on } \partial\Omega. \end{cases} \tag{5.1.4}$$

In our studies we assume that:

(A-I) The triple $(A, \{B_j\}, \Omega)$ forms a *regular elliptic boundary value problem* in the sense of Definition 1.2.1 (i.e. the operator A is *uniformly strongly elliptic* and the *strong complementary condition* is satisfied).

(A-II) The condition

$$\int_\Omega (Av) w dx = \int_\Omega a(v, w) dx$$

holds for all $v \in W^{2m,2}_{\{B_j\}}(\Omega)$, $w \in W^{m,2}_{\{B_j\}}(\Omega)$, whereas the form

$$a(v, w) = \sum_{|\xi|,|\zeta| \le m} a_{\xi,\zeta}(x) D^\xi v D^\zeta w$$

is *symmetric* and *coercive* (see [LI-MA, p. 217]). The latter means that for some $\lambda_0 > 0$, $c > 0$,

$$\int_\Omega a(w, w) dx + \lambda_0 \|w\|^2_{L^2(\Omega)} \ge C_1 \|w\|^2_{W^{m,2}(\Omega)}, \quad w \in W^{m,2}_{\{B_j\}}(\Omega). \tag{5.1.5}$$

Remark 5.1.2. The problem (5.1.1),(5.1.4) has been formulated here in a self-adjoint form which is advantageous from the point of view of applications. However, this particular form is not essential for the considerations appearing further in this chapter. Under the requirements of (A-I) the results reported below will remain true also when A is given by the condition (1.2.60) (see Subsection 1.2.4, Example 1.3.8). Note also that, according to (A-I), we assume here that coefficients $a_{\xi,\zeta}$ ($0 \le |\zeta| \le m$) appearing in (5.1.2) are of class $C^{|\zeta|}(\overline{\Omega})$.

Abstract form of the problem (5.1.1),(5.1.4). Let $\mathcal{A} := A + \lambda_0 I$ and $F(u) := f(x, d^{m_0} u) + \lambda_0 u$. From considerations of Section 1.3 (see Example 1.3.8) the operator $\mathcal{A}$ acting in $L^p(\Omega)$ with the domain $D(A + \lambda_0 I) = W^{2m,p}_{\{B_j\}}(\Omega)$ is sectorial and $Re\,\sigma(A+\lambda_0 I)>0$; therefore the problem (5.1.1),(5.1.4) may be rewritten in the abstract form

$$\begin{cases} \dot{u}+\mathcal{A}u=F(u),\ t>0, \\ u(0)=u_0. \end{cases} \tag{5.1.6}$$

If we choose parameters α and p in such a way that

$$\alpha\in\left(\frac{m_0}{2m},1\right) \quad\text{and}\quad 2m\alpha-m_0>\frac{n}{p}, \tag{5.1.7}$$

then the nonlinearity $F: X^\alpha\to L^p(\Omega)$ will be Lipschitz continuous on bounded sets. Indeed, let $\mathcal{U}\subset X^\alpha$ be bounded. Then, by Proposition 1.3.10 (with $\mu=0$), we have the inclusion $X^\alpha\subset C^{m_0}(\overline{\Omega})$ and hence $\mathcal{U}$ may be treated as a bounded subset of $C^{m_0}(\overline{\Omega})$. In consequence, in so far as we choose the arguments v of F from $\mathcal{U}$, then, according to the Lipschitz assumption on f (see Definition 5.1.1), the composite function $f(\cdot,d^{m_0}v)$ is globally Lipschitz continuous with respect to functional arguments and the Lipschitz constant does not depend on $x\in\Omega$. Note here that since $\mathcal{U}$ is bounded in $C^{m_0}(\overline{\Omega})$, the range of the functional arguments of f is restricted to a compact subset of R^{d_0}.

We thus obtain

$$\begin{aligned} &\|F(v_1)-F(v_2)\|_{L^p(\Omega)} \\ &\le \|f(\cdot,d^{m_0}v_1)-f(\cdot,d^{m_0}v_2)+\lambda_0(v_1-v_2)\|_{L^p(\Omega)} \\ &\le \sum_{|\beta|\le m_0} L_{\mathcal{U},\beta}\|D^\beta(v_1-v_2)\|_{L^p(\Omega)}+\lambda_0\|v_1-v_2\|_{L^p(\Omega)} \\ &\le const._{\mathcal{U},\lambda_0}\|v_1-v_2\|_{X^\alpha}, \quad\text{for all}\quad v_1,v_2\in\mathcal{U}. \end{aligned} \tag{5.1.8}$$

Therefore, the existence of local X^α solutions of the problem (5.1.1),(5.1.4) (or its abstract version (5.1.6)) in the sense of Definition 2.1.1 follows from Theorem 2.1.1.

Remark 5.1.3. It should be emphasized here that it was not important in the considerations connected with (5.1.8) whether Ω was bounded or not and that similar reasoning applies to a nonlinearity in the case of the Cauchy problem in $\Omega=R^n$ considered further in Examples 6.7.1, 6.7.2 and 7.2.1. Consequently, the existence of the local solutions in these examples follows via the above arguments.

Blow up examples. Let us consider the simple examples motivating conditions (3.1.3) and (3.1.4) listed in assumption (A_2). They are connected with difficulties appearing when one considers global solvability of problems like (5.1.1) with the right side term f growing faster than linearly. Generally in such a case the solutions may not exist for all $t\ge 0$. The well known reason is connected with the *blow up phenomena* observed originally for the solutions of ordinary differential equations of the form

$$\begin{cases} y'(t)=y^{1+\varepsilon}(t),\ \varepsilon>0,\ t>0, \\ y(0)=y_0>0, \end{cases} \tag{5.1.9}$$

whose local solutions $y(t) = (y_0^{-\varepsilon} - \varepsilon t)^{\frac{-1}{\varepsilon}}$ are defined only until the time $\tau_{y_0} = \frac{y_0^{-\varepsilon}}{\varepsilon}$.

Such form of the behavior of solutions extends automatically to the second and higher order parabolic Neumann type problems

$$\begin{cases} u_t = \Delta u + u^{1+\varepsilon}, \ \varepsilon > 0, \ x \in \Omega \subset R^n, \ t > 0, \\ \frac{\partial u}{\partial N} = 0 \ \text{ on } \ \partial\Omega, \\ u(0,x) = y_0 \equiv const., \end{cases} \tag{5.1.10}$$

fulfilled by the solutions of (5.1.9), which coincide with the space independent solutions of (5.1.10). Assuming that

$$A(const.) = 0 \text{ and } B_j(const.) = 0 \text{ for } j = 0, \ldots, m-1,$$

the blow up phenomenon may be then easily observed for (5.1.1),(5.1.4) with $f(x, d^{m_0}u) = u^{1+\varepsilon}$, $\varepsilon > 0$. Evidently for (5.1.1),(5.1.4) the problem (5.1.9) also produces space independent solutions which blow up in a finite time.

Actually, for solutions of second order equations with Neumann boundary conditions we can prove more. Not only some but many such solutions blow up in finite time. We include the ingenious estimate of solutions *from below* (see [C-L], [GI 3]) for completeness.

Let the space variable x vary in a bounded, smooth domain $\Omega \in R^n$. Consider the problem

$$\dot{u} = \Delta u + \lambda u|u|^{q-1}, \ \lambda > 0, \ q > 1, \tag{5.1.11}$$

together with the Neumann initial-boundary condition

$$u(0,x) = u_0(x) \ \text{ for } x \in \Omega, \ \frac{\partial u}{\partial N} = 0 \ \text{ for } \ x \in \partial\Omega. \tag{5.1.12}$$

Write

$$E(u(t)) = \|\nabla u\|^2_{L^2(\Omega)} - \frac{2\lambda}{q+1}\|u\|^{q+1}_{L^{q+1}(\Omega)}. \tag{5.1.13}$$

We then have

Lemma 5.1.1. *Every solution of* (5.1.11), (5.1.12) *corresponding to initial data* u_0 *satisfying the condition*

$$\|u_0\|^{q+1}_{L^2(\Omega)} > \frac{E(u_0)c^{q+1}}{\lambda(1 - \frac{2}{q+1})}, \tag{5.1.14}$$

where c *is the inclusion constant of* $L^{q+1}(\Omega) \subset L^2(\Omega)$, *blows up in a finite time.*

Proof. Multiplying (5.1.11) first by u, then by $2\dot{u}$, and integrating over Ω, we get

$$\frac{1}{2}\frac{d}{dt}\|u\|^2_{L^2(\Omega)} = -\|\nabla u\|^2_{L^2(\Omega)} + \lambda\|u\|^{q+1}_{L^{q+1}(\Omega)}, \tag{5.1.15}$$

and

$$2\|\dot{u}\|^2_{L^2(\Omega)} = -\frac{d}{dt}E(u(t)). \tag{5.1.16}$$

The last equation shows that the function $E(u(t))$ is nonincreasing in time, hence

$$\|\nabla u\|^2_{L^2(\Omega)} \le E(u_0) + \frac{2\lambda}{q+1}\|u\|^{q+1}_{L^{q+1}(\Omega)}. \tag{5.1.17}$$

Combining (5.1.15) and (5.1.17), we obtain

$$\frac{1}{2}\frac{d}{dt}\|u\|^2_{L^2(\Omega)} \ge -E(u_0) + C\|u\|^{q+1}_{L^{q+1}(\Omega)}, \tag{5.1.18}$$

with $C = \lambda(1 - \frac{2}{q+1})$ being always positive ($\lambda > 0, q > 1$). Using the embedding estimate

$$\|\phi\|_{L^2(\Omega)} \le c\|\phi\|_{L^{q+1}(\Omega)}, \tag{5.1.19}$$

we find finally a differential inequality for $\|u(t)\|^2_{L^2(\Omega)} = z(t)$:

$$z'(t) \ge -2E(u_0) + \frac{2C}{c^{q+1}}z(t)^{\frac{q+1}{2}}. \tag{5.1.20}$$

It is now easy to see that every solution of this differential inequality satisfying condition (5.1.14) blows up in a finite time; more precisely, its $L^2(\Omega)$ norm becomes unbounded after a finite time that depends on u_0. □

Since we are mostly interested here in a long time behavior of global solutions, for classical results concerning the blow up phenomenon we refer the reader to the considerations due to J. M. Ball, F. Weissler, A. Friedman, and many others. See e.g. [BAL], [WES], [FU-MO]; also [M-Y] for the results and further references concerning equation (5.1.11) in the case when $\Omega = R^n$.

5.2. Global solutions via partial information

A situation is quite different when we have any a priori knowledge about the solutions. It very often happens, especially in the case of equations describing physical or biological processes, that some introductory estimate global in time resulting from the nature of the phenomena described by the equation (like the consequences of the mass conservation law or the behavior of an energy functional) is initially given. With this partial introductory information the proof of the global existence becomes much simpler, and also suitable time independent estimates of the solutions (necessary for dissipativeness of the system) can be derived, very often enabling the construction of an absorbing set and attractor. It is also possible to study growth rates of the nonlinearities for which global solutions exist.

The introductory estimate global in time usually has the form of an $L^r(\Omega)$ a priori estimate of all l-th order partial derivatives of the solution. Our present task will be thus formulated in the form of the following problem.

Admissible growth of nonlinearity. Knowing for some $0 \leq l \leq m_0$ the following a priori estimate for the solution u of the problem (5.1.1),(5.1.4) -

$$\sum_{|\sigma|=l} \|D^\sigma u(t)\|_{L^r(\Omega)} \leq \rho(t),\ t > 0, \tag{5.2.1}$$

with a function $\rho \in C^0\big([0,\infty)\big)$ - find the growth condition for the nonlinear term f in (5.1.1) for which the global solution u exists and, therefore, the semigroup $\{T(t)\}$ is defined by the formula $T(t)u_0 = u(t,u_0)$.

Further, we will also be interested in finding time independent estimates of solutions suitable for study of the dynamics of the system considered.

Remark 5.2.1. When $l > 0$, the estimate (5.2.1) is not sufficient to control the derivatives $D^\alpha u$ with $0 \leq |\alpha| < l$. Using (5.2.1) and the boundary conditions in (5.1.4), we can often estimate these lower derivatives basing on the *generalized Poincaré inequality* [TE 1, p. 50]:

$$\|w\|_{H^{l-1}(\Omega)} \leq C_2 \left\{ \sum_{|\sigma|=l} \|D^\sigma w\|_{L^2(\Omega)} + p(w) \right\},\ w \in H^l(\Omega), \tag{5.2.2}$$

where p is a continuous seminorm on $H^l(\Omega)$ which is a norm on the space $\mathcal{P}^{l-1}$ of polynomials of degree not exceeding $l-1$. Clearly such an estimate is true when (5.1.4) are Dirichlet boundary conditions and

$$p(w) = \sqrt{\sum_{|\sigma|=0}^{l-1} \int_\Gamma |D^\sigma w|^2 dS},$$

where $\Gamma \subset \partial\Omega$, $|\Gamma| > 0$, $l \leq m$. Inequality (5.2.2) then guarantees, in particular, that $\sum_{|\sigma|=l} \|D^\sigma w\|_{L^2(\Omega)}$ is the norm on $H^l_{\{B_j\}}(\Omega)$ equivalent to the standard $H^l(\Omega)$ norm. Further generalizations of the Poincaré inequality, following [ZI], are reported in Chapter 9.

Thus, if $l > 0$ in (5.2.1) we shall assume that for the solution u of (5.1.1),(5.1.4) the full $W^{l,r}(\Omega)$ norm of u is estimated a priori for $t > 0$:

$$\|u(t)\|_{W^{l,r}(\Omega)} \leq \rho(t),\ t > 0. \tag{5.2.3}$$

We assume throughout this section that the conditions (A-I), (A-II) of Section 5.1 are satisfied. Additionally we require that Ω is bounded and the nonlinear term f in (5.1.1) satisfies the following general growth condition:

$$|f(\cdot, d^{m_0}u) + \lambda_0 u| \leq C_3 \left(1 + \sum_{j=0}^{m_0} \sum_{|\alpha|=j} |D^\alpha u|^{\gamma_j}\right), \tag{5.2.4}$$

where $m_0 \leq 2m-1$ and each exponent γ_j is restricted by the conditions:

Restriction 1. $\gamma_j \geq 1$ for $j = 0, \ldots, m_0$ and

$$\text{(a)} \qquad \gamma_j < 1 + \frac{2m - j}{j - l + \frac{n}{r}} \quad \text{if } r(l-j) < n,$$

$$\text{(b)} \qquad \gamma_j \text{ arbitrarily large if } n \leq r(l-j).$$

We shall next prove that under Restriction 1 just stated there exists a global X^α solution of the problem (5.1.1),(5.1.4) provided that (5.2.3) is satisfied.

Lemma 5.2.1. *Let the assumption* (5.1.7) *hold. Fix some* $l \geq 0$, $r \in [1, +\infty)$ *and let* $D(A) = W^{2m,p}_{\{B_j\}}(\Omega) \subset W^{l,r}(\Omega)$. *If* $u(t, u_0)$ *is a local* X^α *solution of* (5.1.1),(5.1.4) *defined on* $[0, \tau]$ *and the condition* (5.2.4) *holds with the exponents* γ_j *as in Restriction* 1, *then the following estimate of the nonlinear term is valid:*

(5.2.5)
$$\|F(u(t,u_0))\|_{L^p(\Omega)} \leq const.k\Big(\|u(t,u_0)\|_{W^{l,r}(\Omega)}\Big)\Big(1 + \|u(t,u_0)\|_{X^{\tilde\alpha}}\Big), \quad 0 \leq t \leq \tau,$$

where $\tilde\alpha$ *is a constant in* $[\alpha, 1)$ *and* $k : [0, +\infty) \to [0, +\infty)$ *is specified in condition* (5.2.14) *below.*

Proof. From the assumption (5.1.7) it follows that

$$p > \frac{n}{2m - (l - \frac{n}{r})} \text{ whenever } l - \frac{n}{r} < m_0. \tag{5.2.6}$$

The further proof rests on the Nirenberg-Gagliardo inequality (1.2.53) of Remark 1.2.1 and Proposition 1.3.10.

From the growth condition (5.2.4), we obtain

$$\|F(u(t,u_0))\|_{L^p(\Omega)} \leq C_3|\Omega|^{\frac{1}{p}} + C_3 \sum_{j=0}^{m_0} \sum_{|\alpha|=j} \|D^\alpha u(t,u_0)\|^{\gamma_j}_{L^{p\gamma_j}(\Omega)}. \tag{5.2.7}$$

Whenever $n \leq r(l-j)$, the Sobolev embedding $W^{l-j,r}(\Omega) \subset L^q(\Omega)$ $(1 \leq q < +\infty)$ gives immediately for any multiindex α, $|\alpha| = j$,

$$\|D^\alpha u(t,u_0)\|^{\gamma_j}_{L^{p\gamma_j}(\Omega)} \leq C_4\|u(t,u_0)\|^{\gamma_j}_{W^{l,r}(\Omega)}, \quad \text{for arbitrarily large } p\gamma_j. \tag{5.2.8}$$

Assume next that $n > r(l-j)$. Using Remark 1.2.1, we find that for $|\alpha| = j$ and $s_j > j$ (s_j is precised in (5.2.10))

(5.2.9)
$$\|D^\alpha u(t,u_0)\|^{\gamma_j}_{L^{p\gamma_j}(\Omega)} \leq C_5\|u(t,u_0)\|^{\gamma_j}_{W^{j,p\gamma_j}(\Omega)} \leq C_6\|u(t,u_0)\|^{\theta_j\gamma_j}_{W^{s_j,p}(\Omega)}\|u(t,u_0)\|^{(1-\theta_j)\gamma_j}_{W^{l,r}(\Omega)},$$

provided that $\theta_j \in (0,1)$ and (1.2.54) is satisfied:

$$\begin{cases} (j - l + \frac{n}{r})\gamma_j < \gamma_j\theta_j(s_j - \frac{n}{p} - l + \frac{n}{r}) + \frac{n}{p}, \\ \frac{1}{p\gamma_j} - \frac{1}{r} \leq \theta_j(\frac{1}{p} - \frac{1}{r}). \end{cases} \tag{5.2.10}$$

We need additionally to assume that

$$\gamma_j\theta_j < 1, \tag{5.2.11}$$

then the first condition in (5.2.10) reduces to Restriction 1(a) while the value s_j there will grow arbitrarily close to $2m$.

We shall now explain whether simultaneous occurrence of conditions (5.2.10) and (5.2.11) allows exponents γ_j to approach the maximal values expected in Restriction 1(a). For this we note the properties of a function $\mathcal{H} : [1, +\infty) \to R$:

$$\mathcal{H}(x) := \frac{(j - l + \frac{n}{r}) - \frac{n}{p}\frac{1}{x}}{s_j - \frac{n}{p} - l + \frac{n}{r}}.$$

These are:

- $\mathcal{H}(1) < 1$,
- $\mathcal{H}(x)$ is strictly increasing,
- $\mathcal{H}(x) < \frac{1}{x}$ for $x \in [1, x_0)$,

where $x_0 = 1 + \frac{s_j - j}{j-(l-\frac{n}{r})}$ is the only root of the equation $\mathcal{H}(x) = \frac{1}{x}$.

For a given $\gamma_j \in (1, x_0)$ the above properties allow us to choose $\theta_j = \theta_j(\gamma_j)$ in such a way that

$$\frac{j - l + \frac{n}{r} - \frac{n}{p}\frac{1}{\gamma_j}}{s_j - \frac{n}{p} - l + \frac{n}{r}} < \theta_j < \frac{1}{\gamma_j}. \tag{5.2.12}$$

Such a choice justifies the first inequality in (5.2.10). Since, in addition, the value θ_j in (5.2.12) may be taken arbitrarily close to $\frac{1}{\gamma_j}$, the second inequality in (5.2.10) will also be satisfied. Finally, fulfillment of (5.2.11) is obvious from (5.2.12).

In the above considerations the value $\gamma_j = 1$ is admissible provided that $p \geq r$.

If we take $\gamma_j = 1$ without the additional assumptions just mentioned, then the inequality in (5.2.11) would be weak.

Inserting the estimate (5.2.9) in the right side of (5.2.7) and applying condition (5.2.3) and Proposition 1.3.10, for

$$t = s := \max_{s<j\leq m_0} s_j, \ \alpha = \tilde{\alpha} = \frac{s}{2m}, \tag{5.2.13}$$

we find that

(5.2.14)

$$\begin{aligned}
\|F(u(t,u_0))\|_{L^p(\Omega)} &\leq \Big(C_3|\Omega|^{\frac{1}{p}} + C_3C_4 \sum_{\{j:n\leq r(l-j)\}} \sum_{|\alpha|=j} \|u(t,u_0)\|^{\gamma_j}_{W^{l,r}(\Omega)} \\
&\quad + C_3C_6 \sum_{\{j:n>r(l-j)\}} \sum_{|\alpha|=j} \|u(t,u_0)\|^{\gamma_j-\gamma_j\theta_j}_{W^{l,r}(\Omega)}\Big)\Big(1 + \|u(t,u_0)\|^{\gamma_j\theta_j}_{W^{s,p}(\Omega)}\Big) \\
&=: k\Big(\|u(t,u_0)\|_{W^{l,r}(\Omega)}\Big)\Big(1 + \|u(t,u_0)\|^{\gamma_j\theta_j}_{W^{s,p}(\Omega)}\Big) \\
&\leq const.k\Big(\|u(t,u_0)\|_{W^{l,r}(\Omega)}\Big)\left(1 + \|u(t,u_0)\|_{X^{\tilde{\alpha}}}\right), \ 0 \leq t \leq \tau,
\end{aligned}$$

where the Sobolev and Young inequalities have been used in the last estimate. The proof of Lemma 5.2.1 is thus finished. □

Remark 5.2.2. As seen from the proof of Lemma 5.2.1, under either of the additional assumptions

- $\gamma_j > 1$ if $n > r(l-j)$,
- $p \geq r$,

condition (5.2.5) may be strengthened to (3.1.4) having the form

(5.2.15)
$$\|F(u(t,u_0))\|_{L^p(\Omega)} \leq g\left(\|u(t,u_0)\|_{W^{l,r}(\Omega)}\right)\left(1+\|u(t,u_0)\|^{\eta}_{X^\beta}\right),\ t \geq 0,$$

with any $\beta \in (\tilde{\alpha},1)$ and a certain constant $\eta(\beta) \in [0,1)$. However, the estimate (5.2.5) ensures *not faster than linear* growth of F on local solutions of (5.1.1),(5.1.4). According to the sublinear growth restriction of Chapter 3 this suffices for X^β solvability of this problem global in time.

It is also evident that without loss of generality the exponents γ_j in Restriction 1 may be taken strictly greater than 1 (which is a consequence of the Young inequality).

We thus have

Proposition 5.2.1. *Let the introductory estimate* (5.2.3) *be satisfied and the assumptions of Lemma* 5.2.1 *hold. Then there exists a global X^β solution of the problem* (5.1.6) *for any $\beta \in [\tilde{\alpha},1)$.*

Remark 5.2.3. In fact β in the above proposition may be taken from a larger interval $[\alpha,1)$ with α given in (5.1.7). Indeed, under condition (5.1.7) we have the existence of the local X^α solutions. Next, by definition, we know that these solutions enter $D(A)$ immediately. Therefore, for $t>0$, they may be considered as the $X^{\tilde{\alpha}}$ solutions that exist globally in time.

5.3. Existence of a global attractor

It was observed in [C-D 3] (see Theorem 4.1.1) that by strengthening slightly the assumptions of Lemma 5.2.1 a necessary and sufficient condition for the existence of a global attractor for the semigroup generated by (5.1.1),(5.1.4) will be obtained. For convenience of further use we will collect below all the required assumptions and formulate a suitable theorem, being a consequence of Theorem 4.2.1.

Consider the problem (5.1.1),(5.1.4) under the assumptions (A-I) and (A-II). Let the function f be continuous and locally Lipschitz continuous with respect to its functional arguments uniformly for $x \in \Omega$. Choose parameters α and p satisfying condition (5.1.7), i.e.

$$\alpha \in \left(\frac{m_0}{2m},1\right) \quad \text{and} \quad 2m\alpha - m_0 > \frac{n}{p}$$

to justify local X^α solvability of (5.1.6).

We have

Theorem 5.3.1. *Let the assumptions of Lemma* 5.2.1 *hold. Assume the condition* (5.2.3) *is satisfied in the form*

$$\|u(t,u_0)\|_{W^{1,r}(\Omega)} \leq c(\|u_0\|_{X^\alpha}),\ t \geq 0,$$

and there exists a closed and positively invariant (conditions (i) and (iii) in Theorem 4.2.1*) subset V of X^α, such that*

$$\exists_{const.>0}\ \forall_{u_0\in V}\ \limsup_{t\to+\infty} \|u(t,u_0)\|_{W^{1,r}(\Omega)} \leq const., \tag{5.3.1}$$

where $u(t,u_0)$ denotes the global X^α solution given in Proposition 5.2.1. *Then there exists a global attractor for the restriction to V of the semigroup of X^β solutions of* (5.1.6)*, where $\beta \in [\tilde{\alpha}, 1)$.*

Proof. We need to check that the assumptions of Theorem 4.2.1 are satisfied. As was stated in Example 1.3.8 the elliptic operator $\mathcal{A} = A + \lambda_0 I$ is sectorial and positive in $L^p(\Omega)$. Also, the resolvent of $\mathcal{A}$ is compact (see Proposition 1.2.3).

We shall further recall that the condition (5.2.5) may be strengthened to (3.1.4), having actually the form

$$\begin{aligned}&\|F(u(t,u_0))\|_{L^p(\Omega)}\\ &\leq g\left(\|u(t,u_0)\|_{W^{1,r}(\Omega)}\right)\left(1 + \|u(t,u_0)\|^\eta_{X^\beta}\right),\ t \geq 0,\end{aligned} \tag{5.3.2}$$

with a certain constant $\eta \in [0,1)$ (see Remark 5.2.2). Therefore, the existence of a global attractor for the semigroup generated by (5.1.6) on X^β follows directly from Theorem 4.2.1. □

Remark 5.3.1. Similarly as in Remark 5.2.3 the parameter β in the above theorem may vary in the interval $[\alpha, 1)$ with α defined in (5.1.7).

Remark 5.3.2. The sublinear growth restriction (condition (3.0.1) or (5.2.5)) is not sufficient to bound a global solution $u(t,u_0)$ on the half line $[0,+\infty)$. Moreover, it is well known that solutions of the sample problem

$$u_t = \Delta u + \lambda u,\ x \in \Omega \subset R^n,\ t > 0, \tag{5.3.3}$$

with homogeneous Dirichlet boundary condition and $\lambda > \lambda_1$, λ_1 being the first positive eigenvalue of $(-\Delta, I)$ on Ω, will satisfy the sublinear growth condition; however they become unbounded when $t \to +\infty$.

Indeed, multiplying (5.3.3) in $L^2(\Omega)$ by the eigenfunction v_1 corresponding to λ_1, we find that

$$\frac{d}{dt}\int_\Omega uv_1dx = \int_\Omega u\Delta v_1dx + \lambda\int_\Omega uv_1dx = (\lambda - \lambda_1)\int_\Omega uv_1dx,$$

which shows that $c_1(t) = \int_\Omega u(t)v_1dx$ grows exponentially if only the Fourier coefficient $c_1(0) \neq 0$. Now, by the Parseval identity, the $L^2(\Omega)$ norm of $u(t)$ is forced to increase at least as fast as $c_1(t)$.

If (5.3.3) is considered with homogeneous Neumann condition, we simply integrate (5.3.3) over Ω, to obtain

$$\frac{d}{dt}\int_\Omega u dx = \lambda \int_\Omega u dx,$$

which ensures that the integral mean value of $u(t)$ over Ω grows exponentially when $\lambda > 0$.

In comparison with the situation just described, the strengthened condition (3.1.4) (or (5.3.2)) gives the estimate on $[0, +\infty)$ uniform in time as shown in Theorem 3.1.1.

We close this chapter showing how the theory just described works in practical examples.

Example 5.3.1. Consider the single second order equation

$$u_t = \sum_{i,j=1}^{n} (a_{ij}(x) u_{x_i})_{x_j} + f(x, u, \nabla u), \tag{5.3.4}$$

$x \in \Omega \subset R^n$, Ω bounded, smooth, $\nabla u = (\frac{\partial u}{\partial x_1}, \ldots, \frac{\partial u}{\partial x_n})$. We assume that a_{ij} are $C^1(\overline{\Omega})$ smooth, the main part is uniformly elliptic and the nonlinear term f is continuous and also locally Lipschitz continuous with respect to functional arguments uniformly for $x \in \Omega$. In the case of the homogeneous Dirichlet boundary condition we set

$$Au := -\sum_{i,j=1}^{n} (a_{ij}(x) u_{x_i})_{x_j}$$

and consider A on the domain $D(A) = W^{2,p}(\Omega) \cap W_0^{1,p}(\Omega)$ with $p > n$. Such an operator is sectorial (see Example 1.3.10). Also, as a consequence of the inclusion $X^\alpha \subset C^{1+\nu}(\overline{\Omega})$ with $0 < \nu < 2\alpha - \frac{n}{p} - 1$ (see Remark 1.3.8), the nonlinear term acting from X^α to $L^p(\Omega)$ is for $\alpha \in (\frac{1}{2} + \frac{n}{2p}, 1)$ Lipschitz continuous on bounded sets. Indeed, for a bounded set $\mathcal{U} \subset X^\alpha$ condition (5.1.8) now reads

$$\begin{aligned} &\|f(\cdot, w, \nabla w) - f(\cdot, v, \nabla v)\|_{L^p(\Omega)} \\ &\leq L_w(\mathcal{U}) \|w - v\|_{L^p(\Omega)} + L_\nabla(\mathcal{U}) \|\nabla w - \nabla v\|_{L^p(\Omega)} \\ &\leq const. \|w - v\|_{W^{1,p}(\Omega)} \\ &\leq const.' \|w - v\|_{X^\alpha}, \quad w, v \in \mathcal{U}, \end{aligned} \tag{5.3.5}$$

where $L_w(\mathcal{U})$, $L_\nabla(\mathcal{U})$ are Lipschitz constants with respect to functional arguments for f restricted to $\mathcal{U}$ ($\mathcal{U}$ is bounded in $C^{1+\nu}(\overline{\Omega})$).

Assume now that $Y = L^\infty(\Omega)$ and we have an a priori estimate of u of the form

$$\|u(t, u_0)\|_{L^\infty(\Omega)} \leq \|u_0\|_{L^\infty(\Omega)}, \tag{5.3.6}$$

which is often a consequence of the maximum principle (e.g. when $f(x, u, 0) = 0$, see [WAL, p. 199]). Restriction 1(a) then reduces to the well known 'subquadratic growth condition' ($\gamma_1 < 2$):

(5.3.7) $$|f(x,u,\nabla u)| \le h(|u|)(1+|\nabla u|^{2-\varepsilon}),$$

where $h : [0,+\infty) \longrightarrow [0,+\infty)$ is locally Lipschitz continuous and $\varepsilon \in (0,2)$ is arbitrary. Therefore the estimate (5.3.6) together with the growth restriction (5.3.7) guarantees global solvability of the Dirichlet problem for (5.3.4).

The existence of a global attractor for a particular version of equation (5.3.4) (with f independent of ∇u) will be studied in Chapter 6.

Remark 5.3.3. Calling the values γ_j fulfilling the equalities in Restriction 1(a),

$$\gamma_j = 1 + \frac{2m-j}{j-l+\frac{n}{r}},\ r(l-j) < n,$$

critical exponents, an interesting problem would be to select classes of parabolic equations with nonlinear terms having *critical growth* (satisfying condition (5.2.4) with critical exponents) possessing only solutions global in time and, eventually, also a global attractor. Some partial results in this direction were reported in [WA 1], [KI]. Note, that the critical growth is defined in relation to the a priori $W^{l,r}(\Omega)$ estimate. The critical exponents γ_j will grow together with the net smoothness $(l-\frac{n}{r})$ of such estimates. For the problem considered in Example 5.3.1 under the $L^\infty(\Omega)$ a priori estimate (5.3.6) the critical exponent γ_1 will be equal to 2. Another example of the problem with nonlinearity having critical growth will be presented next.

Example 5.3.2. Consider the three-dimensional Cahn-Hilliard equation (see Chapter 6 for more complete presentation) with special nonlinearity $f(u) = u^5 - u$:

$$\begin{cases} u_t = -\varepsilon^2\Delta^2 u + \Delta(u^5-u), & t>0,\ x\in\Omega\subset R^3, \\ u(0,x) = u_0(x), & x\in\Omega, \\ \frac{\partial u}{\partial N} = \frac{\partial(\Delta u)}{\partial N} = 0,\ x\in\partial\Omega, & t>0. \end{cases}$$

Since

$$\Delta(u^5-u) = (5u^4-1)\Delta u + 20u^3|\nabla u|^2,$$

we get, based on the Young inequality (1.2.2), an estimate:

(5.3.8) $$|\Delta(u^5-u)| \le const.(|\Delta u|^{\frac{9}{5}} + |\nabla u|^3 + |u|^9 + 1).$$

As shown in the formula (6.4.13) below, for the solution of the Cahn-Hilliard equation an a priori estimate in the $H^1(\Omega)$ norm (being the $W^{l,r}(\Omega)$ estimate in this example) holds. Calculating the values of critical exponents relative to these a priori $H^1(\Omega)$ estimates, we find

$$\gamma_0 = 9,\ \gamma_1 = 3,\ \gamma_2 = \frac{9}{5},$$

which are precisely the exponents specified in the estimate (5.3.8). Therefore, the nonlinear term $\Delta(u^5-u)$ in this example has critical growth relative to the $H^1(\Omega)$ a priori estimate.

Remark 5.3.4. So far, we have considered problems of type (2.1.1) with nonlinearity F acting from some fractional power space X^α, $\alpha \in [0, 1)$, into X. If we allow $\alpha = 1$ and understand the critical growth of F in the sense that $\|F(u(t))\|_X$ has the same *'order of magnitude'* as $\|Au(t)\|_X$ (see [WA 1]) then, in particular, the known *non-well-posed equation* (see [FR 2, p. 172], [PA])

$$u_t - \Delta u = -c\Delta u, \ c > 1,$$

falls into a class of problems with critical growth.

Bibliographical notes.

- **Section 5.1.** A variational approach to elliptic boundary value problems was used in [LI-MA]. The abstract formulation considered next of such problems follows [HE 1]. From the huge literature devoted to blow up phenomena few items were listed at the end of the section.
- **Section 5.2.** Here the results of [AM 1], [C-D 1] are reported. We remark that the conditions of type (5.2.4) (and the corresponding Restriction 1) were originally found in the monograph [FR 1].
- **Section 5.3.** Existence of the global attractor is treated as in [C-D 3] and remains in a vein of the *dissipative systems* theory [HA 2].
- **Miscellaneous.** The *Poincaré inequality* is one of the main tools in obtaining various estimates of solutions of partial differential equations in Sobolev spaces. We thus quote some references in which more or less general versions of the Poincaré inequality can be found. These are, in particular, [GI-TR, p. 164], [AD, p. 158], [MO, pp. 64, 82], [NE, p. 34], [SM, p. 112]; or [ZI, Chapter 4], where more sophisticated versions are considered. The general results of [ZI] are reported briefly in Section 9.1. We also point to the *Nirenberg-Gagliardo* (*interpolation*) *inequality*, which is now a standard tool in the estimates mentioned above; one may refer to [FR 1, Part 1], [HE 1, Section 1.6], [HE 2], [AD, Chapter IV], or [NI 2] for the proofs of this inequality. The general version which we use in this book, is proved within the *theory of interpolation* and reported in [TR]; see also [AM 1].

CHAPTER 6

Examples of global attractors in parabolic problems

Based on considerations of Chapters 3 and 4 we shall study in this chapter a number of examples of second and higher order parabolic problems. Most of these problems have their origin in applications. The central examples here will be the Cahn-Hilliard pattern formation equation and the Navier-Stokes equation from hydrodynamics.

6.1. Introductory example

We will start with the simple source second order parabolic equation of the form

$$u_t = \Delta u - u^q + u, \quad \text{for} \quad x \in \Omega \subset R^n, \tag{6.1.1}$$

Ω bounded, $\partial\Omega \in C^2$, subject to the Dirichlet boundary conditions

$$u(t,x) = 0 \quad \text{for} \quad x \in \partial\Omega, \tag{6.1.2}$$

$$u(0,x) = u_0(x) \quad \text{for} \quad x \in \Omega. \tag{6.1.3}$$

Let $q \in N$ be an arbitrary large odd number. Consider the sectorial operator $A = -\Delta$ on Ω with zero Dirichlet condition. Fix $p > n$ and set $X = L^p(\Omega)$. The problem (6.1.1)-(6.1.3) will be studied abstractly on X^α with $\alpha \geq \frac{1}{2}$ since, for such values of α (see Remark 1.3.8),

$$X^\alpha \subset L^\infty(\Omega), \tag{6.1.4}$$

so that the nonlinear term $f(u) = -u^q + u$ defines a nonlinear substitution operator $F : W_0^{1,p}(\Omega) \to L^p(\Omega)$ by the formula

$$F(\phi) = -\phi^q + \phi, \quad \text{for} \quad \phi \in W_0^{1,p}(\Omega).$$

It is easy to check, because of the inclusion (6.1.4), that such an F is also Lipschitz continuous on bounded subsets of $X^{\frac{1}{2}}$. Indeed, for any bounded subset $\mathcal{U}$ of $X^{\frac{1}{2}}$, we obtain

$$\begin{aligned}\|F(\phi)-F(\psi)\|_{L^p(\Omega)} &\le \|\phi^q-\psi^q\|_{L^p(\Omega)}+\|\phi-\psi\|_{L^p(\Omega)}\\ &\le \|\phi-\psi\|_{L^p(\Omega)}\left(\|\phi^{q-1}+\phi^{q-2}\psi+\ldots+\phi\psi^{q-2}+\psi^{q-1}\|_{L^\infty(\Omega)}+1\right)\\ &\le \|\phi-\psi\|_{L^p(\Omega)}\left(\sup_{\eta\in\mathcal{U}}\|\eta\|^{q-1}_{L^\infty(\Omega)}+1\right)\\ &\le \|\phi-\psi\|_{L^p(\Omega)}\left(\sup_{\eta\in\mathcal{U}}(c\|\eta\|_{X^{\frac12}})^{q-1}+1\right),\ \phi,\psi\in\mathcal{U},\end{aligned}$$

where c is an inclusion constant between the $L^\infty(\Omega)$ and $X^{\frac12}$ norms. Therefore, local solvability of the abstract problem

$$\dot u + Au = F(u),\ u(0)=u_0, \tag{6.1.5}$$

corresponding to (6.1.1)-(6.1.3) follows. To justify global solvability of (6.1.5) and the existence of the global attractor for the semigroup generated by (6.1.5) on X^α, $\alpha\in[\frac12,1)$, we need to find, according to Theorem 3.1.1, an a priori estimate global in time and asymptotically independent of u_0. For the present problem this will be the $L^\infty(\Omega)$ estimate.

We are able to formulate

Lemma 6.1.1. *The two following estimates of X^α solutions of* (6.1.5), $\alpha\in[\frac12,1)$, *hold:*

$$\|u(t,u_0)\|_{L^\infty(\Omega)}\le\max\{\|u_0\|_{L^\infty(\Omega)},1\}, \tag{6.1.6}$$

$$\limsup_{t\to+\infty}\|u(t,u_0)\|_{L^\infty(\Omega)}\le 1. \tag{6.1.7}$$

Proof. Multiplying equation (6.1.1) by u^{2m-1}, $m=1,2,\ldots$, we get

$$\int_\Omega u_t\,u^{2m-1}dx=\int_\Omega \Delta u\,u^{2m-1}dx-\int_\Omega u^{q+2m-1}dx+\int_\Omega u^{2m}dx.$$

Then integrating by parts and using the Hölder estimate

$$\int_\Omega u^{2m}dx\le|\Omega|^{\frac{q-1}{q+2m-1}}\left(\int_\Omega u^{q+2m-1}dx\right)^{\frac{2m}{q+2m-1}},$$

we have

$$\begin{aligned}\frac{1}{2m}\frac{d}{dt}\int_\Omega u^{2m}dx&=-\frac{2m-1}{m^2}\int_\Omega|\nabla(u^m)|^2dx-\int_\Omega u^{q+2m-1}dx+\int_\Omega u^{2m}dx\\ &\le-|\Omega|^{-\frac{q-1}{2m}}\left(\int_\Omega u^{2m}dx\right)^{\frac{2m+q-1}{2m}}+\int_\Omega u^{2m}dx.\end{aligned}$$

For $y(t):=\int_\Omega u^{2m}(t,u_0)dx$ the above differential inequality reads

$$y'(t)\le-2m\,|\Omega|^{-\frac{q-1}{2m}}[y(t)]^{\frac{2m+q-1}{2m}}+2m\,y(t). \tag{6.1.8}$$

According to Lemma 1.2.4 solutions of (6.1.8) satisfy

$$y(t) \leq \max\{y(0), |\Omega|\}$$
$$\limsup_{t\to+\infty} y(t) \leq |\Omega|.$$

Taking the $2m$-th roots of both sides of the above estimates, using monotonicity of these roots and then letting m tend to infinity, we finally get the formulas (6.1.6), (6.1.7). The proof is complete. □

Estimates of Lemma 6.1.1 and the simple nonlinear term F allow us to check easily the validity of assumptions of the abstract Theorem 4.2.1. With $Y = L^\infty(\Omega)$ and $\alpha_1 = \frac{1}{2}$, together with estimates (6.1.6), (6.1.7), we have

$$\begin{aligned}\|F(u(t))\|_{L^p(\Omega)} &= \|-u^q + u\|_{L^p(\Omega)} \\ &\leq |\Omega|^{\frac{1}{p}}(\|u\|^q_{L^\infty(\Omega)} + \|u\|_{L^\infty(\Omega)}) =: g(\|u\|_{L^\infty(\Omega)})\end{aligned}$$

which is a simple form of the estimate (3.1.4). Now the general Theorem 4.2.1 justifies the existence of the global attractor in X^α, $\alpha \in [\frac{1}{2}, 1)$, for the semigroup generated by the problem (6.1.1)-(6.1.3).

6.2. Single second order dissipative equation

Consider a generalization of the previous introductory example to the second order equation

$$u_t = \sum_{j,k=1}^{n} \frac{\partial}{\partial x_j}\left(a_{jk}(x)\frac{\partial u}{\partial x_k}\right) + f(x,u),\ t>0,\ x \in \Omega, \tag{6.2.1}$$

where Ω is a bounded domain in R^n with $\partial\Omega$ of class C^2. Let $a_{jk} : \Omega \to R$ be $C^1(\overline{\Omega})$ smooth and satisfy the symmetry condition $a_{jk} = a_{kj}$ for $j,k = 1,\dots,n$ and f be continuous in $\overline{\Omega} \times R$ and locally Lipschitz continuous with respect to its functional argument uniformly for $x \in \Omega$. First we shall consider (6.2.1) with the homogeneous Dirichlet condition

$$u = 0, \quad \text{for } x \in \partial\Omega, \tag{6.2.2}$$

and the initial condition

$$u(0,x) = u_0(x),\ x \in \Omega. \tag{6.2.3}$$

We need to assume the ellipticity condition

$$\exists_{C>0}\ \forall_{x\in\overline{\Omega}}\ \forall_{\xi\in R^n}\ \sum_{j,k=1}^{n} a_{jk}(x)\xi_j\xi_k \geq C|\xi|^2, \tag{6.2.4}$$

and the *dissipativeness condition*

$$\limsup_{|v|\to\infty} \frac{f(x,v)}{v} \leq 0 \ \text{ uniformly in } x \in \overline{\Omega}. \tag{6.2.5}$$

For $p > n$ we consider the sectorial operator $A = -\sum_{j,k=1}^{n} \frac{\partial}{\partial x_j}(a_{jk}(x)\frac{\partial u}{\partial x_k})$ in $L^p(\Omega)$ with domain $D(A) = W^{2,p}(\Omega) \cap W_0^{1,p}(\Omega)$. Thanks to the inclusion $X^{\frac{1}{2}} \subset L^\infty(\Omega)$, for $\alpha \in [\frac{1}{2}, 1)$, Lipschitz continuity of the nonlinear term $F(u) = f(\cdot, u)$ on bounded subsets of X^α follows as in the previous example, which ensures local X^α solvability of (6.2.1)-(6.2.3). To justify the global X^α solvability we need, according to Theorem 3.1.1, to indicate an a priori estimate global in time. For (6.2.1)-(6.2.3) that will be the $L^\infty(\Omega)$ estimate. We start with the following observation.

Observation 6.2.1. *It follows from* (6.2.5) *that*

(6.2.6)
$$\forall_{\varepsilon>0}\ \exists_{M_\varepsilon}\ \forall_{|v|\geq\varepsilon}\ \forall_{x\in\overline{\Omega}}\ \frac{f(x,v)}{v} \leq \varepsilon$$

or, with the same quantification, $vf(x,v) \leq \varepsilon v^2$. *Since on the bounded subset* $\{x \in \overline{\Omega} : |v| \leq M_\varepsilon\}$ *of* R^{n+1} *we have* $|vf(x,v)| \leq m_\varepsilon$, *also*

(6.2.7)
$$\forall_{\varepsilon>0}\ \exists_{m_\varepsilon>0}\ \forall_{v\in R}\ vf(x,v) \leq m_\varepsilon + \varepsilon v^2.$$

We are now able to formulate

Lemma 6.2.1. *For every local* X^α *solution to* (6.2.1)-(6.2.3) *the following estimate holds:*

(6.2.8)
$$\sup_{t\geq0} \|u(t,u_0)\|_{L^\infty(\Omega)} \leq const. \max\left\{\sup_{t\geq0} \|u(t,u_0)\|_{L^2(\Omega)}, 1\right\},$$

with const. depending on n, m_ε, $\overline{\Omega}$ *and* $\|u_0\|_{L^\infty(\Omega)}$.

Proof. We start with an a priori $L^2(\Omega)$ estimate. Multiplying (6.2.1) by u and integrating over Ω, we get

(6.2.9)
$$\begin{aligned}\frac{1}{2}\frac{d}{dt}\int_\Omega u^2dx &= -\int_\Omega \sum_{j,k=1}^{n} a_{jk}(x)\frac{\partial u}{\partial x_k}\frac{\partial u}{\partial x_j}dx + \int_\Omega f(x,u)udx \\ &\leq -C\int_\Omega \sum_{j=1}^{n}\left(\frac{\partial u}{\partial x_j}\right)^2 dx + \varepsilon\int_\Omega u^2dx + m_\varepsilon|\Omega| \\ &\leq -c_\Omega C\int_\Omega u^2dx + \varepsilon\int_\Omega u^2dx + m_\varepsilon|\Omega|,\end{aligned}$$

where (6.2.4), (6.2.7) and the Poincaré inequality were used. Next choosing $\varepsilon = \varepsilon_0 = \frac{1}{2}c_\Omega C$ we obtain the estimate

(6.2.10)
$$\frac{d}{dt}\int_\Omega u^2dx \leq -c_\Omega C\int_\Omega u^2dx + 2m_{\varepsilon_0}|\Omega|$$

which, through Lemma 1.2.4, leads to the condition

(6.2.11)
$$\|u(t,u_0)\|_{L^2(\Omega)} \leq \max\left\{\|u_0\|_{L^2(\Omega)}, \frac{2m_{\varepsilon_0}|\Omega|}{c_\Omega C}\right\}.$$

Given the estimate (6.2.7), (6.2.11) an iteration procedure of Alikakos (see [AL 1], [DL 3, p. 37]) provides the $L^\infty(\Omega)$ a priori estimate required in (6.2.8). We omit here the technical proof of this estimate (which is based on a version of the Nirenberg-Gagliardo estimate; see Section 9.3). □

Estimate (6.2.8) is a version of (3.1.3) with $Y = L^\infty(\Omega)$, $\alpha \in [\frac{1}{2}, 1)$, and provides for the problem (6.2.1)-(6.2.3) considered the following simple form of (3.1.4):

(6.2.12)

$$\|F(\cdot, u(t, u_0))\|_{L^p(\Omega)} \leq |\Omega|^{\frac{1}{p}} \sup_{\substack{x\in\Omega \\ \{y\in R:\ |y|\leq \|u(t,u_0)\|_{L^\infty(\Omega)}\}}} |f(x,y)| =: g(\|u(t, u_0)\|_{L^\infty(\Omega)}).$$

Global solvability of (6.2.1)-(6.2.3) follows now from Theorem 3.1.1.

To justify the existence of a global attractor we will show point dissipativeness of the semigroup generated by (6.2.1)-(6.2.3) on X^α using the properties of the Lyapunov function and applying Theorem 4.2.3. Multiplying by u_t and integrating over Ω we find

(6.2.13)

$$\begin{aligned}
\int_\Omega u_t^2 dx &= \int_\Omega \left(\sum_{j,k=1}^n \frac{\partial}{\partial x_j} \left(a_{jk}(x) \frac{\partial u}{\partial x_k} \right) + f(x,u) \right) u_t dx \\
&= \int_\Omega \left(-a_{jk}(x) \frac{\partial u}{\partial x_k} \frac{\partial u_t}{\partial x_j} + \frac{\partial}{\partial t} \int_0^{u(t,x)} f(x,s) ds \right) dx \\
&= \frac{d}{dt} \int_\Omega \left(-a_{jk}(x) \frac{\partial u}{\partial x_k} \frac{\partial u}{\partial x_j} + \int_0^{u(t,x)} f(x,s) ds \right) dx.
\end{aligned}$$

It is thus seen that the function

$$\mathcal{L}(\phi) = \frac{1}{2} \int_\Omega \sum_{j,k=1}^n a_{jk}(x) \frac{\partial \phi}{\partial x_k} \frac{\partial \phi}{\partial x_j} - \int_\Omega \int_0^{\phi(x)} f(x,s) ds dx. \tag{6.2.14}$$

decreases along trajectories of (6.2.1)-(6.2.3). It is now easy to check that $\mathcal{L}$ satisfies all the requirements imposed in Definition 1.1.7. First, continuity of $\mathcal{L} : X^\alpha \to R$, $\alpha \in [\frac{1}{2}, 1)$, is a result of the inclusions $X^\alpha \subset L^\infty(\Omega)$, $X^\alpha \subset H^1(\Omega)$ and smoothness of a_{jk}, f. Assume now that $\mathcal{L}$ is constant on some X^α solution $u(\cdot, u_0)$ of (6.2.1)-(6.2.3). Then by (6.2.13)

$$\int_\Omega u_t^2(t, u_0) dx = 0 \text{ for all } t > 0$$

so that at each $t > 0$, $u_t(t, u_0)$ is a zero element in $L^p(\Omega)$. As a consequence $u(\cdot, u_0)$, acting from $[0, \infty)$ to $L^p(\Omega)$, is constant in time and hence, from continuity,

$$u(t, u_0) = u_0 \text{ for all } t > 0,$$

which proves that u is a time independent solution to that problem.

According to Proposition 1.1.2 point dissipativeness of the semigroup generated by (6.2.1)-(6.2.3) is now a consequence of boundedness of the set S of its stationary solutions. Boundedness of S is thus the last information required in Theorem 4.2.3 for the existence of a global attractor.

Let us show now that the set of stationary solutions of (6.2.1)-(6.2.3) is bounded in $W^{2,p}(\Omega)$. Recall that stationary solutions of (6.2.1)-(6.2.3) are simultaneously solutions of the elliptic problem

$$\begin{cases} \sum_{j,k=1}^{n} \frac{\partial}{\partial x_j}(a_{jk}(x)\frac{\partial v}{\partial x_k}) + f(x,v) = 0, \ x \in \Omega, \\ v = 0 \ \text{ on } \ \partial\Omega. \end{cases} \tag{6.2.15}$$

We have

Lemma 6.2.2. *The solutions of* (6.2.15) *are bounded in* $W^{2,p}(\Omega)$ *in terms of* p, $|\Omega|$, C, c_Ω *and* m_ε.

Proof. We start with the $L^\infty(\Omega)$ estimate. Since such an estimate is very similar to the $L^\infty(\Omega)$ bound of solutions of the parabolic problem (6.2.1) reported in Lemma 9.3.1 (see Appendix 9.3), we recapitulate here only the method of the proof, underlining the natural differences.

As in the parabolic case we multiply (6.2.15) by v^{2^k-1}, integrate over Ω to get

(6.2.16)

$$0 = -\frac{2^k-1}{2^{2k-2}}\int_\Omega \sum_{j,k=1}^{n} a_{jk}(x)\frac{\partial(v^{2^{k-1}})}{\partial x_k}\frac{\partial(v^{2^{k-1}})}{\partial x_j}dx + \int_\Omega f(x,v)v^{2^k-1}dx.$$

We follow further the estimates of Lemma 9.3.1 (remembering that now the time derivative vanishes) until we obtain

$$0 \le -2^k\int_\Omega v^{2^k}dx + const.(2^k)^{\frac{n}{2}+1}\left(\int_\Omega v^{2^{k-1}}dx\right)^2 + 2^k m_\varepsilon|\Omega|.$$

Writing $\mu_k := (\int_\Omega v^{2^k}(x)dx)^{\frac{1}{2^k}}$ we thus have a recurrent estimate

$$\mu_k \le (const.(2^k)^{\frac{n}{2}}(\mu_{k-1})^{2^k} + m_\varepsilon|\Omega|)^{\frac{1}{2^k}}. \tag{6.2.17}$$

Enlarging *const.* to the value $c' = \max\{const., 1, m_\varepsilon|\Omega|\}$ and enlarging μ_0 (which was equal to $\|v\|_{L^1(\Omega)}$) to the value $x_0 = \max\{\mu_0, 1\}$ we observe that the numbers μ_k are dominated by the corresponding numbers x_k given by the recurrence

$$x_k = \left(c'\left(2^k\right)^{\frac{n}{2}}x_{k-1}^{2^k} + c'\right)^{\frac{1}{2^k}}.$$

Furthermore, elements of the sequence $\{x_k\}$ are dominated, respectively, by y_k given by

$$\begin{cases} y_0 = x_0, \\ y_k = (2c'(2^k)^{\frac{n}{2}}y_{k-1}^{2^k})^{\frac{1}{2^k}} = (2c'(2^k)^{\frac{n}{2}})^{\frac{1}{2^k}}y_{k-1}, \ k = 1, 2, \ldots. \end{cases}$$

We are now able to put these estimates together, obtaining

(6.2.18)
$$\|v\|_{L^\infty(\Omega)} \le y_\infty = \lim_{k\to\infty} y_k$$
$$= x_0 \prod_{k=1}^{\infty} (2c'(2^k)^{\frac{n}{2}})^{\frac{1}{2^k}} = 2c'P^{\frac{n}{2}} \max\{\|v\|_{L^1(\Omega)}, 1\},$$

with $P = \prod_{k=1}^{\infty} 2^{\frac{k}{2^k}} = 4$. To derive the $L^2(\Omega)$ estimate of v we use (6.2.16) with $k = 1$ and the Poincaré inequality (see similar calculations (6.2.9)-(6.2.10)) which leads to the condition

$$c_\Omega C \int_\Omega v^2 dx \le \varepsilon \int_\Omega v^2 dx + m_\varepsilon |\Omega|, \tag{6.2.19}$$

where $\varepsilon = \varepsilon_0 = \frac{1}{2} c_\Omega C$. Therefore, the required $L^\infty(\Omega)$ bound for solutions of (6.2.15) follows as a result of the estimates (6.2.18), (6.2.19).

Set $-Av := \sum_{j,k=1}^{n} \frac{\partial}{\partial x_j}(a_{jk}(x)\frac{\partial v}{\partial x_k})$. The easiest way to obtain the $W^{2,p}(\Omega)$ estimate of the solutions to (6.2.15) is to use the Calderón-Zygmund type estimate ([GI-TR, Theorem 9.14]). We only need to note that given (6.2.18) $f(\cdot, v(\cdot))$ is an element of $L^p(\Omega)$ with arbitrary $p \in [1, +\infty]$. Therefore, for sufficiently large constant $\sigma > 0$ we obtain the inequality (see [GI-TR, p. 240])

(6.2.20)
$$\|v\|_{W^{2,p}(\Omega)} \le const.' \| - Av - \sigma v\|_{L^p(\Omega)}$$
$$= const.' \|f(\cdot, v) - \sigma v\|_{L^p(\Omega)}, \; 1 < p < \infty,$$

whose right side is bounded, due to the estimate (6.2.18). Lemma 6.2.2 is thus proved. □

6.3. The method of invariant regions

For simplicity of the presentation we shall restrict the considerations of this section to the case of divergence systems of equations for the unknown $u = (u_1, \dots, u_m)$, $u_i = u_i(t, x)$. Consider

$$\frac{\partial u_i}{\partial t} = \sum_{j,k=1}^{n} \frac{\partial}{\partial x_j}\left(a^i_{jk}(x)\frac{\partial u_i}{\partial x_k}\right) + f_i(x, u), \; t > 0, \; x \in \Omega \subset R^n. \tag{6.3.1}$$

with homogeneous Dirichlet initial-boundary conditions

$$u_i(0, x) = u_{i0}(x) \text{ for } x \in \Omega, \; u_i(t, x) = 0 \text{ for } t > 0, \; x \in \partial\Omega. \tag{6.3.2}$$

We assume that Ω is bounded with $\partial\Omega \in C^2$ and the ellipticity condition is satisfied:

$$\exists_{C>0} \forall_{\xi \in R^n} \forall_{x\in\Omega} \sum_{j,k=1}^{n} a^i_{j,k}(x)\xi_j\xi_k \ge C|\xi|^2 \tag{6.3.3}$$

where $1 \leq i \leq m$ is arbitrary. Moreover we need the following *local dissipativeness condition* to hold:

$$\exists_{M_1,\dots,M_m>0} \forall_{i=1,\dots,m} \forall_{x\in\overline{\Omega}} \forall_{v\in\{w\in R^m:\ |w_i|=M_i,\ |w_j|\leq M_j,\ \text{for}\ j\neq i\}}\ v_i f_i(x,v) < 0. \tag{6.3.4}$$

Write further

$$R_0 = [-M_1, M_1] \times \ldots \times [-M_m, M_m]. \tag{6.3.5}$$

We assume that the components $f_i : \overline{\Omega} \times R^m \to R$ are continuous for $x \in \overline{\Omega}$ and locally Lipschitz with respect to functional arguments uniformly for $x \in \Omega$. Moreover, $a^i_{jk} \in C^1(\overline{\Omega})$ for the whole range of i, j, k.

Remark 6.3.1. Usually, condition (6.3.4) is expressed in a slightly more general form; assume that for some nonempty open bounded convex set $S \subset R^m$ containing 0 and having outer normal $N(v)$ at each point $v \in \partial S$

$$\forall_{v\in\partial S} \forall_{x\in\overline{\Omega}} \langle N(v), f(x,v)\rangle < 0.$$

It is then shown in the literature (see e.g. [SM, p. 198], [C-C-S]) that such an S is positively invariant with respect to the system (6.3.1), (6.3.2) (as a phase space, $[C(\overline{\Omega})]^m$ is chosen there). We shall show a similar result below, but using a wholly different technique of estimates. Also, our assumptions and formulations seem to be more precise than in [SM].

Note also that, for $p > n$, the problem (6.3.1), (6.3.2) is a special form of a system considered in [CO] (see Example 1.3.10). In particular $A : D(A) \to [L^p(\Omega)]^m$, where

$$Au = \left(-\sum_{j,k=1}^{n} \frac{\partial}{\partial x_j}\left(a^1_{jk}(x)\frac{\partial u_1}{\partial x_k}\right), \dots, -\sum_{j,k=1}^{n} \frac{\partial}{\partial x_j}\left(a^m_{jk}(x)\frac{\partial u_m}{\partial x_k}\right)\right),$$

and

$$D(A) = [W^{2,p}(\Omega) \cap W^{1,p}_0(\Omega)]^m$$

is a sectorial and positive operator in $[L^p(\Omega)]^m$ whereas, corresponding to the nonlinear term f, the substitution operator acting from $X^{\frac{1}{2}}$ into $[L^p(\Omega)]^m$ is Lipschitz continuous on bounded sets. Hence, for $\alpha \in [\frac{1}{2}, 1)$, local X^α solvability of (6.3.1), (6.3.2) follows.

Global solutions of (6.3.1)-(6.3.2). To ensure global solvability and the existence of a global attractor we will need an a priori $[L^\infty(\Omega)]^m$ estimate of solutions given below (note, that the $[L^\infty(\Omega)]^m$ and $[C(\overline{\Omega})]^m$ norms are equivalent for smooth solutions under considerations).

Lemma 6.3.1. *For $p > n$ and $\alpha \in [\frac{1}{2}, 1)$ consider a complete metric space $V^\alpha_{R_0}$:*

$$V^\alpha_{R_0} := \{v \in X^\alpha : v(x) \in R_0 \text{ for all } x \in \Omega\}, \tag{6.3.6}$$

$R_0 = [-M_1, M_1] \times \ldots \times [-M_m, M_m]$. Local X^α solutions $u(t, u_0)$ of (6.3.1), (6.3.2) corresponding to $u_0 \in V^\alpha_{R_0}$ ($\alpha \in [\frac{1}{2}, 1)$) exist for all $t \geq 0$. In addition, $u(t, u_0) \in V^\alpha_{R_0}$ for each $t \geq 0$ and every $u_0 \in V^\alpha_{R_0}$.

Proof. Denote by τ_{u_0} the lifetime of a local X^α solution $u(t, u_0)$. If $\tau \in (0, \tau_{u_0})$ then, from Sobolev embeddings, $u(\cdot, u_0) \in C\big([0, \tau], [C(\overline{\Omega})]^m\big)$. We shall prove that for each $u_0 \in V^\alpha_{R_0}$ a solution $u(t, u_0)$ is estimated uniformly in the $[L^\infty(\Omega)]^m$ norm as long as it exists; more precisely,

$$\|u_i(t, u_0))\|_{L^\infty(\Omega)} \leq M_i, \ t \in [0, \tau_{u_0}), \ i = 1, \ldots, m. \tag{6.3.7}$$

Since the components of f are continuous, then the assumption (6.3.4) implies that for some $\varepsilon > 0$ and each $i = 1, \ldots, m$,

$$v_i f_i(x, v) < 0 \tag{6.3.8}$$

whenever $(x, v) \in \overline{\Omega} \times \mathcal{W}_\varepsilon$:

$$\begin{aligned} \mathcal{W}_\varepsilon =& \{w \in R^m : |w_i - M_i| < \varepsilon, \ |w_j| < M_j + \varepsilon \ \text{ for } \ j \neq i\} \\ & \cup \{w \in R^m : |w_i + M_i| < \varepsilon, \ |w_j| < M_j + \varepsilon \ \text{ for } \ j \neq i\}. \end{aligned}$$

Fix ε from now on. For arbitrary $\tau \in (0, \tau_{u_0})$ and for $i \in \{1, \ldots, m\}$ write

$$F_i(\tau) := \sup_{t \in [0,\tau]} \sup_{x \in \overline{\Omega}} |f_i(x, u(t, u_0))|, \tag{6.3.9}$$

which is finite because of the inclusion $X^\alpha \subset [L^\infty(\Omega)]^m$ and continuity of f_i. Since $u(\cdot, u_0) \in C\big([0, \tau], [C(\overline{\Omega})]^m\big)$ with any $\tau \in (0, \tau_{u_0})$ and $u_0 \in V^\alpha_{R_0}$, either (6.3.7) holds or there exists $\tau_1 \in (0, \tau_{u_0})$ such that

$$M_{i_0} < \sup_{t \in [0, \tau_1]} \|u_{i_0}(t, u_0)\|_{L^\infty(\Omega)} < M_{i_0} + \varepsilon, \tag{6.3.10}$$

for some $i_0 \in \{1, \ldots, m\}$ and $\varepsilon > 0$ fixed above.

Multiply the i_0-th equation of the system (6.3.1) by $u_{i_0}^{2l-1}$, $l = 1, 2, \ldots$, and integrate the result over Ω to get

$$\int_\Omega \frac{\partial u_{i_0}}{\partial t} u_{i_0}^{2l-1} dx = \int_\Omega u_{i_0}^{2l-1} \sum_{j,k=1}^n \frac{\partial}{\partial x_j} \left(a^{i_0}_{jk}(x) \frac{\partial u_{i_0}}{\partial x_k} \right) dx + \int_\Omega u_{i_0}^{2l-1} f_{i_0}(x, u) dx.$$

Let us assume that (6.3.10) holds. Then, for $t \in (0, \tau_1]$ and $l = 1, 2, \ldots$, we estimate as follows:

(6.3.11)
$$\begin{aligned}\frac{1}{2l}\frac{d}{dt}\int_\Omega u_{i_0}^{2l}dx &\leq -(2l-1)\int_\Omega u_{i_0}^{2l-2}\sum_{j,k=1}^n a_{jk}^{i_0}(x)\frac{\partial u_{i_0}}{\partial x_k}\frac{\partial u_{i_0}}{\partial x_j}dx \\ &\quad + \int_{\{x\in\Omega:\ |u_{i_0}(x)|\leq M_{i_0}-\varepsilon\}} u_{i_0}^{2l-1}f_{i_0}(x,u)dx \\ &\leq -\frac{2l-1}{l^2}C\int_\Omega\sum_{j=1}^n\left[\frac{\partial(u_{i_0}^l)}{\partial x_j}\right]^2dx + \int_\Omega (M_{i_0}-\varepsilon)^{2l-1}F_{i_0}(\tau_1)dx \\ &\leq -\frac{2l-1}{l^2}c_\Omega C\int_\Omega u_{i_0}^{2l}dx + F_{i_0}(\tau_1)(M_{i_0}-\varepsilon)^{2l-1}|\Omega|,\end{aligned}$$

where the Poincaré inequality with constant c_Ω was used. As a consequence of Lemma 1.2.4, for $t\in[0,\tau_1]$, we thus have

$$\int_\Omega u_{i_0}^{2l}dx \leq \max\left\{\int_\Omega u_{i_0}^{2l}(0,x)dx,\ \frac{F_{i_0}(\tau_1)(M_{i_0}-\varepsilon)^{2l-1}|\Omega|}{\frac{2l-1}{l^2}c_\Omega C}\right\}, \tag{6.3.12}$$

or, taking the $2l$-th roots and letting l tend to infinity,

(6.3.13)
$$\begin{aligned}\|u_{i_0}(t,u_0)\|_{L^\infty(\Omega)} &\leq \max\{\|u_{i_0}(0,\cdot)\|_{L^\infty(\Omega)}, M_{i_0}-\varepsilon\} \\ &\leq \max\{M_{i_0}, M_{i_0}-\varepsilon\}\leq M_{i_0},\ t\in[0,\tau_1],\end{aligned}$$

which contradicts (6.3.10) and therefore ensures validity of (6.3.7).

For all $u_0\in V_{R_0}^\alpha$ the assumptions on f_i imply immediately that

$$\begin{aligned}\|f(\cdot,u(t,u_0))\|_{[L^p(\Omega)]^m} &\leq |\Omega|^{\frac{1}{p}}\max_{i=1,\dots,m}\left(\sup_{\substack{x\in\Omega \\ \{y\in R^m:\ |y|\leq\|u(t,u_0)\|_{[L^\infty(\Omega)]^m}\}}}|f_i(x,y)|\right) \\ &=: g(\|u(t,u_0)\|_{[L^\infty(\Omega)]^m}),\ t\in(0,\tau_{u_0}).\end{aligned}$$

Hence the requirements (A_2) of Chapter 3 are satisfied and global solvability of the problem (6.3.1), (6.3.2) with $u_0\in V_{R_0}^\alpha$ follows from Corollary 3.1.2. The proof is complete. □

Corollary 6.3.1. *The problem* (6.3.1), (6.3.2) *generates a C^0 semigroup of global X^α solutions ($\alpha\in[\frac{1}{2},1)$) on a complete metric space $V_{R_0}^\alpha$ and has a global attractor in $V_{R_0}^\alpha$.*

Proof. As seen from Lemma 6.3.1 all the requirements of (A_2) are satisfied provided that the initial functions u_0 belong to $V_{R_0}^\alpha$. Also $u(t,u_0)\in V_{R_0}^\alpha$ for all $t\geq 0$. Therefore, Theorem 4.2.1 is applicable, ensuring the existence of a global attractor in $V_{R_0}^\alpha$. □

Global attractor in the whole space X^α. In the above considerations we have used essentially the assumption (6.3.4) to prove that R_0 is a positively invariant region for local X^α solutions of (6.3.1), (6.3.2) (see Lemma 6.3.1) and to derive next Corollary 6.3.1. In order to extend these results to the whole space X^α we shall adapt the idea of [SM, pp. 200, 227], strengthening condition (6.3.4) to the following global dissipativeness assumption:

$$\text{(6.3.14)} \qquad \exists_{M_1,\dots,M_m>0}\ \forall_{s\geq 1}\ \forall_{i=1,\dots,m}\ \forall_{x\in\overline{\Omega}}$$
$$\forall_{v\in\{w\in R^m:\ |w_i|=sM_i,\ |w_j|\leq sM_j\ \text{for}\ j\neq i\}}\ v_i f_i(x,v) < 0.$$

Remark 6.3.2. Note that (6.3.14) coincides with (6.3.4) for $s = 1$. Generally, (6.3.14) guarantees that there exists a family $\{sR_0 : s \geq 1\}$ of *contracting rectangles* covering the whole of R^m. It is also seen from the Sobolev embedding $X^\alpha \subset [C(\overline{\Omega})]^m$ ($\alpha \in [\frac{1}{2}, 1)$) that for each $u_0 \in X^\alpha$ there is an $s \geq 1$ such that $u_0 \in V^\alpha_{sR_0}$, where $V^\alpha_{sR_0}$ is defined similarly to (6.3.6) as

$$V^\alpha_{sR_0} := \{v \in X^\alpha : v(x) \in sR_0 \ \text{ for all }\ x \in \Omega\}.$$

Considerations analogous to those of Lemma 6.3.1 lead to the following conclusion.

Corollary 6.3.2. *Let the strengthened assumption* (6.3.14) *be satisfied and* $\alpha \in [\frac{1}{2}, 1)$. *Then local* X^α *solutions of* (6.3.1), (6.3.2) *satisfy the estimate*

$$\text{(6.3.15)} \qquad \forall_{r>0}\ \exists_{s\geq 1}\ \forall_{u_0\in B_{X^\alpha}(r)}\ \forall_{i=1,\dots,m}\ \sup_{t\geq 0} \|u_i(t,u_0)\|_{L^\infty(\Omega)} \leq sM_i,$$

where $s \geq 1$ *is such that* $B_{X^\alpha}(r) \subset V^\alpha_{sR_0}$. *Therefore, the problem* (6.3.1), (6.3.2) *generates on* X^α *a* C^0 *semigroup* $T(t) : X^\alpha \to X^\alpha$, $t \geq 0$, *of global* X^α *solutions such that* $\{T(t)\}$ *keeps orbits of bounded sets bounded and* $T(t)V^\alpha_{sR_0} \subset V^\alpha_{sR_0}$ *for all* $t \geq 0$, $s \geq 1$.

As a consequence of Theorem 4.2.1, to prove the existence of a global attractor for $\{T(t)\}$ on X^α ($\alpha \in [\frac{1}{2}, 1)$) it is necessary to show that the estimate (6.3.15) will be strengthened to a similar one, asymptotically independent of initial condition.

Lemma 6.3.2. *If the strengthened assumption* (6.3.14) *is satisfied, then*

$$\text{(6.3.16)} \qquad \forall_{i=1,\dots,m}\ \forall_{u_0\in X^\alpha}\ \forall_{\alpha\in[\frac{1}{2},1)}\ \limsup_{t\to+\infty} \|u_i(t,u_0)\|_{L^\infty(\Omega)} \leq M_i.$$

Proof. We shall prove first that

$$\text{(6.3.17)} \qquad \forall_{s\geq 1}\ \exists_{\delta>0}\ \forall_{u_0\in V^\alpha_{(s+\delta)R_0}}\ \exists_{t_0\geq 0}\ \forall_{t\geq t_0}\ u(t,u_0) \in V^\alpha_{(s-\frac{\delta}{2})R_0}.$$

Choose arbitrary $s \geq 1$. Since the components of f are continuous, (6.3.14) implies that for some $\varepsilon > 0$ and each $i = 1, \dots, m$

$$\text{(6.3.18)} \qquad v_i f_i(x,v) < 0$$

on the set $\overline{\Omega} \times \mathcal{W}^s_\varepsilon$:

$$\begin{aligned}\mathcal{W}^s_\varepsilon =&\{w \in R^m : |w_i - sM_i| < \varepsilon,\ |w_j| < sM_j + \varepsilon \text{ for } j \neq i\}\\ &\cup \{w \in R^m : |w_i + sM_i| < \varepsilon,\ |w_j| < sM_j + \varepsilon \text{ for } j \neq i\}.\end{aligned}$$

Fix $\varepsilon > 0$ as above and choose any $u_0 \in V^\alpha_{(s+\frac{\varepsilon}{M})R_0}$, where $M := \max_{i=1,\dots,m} M_i$. We know from Corollary 6.3.2 that the global solution u starting from $u_0 \in V^\alpha_{(s+\frac{\varepsilon}{M})R_0}$ remains in $V^\alpha_{(s+\frac{\varepsilon}{M})R_0}$ for all $t \geq 0$. Define $\overline{u}_i(t,x) := \max\{u_i(t,x) - sM_i + \varepsilon, 0\}$ and $\Omega_t := \{x \in \Omega : \overline{u}_i(t,x) > 0\}$. Multiplying the i-th equation of (6.3.1) by $\overline{u}_i$ and integrating over $x \in \Omega$ and $t \in (t_1, t_2)$ we obtain that

(6.3.19)

$$\begin{aligned}\int_{t_1}^{t_2}\int_\Omega u_{it}\overline{u}_i dx dt =& \int_{t_1}^{t_2}\int_\Omega \sum_{j,k=1}^n \frac{\partial}{\partial x_j}\left(a^i_{jk}(x)\frac{\partial u_i}{\partial x_k}\right)\overline{u}_i dx dt\\ &+ \int_{t_1}^{t_2}\int_\Omega f_i(x,u)\overline{u}_i dx dt.\end{aligned}$$

Based on regularity of u_t [GI-TR, Lemma 7.6] (here, locally, u_t is understood as the Sobolev derivative; see Section 9.4, also [L-S-U, §4] for formal justification of the calculations below), we have

$$\begin{aligned}\int_{t_1}^{t_2}\int_\Omega u_{it}\overline{u}_i dx dt &= \int_{t_1}^{t_2}\frac{1}{2}\int_\Omega \frac{d}{dt}(\overline{u}_i)^2 dx dt\\ &= \frac{1}{2}\int_\Omega \overline{u}_i^2(t_2)dx - \frac{1}{2}\int_\Omega \overline{u}_i^2(t_1)dx.\end{aligned} \tag{6.3.20}$$

Note further that $\Omega_t \subset \Omega$, $\frac{\partial u_i}{\partial x_j}(t) = \frac{\partial \overline{u}_i}{\partial x_j}(t)$ on Ω_t and $\overline{u}(t) = 0$ on $\Omega \backslash \Omega_t$. In addition $\overline{u}_i \in W^{1,p}(\Omega)$ (see [GI-TR, Theorem 7.8]). Hence integrating by parts, using the ellipticity condition (6.3.3) and the Poincaré inequality, we get

(6.3.21)

$$\begin{aligned}&\int_{t_1}^{t_2}\int_\Omega \sum_{j,k=1}^n \frac{\partial}{\partial x_j}\left(a^i_{jk}(x)\frac{\partial u_i}{\partial x_k}\right)\overline{u}_i dx dt = -\int_{t_1}^{t_2}\int_\Omega \sum_{j,k=1}^n a^i_{jk}(x)\frac{\partial u_i}{\partial x_k}\frac{\partial \overline{u}_i}{\partial x_j} dx dt\\ &= -\int_{t_1}^{t_2}\int_{\Omega_t} \sum_{j,k=1}^n a^i_{jk}(x)\frac{\partial u_i}{\partial x_k}\frac{\partial \overline{u}_i}{\partial x_j} dx dt = -\int_{t_1}^{t_2}\int_{\Omega_t} \sum_{j,k=1}^n a^i_{jk}(x)\frac{\partial \overline{u}_i}{\partial x_k}\frac{\partial \overline{u}_i}{\partial x_j} dx dt\\ &\leq -C\int_{t_1}^{t_2}\int_{\Omega_t} \sum_{j=1}^n \left(\frac{\partial \overline{u}_i}{\partial x_j}\right)^2 dx dt = -C\int_{t_1}^{t_2}\int_\Omega \sum_{j=1}^n \left(\frac{\partial \overline{u}_i}{\partial x_j}\right)^2 dx dt\\ &\leq -c_\Omega C\int_{t_1}^{t_2}\int_\Omega \overline{u}_i^2 dx dt\end{aligned}$$

Since, as a consequence of (6.3.18) and Corollary 6.3.2, $f_i(x, u(t,x)) < 0$ for $x \in \Omega_t$ we have

$$\int_{t_1}^{t_2}\int_\Omega f_i(x,u)\overline{u}_i dx dt = \int_{t_1}^{t_2}\int_{\Omega_t} f_i(x,u)\overline{u}_i dx dt \leq 0. \tag{6.3.22}$$

Connecting now (6.3.19)-(6.3.22) we come to the condition

$$\frac{1}{2}\int_\Omega \overline{u}_i^2(t_2)dx \le \frac{1}{2}\int_\Omega \overline{u}_i^2(t_1)dx - c_\Omega C\int_{t_1}^{t_2}\int_\Omega \overline{u}_i^2 dxdt. \tag{6.3.23}$$

Inequality (6.3.23) holds for all $t_2 > t_1 > 0$ so that the function $(0,+\infty) \ni t \to \frac{1}{2}\int_\Omega \overline{u}_i^2(t)dx \in R$ is nonincreasing. Since it is also bounded below, (6.3.23) ensures further that

$$\lim_{t\to+\infty}\int_\Omega \overline{u}_i^2(t)dx = 0; \tag{6.3.24}$$

moreover, similar calculations show that, $\underline{u}_i(t,x) := \min\{u_i(t,x) + sM_i - \varepsilon, 0\}$, satisfies

$$\lim_{t\to+\infty}\int_\Omega \underline{u}_i^2(t)dx = 0. \tag{6.3.25}$$

Since $X^\alpha \subset [C^\nu(\overline{\Omega})]^m$ for $p > n$, $\alpha \in [\frac{1}{2}, 1)$, $0 < \nu < 2\alpha - \frac{n}{p}$ (see Remark 1.3.8) and the X^α solution u is bounded in the X^α norm uniformly for $t \in [0,+\infty)$, then $\overline{u}_i$ and $\underline{u}_i$ are also bounded in $C^\nu(\overline{\Omega})$ uniformly for $t \in [0,+\infty)$ (the maximum of two Hölder functions is Hölder itself). As a result of the Ascoli-Arzelà theorem and (6.3.24), (6.3.25) we thus conclude that

$$\lim_{t\to+\infty}\|\overline{u}_i(t)\|_{C(\overline{\Omega})} = 0 = \lim_{t\to+\infty}\|\underline{u}_i(t)\|_{C(\overline{\Omega})}$$

so that

$$\limsup_{t\to+\infty}\|u_i(t,u_0)\|_{L^\infty(\Omega)} \le sM_i - \varepsilon, \tag{6.3.26}$$

and, by continuity of u in X^α, there exists $t_0 > 0$ for which

(6.3.27)

$$\begin{aligned}\|u_i(t,u_0)\|_{L^\infty(\Omega)} &\le sM_i - \frac{\varepsilon}{2}\\ &= \left(s - \frac{\varepsilon}{2M_i}\right)M_i \le \left(s - \frac{\varepsilon}{2M}\right)M_i,\ t \ge 0.\end{aligned}$$

We have thus shown that the condition (6.3.17) holds with $\delta = \frac{\varepsilon}{M}$.

Condition (6.3.17) is crucial for the proof of (6.3.16). Choose $u_0 \in X^\alpha$ and define the set $I_{u_0} \subset R^+$:

$$I_{u_0} := \{s \ge 1 :\ \exists_{t\ge 0}\ u(t,u_0) \in V^\alpha_{sR_0}\}. \tag{6.3.28}$$

We shall show that $I_{u_0} = [1,+\infty)$. Note first that I_{u_0} is nonempty, since u_0 belongs to some $V^\alpha_{\overline{s}R_0}$ as a consequence of the Sobolev embedding $X^\alpha \subset [L^\infty(\Omega)]^m$. Also, directly from the definition (6.3.28) and Corollary 6.3.2, all $s' \ge \overline{s}$ must belong to I_{u_0}. Further, it follows from (6.3.17) that $\inf I_{u_0} =: s_0$ belongs to I_{u_0}. Indeed, if $\{s_n\} \subset I_{u_0}$ is such that $s_n \to s_0$, then

$$\forall_{n\in N}\ \forall_{i=1,\dots,m}\ \|u_i(t_{s_n},u_0)\|_{L^\infty(\Omega)} \le s_n M_i$$

and consequently, since $s_n < s_0 + \delta_{s_0}$ for all $n \ge n_0$,

$$\forall_{n\ge n_0}\ \forall_{i=1,\dots,m}\ \|u_i(t_{s_n},u_0)\|_{L^\infty(\Omega)} \le (s_0 + \delta_{s_0})M_i$$

where δ_{s_0} is chosen to s_0 according to the condition (6.3.17). In particular

$$u(t_{s_{n_0}}, u_0) \in V^{\alpha}_{(s_0+\delta_{s_0})R_0}$$

and (6.3.17) ensures the existence of $t_0 \geq 0$ such that

$$u(t_0, u(t_{s_{n_0}}, u_0)) = u(t_0 + t_{s_{n_0}}, u_0) \in V^{\alpha}_{(s_0-\frac{\delta_{s_0}}{2})R_0} \subset V^{\alpha}_{s_0 R_0}, \tag{6.3.29}$$

which shows that $s_0 \in I_{u_0}$.

The condition (6.3.17) shows next that $1 \in I_{u_0}$. Indeed, assume to the contrary that $s_0 > 1$. Then applying (6.3.17) to s_0 we find some $\delta_0 > 0$ such that $s_0 - \frac{\delta_0}{2} \in I_{u_0}$, which contradicts minimality of s_0. Since $1 \in I_{u_0}$, $u(t, u_0) \in V^{\alpha}_{R_0}$ for some $t \geq 0$. But the set $V^{\alpha}_{R_0}$ is positively invariant as a consequence of Corollary 6.3.2 (with $s = 1$). This shows that (6.3.16) is satisfied. The proof is complete. □

Corollary 6.3.3. *There exists a global attractor for the semigroup generated by* (6.3.1)-(6.3.2) *on* X^{α} $(\alpha \in [\frac{1}{2}, 1))$.

Proof. It suffices to note that Lemmas 6.3.1, 6.3.2 allow us to fulfill all the requirements of Theorem 4.2.1 with $V = X^{\alpha}$ and hence the global attractor in X^{α} exists. □

We are now able to present examples of reaction-diffusion systems that fall into the class described.

6.3.1. The Fitzhugh-Nagumo Equation. Consider the system

$$\begin{cases} u_t = \Delta u + f(u) - v, \\ v_t = \Delta v + \sigma u - \gamma v, \ x \in \Omega, \ t > 0, \end{cases} \tag{6.3.30}$$

under the Dirichlet initial-boundary condition (6.3.2). Here σ, γ are positive constants and f has the form of a cubic polynomial:

$$f(u) = -u(u-\beta)(u-1), \ 0 < \beta < \frac{1}{2}.$$

This system provides a mathematical model of transmission of nerve impulses in axons.

Choose a rectangle R_0 such that $\sigma u - \gamma v > 0$ in the lower edge, $\sigma u - \gamma v < 0$ in the upper edge, $f(u) - v > 0$ in the left edge and $f(u) - v < 0$ in the right edge. The more than linear growth of f ensures that it is possible to find such a rectangle. Then condition (6.3.14) will be satisfied and our abstract theory will work.

6.3.2. Superconductivity of liquids. Consider a system of the form

$$u_t = D\Delta u + (1 - |u|^2)u, \ x \in \Omega, \ t > 0, \tag{6.3.31}$$

where $u = (u_1, \dots, u_m)$ and D is a diagonal positive definite matrix. This system arises in the study of superconductivity of liquids. Choose $R_0 = [-1, 1]^m$; then the condition (6.3.14) will be satisfied and we easily justify the existence of a global attractor.

6.4. The Cahn-Hilliard pattern formation model

The Cahn-Hilliard equation was studied under various assumptions by many authors (see [TE 1], [DL 1], [C-D 2], [L-Z]). The following considerations are based on the most recent approach of Li and Zhong, which allows one to weaken the growth restrictions imposed in this equation on the nonlinear term f.

The Cahn-Hilliard equation has the form

$$u_t = \Delta(-\Delta u + f(u)),\ x \in \Omega \subset R^n,\ t > 0, \tag{6.4.1}$$

where $\Omega \subset R^n$, $n \leq 3$, is a bounded domain with $\partial\Omega \in C^{4+\varepsilon}$ ($\varepsilon > 0$). The equation is considered with initial-boundary conditions

$$u(0,x) = u_0(x),\ x \in \Omega, \tag{6.4.2}$$

$$\frac{\partial u}{\partial N} = \frac{\partial(\Delta u)}{\partial N} = 0,\ x \in \partial\Omega,\ t > 0. \tag{6.4.3}$$

For further use we write

$$k(u) := -\Delta u + f(u). \tag{6.4.4}$$

The following conditions are imposed on the nonlinear term $f \in C^{2+\text{Lipschitz}}(R)$:

$$\exists_{M>0}\ \forall_{v\in R}\ F(v) := \int_0^v f(s)ds \geq -M, \tag{6.4.5}$$

$$\exists_{\lambda>0}\ \forall_{v\in R}\ -f'(v) \leq \lambda, \tag{6.4.6}$$

$$\forall_{M'\geq 0}\ \exists_{\beta>0}\ \forall_{|m|\leq M'}\ \forall_{v\in R}\ vf(v+m) \geq -\beta. \tag{6.4.7}$$

The original assumption [L-Z, condition (1.6)] is more general than (6.4.6) (see [L-Z, Remark 1.1]). However the above conditions are still not very restrictive. One can easily check that they are fulfilled by arbitrary polynomial of the form

$$f(v) = \sum_{k=1}^{2p-1} a_k v^k, \tag{6.4.8}$$

with the leading coefficient a_{2p-1} positive. There is no restriction on the order $2p-1$. We will start the presentation with two simple properties of solutions of (6.4.1)-(6.4.3). Integrating equation (6.4.1) in the presence of boundary conditions (6.4.3) we get

$$\frac{d}{dt}\int_\Omega u(t,x)dx = 0,\ t > 0,$$

or, equivalently, for integral mean values $\overline{u}(t) = |\Omega|^{-1}\int_\Omega u(t,x)dx$,

$$\overline{u}(t) = const. = \overline{u}_0 \ \text{ for }\ t > 0. \tag{6.4.9}$$

Therefore, the problem (6.4.1)-(6.4.3) preserves spatial average in time. Also, multiplying the equation (6.4.1) by $-\Delta u + f(u)$, integrating by parts and using (6.4.3), we find that

$$\int_\Omega u_t[-\Delta u + f(u)]dx = -\int_\Omega |\nabla[-\Delta u + f(u)]|^2 dx \leq 0$$

or, further, that

$$\frac{d}{dt}\left[\frac{1}{2}\int_\Omega |\nabla u|^2 dx + \int_\Omega \int_0^u f(s)ds dx\right] \leq 0. \tag{6.4.10}$$

The last inequality defines a *Lyapunov function* for the Cahn-Hilliard equation,

$$\mathcal{L}(u(t)) := \frac{1}{2}\int_\Omega |\nabla u|^2 dx + \int_\Omega F(u(t))dx, \tag{6.4.11}$$

where $F(v) = \int_0^v f(s)ds$ is a primitive of f. For sufficiently smooth solutions of (6.4.1)-(6.4.3) (less smooth than constructed below) the existence of the Lyapunov function (6.4.11) together with the assumptions (6.4.5) and (6.4.9) guarantees an a priori $H^1(\Omega)$ estimate,

$$\begin{aligned} \frac{1}{2}\int_\Omega |\nabla u|^2 dx &\leq \mathcal{L}(u_0) - \int_\Omega F(u(t))dx \\ &\leq \mathcal{L}(u_0) + M|\Omega|, \end{aligned} \tag{6.4.12}$$

or, further,

$$\begin{aligned} \|u(t)\|_{H^1(\Omega)} &:= (\|\nabla u\|^2_{L^2(\Omega)} + |\overline{u}(t)|^2)^{\frac{1}{2}} \\ &\leq [2(\mathcal{L}(u_0) + M|\Omega|) + |\overline{u}_0|^2]^{\frac{1}{2}}, \end{aligned} \tag{6.4.13}$$

where an equivalent norm in $H^1(\Omega)$ was used. These properties are natural and well known and now we proceed to systematic construction of the X^α solution to the problem.

The operator $A = \Delta^2 + \rho I$, $\rho > 0$, with the Neumann type boundary conditions (6.4.3) defined on the domain

$$D(A) = \left\{\phi \in H^4(\Omega) : \frac{\partial \phi}{\partial N} = \frac{\partial(\Delta\phi)}{\partial N} = 0 \text{ on } \partial\Omega\right\}$$

is closable in a Hilbert space $X = L^2(\Omega)$. Then A will be easily seen to be positive definite (see Example 1.3.9),

$$\forall_{\phi \in D(A)} \langle A\phi, \phi\rangle_{L^2(\Omega)} = \|\Delta\phi\|^2_{L^2(\Omega)} + \rho\|\phi\|^2_{L^2(\Omega)},$$

and self-adjoint on $L^2(\Omega)$. From Proposition 1.3.3, A is thus sectorial on $L^2(\Omega)$.

Also, the following Sobolev type embeddings hold:

$$\begin{cases} H^2(\Omega) \subset L^\infty(\Omega), & n \leq 3, \\ H^2(\Omega) \subset W^{1,4}(\Omega), & n \leq 3, \\ X^{\frac{1}{2}} \subset H^2(\Omega). \end{cases} \tag{6.4.14}$$

Since boundedness of $\mathcal{U}$ in X^α implies its boundedness in $L^\infty(\Omega)$, the nonlinear term $\Delta f : X^\alpha \to L^2(\Omega)$, $\alpha \in [\frac{1}{2}, 1)$, is Lipschitz continuous on bounded sets; i.e.

(6.4.15)

$$\begin{aligned}
&\|\Delta f(\phi) - \Delta f(\psi)\|_{L^2(\Omega)} \\
&\leq \left\| [f'(\phi)\Delta\phi - f'(\psi)\Delta\psi] + \left[f''(\phi)|\nabla\phi|^2 - f''(\psi)|\nabla\psi|^2\right] \right\|_{L^2(\Omega)} \\
&\quad \text{(adding and subtracting, using (6.4.14))} \\
&\leq C_{\mathcal{U}} \|\phi - \psi\|_{X^\alpha}, \ \phi, \psi \in \mathcal{U} \subset X^\alpha.
\end{aligned}$$

This justifies local solvability of (6.4.1)-(6.4.3) in X^α.

There are different a priori estimates global in time leading to global solvability of (6.4.1)-(6.4.3). Under some growth restrictions imposed on the nonlinear term f (which should grow like a fourth order polynomial when $n = 3$) such estimates were presented in [TE 1], [C-D 2], [DL 1]. With our present weaker assumptions (6.4.5)-(6.4.7) we study first the case of smoother X^α solutions with $\alpha \in [\frac{3}{4}, 1)$. Then we complete the case of $\alpha \in [\frac{1}{2}, \frac{3}{4})$ which requires additional care.

We have

Proposition 6.4.1. *Under the assumptions* (6.4.5)-(6.4.7), *for* $n \leq 3$ *and* $\alpha \in [\frac{3}{4}, 1)$ *there exists an* X^α *solution global in time to the problem* (6.4.1)-(6.4.3).

Proof. The proof, according to the 'continuation theorem', is reduced to justification of an a priori X^α estimate, global in time, of solutions. Such an estimate will be given in a number of steps.

Step 1. Because of the boundary conditions (6.4.3) and the equation (6.4.1) the space average of time derivative u_t is zero:

$$\overline{u_t}(t) = |\Omega|^{-1} \int_\Omega u_t(t, x)dx = 0.$$

Applying to (6.4.1) the inverse operator to $-\Delta_{\frac{\partial}{\partial N}}$, multiplying the result by u_t and using the equality (see [N-S-T])

$$\left\langle u_t, (-\Delta)^{-1} u_t \right\rangle_{L^2(\Omega)} = \|u_t\|_{H^{-1}(\Omega)},$$

we get

(6.4.16)

$$\begin{aligned}
-\|u_t\|^2_{H^{-1}(\Omega)} &= \int_\Omega k(u) u_t dx = \int_\Omega \nabla u \nabla u_t dx + \int_\Omega f(u) u_t dx \\
&= \frac{d}{dt}\left[\frac{1}{2}\int_\Omega |\nabla u|^2 dx + \int_\Omega \int_0^{u(t)} f(s) ds dx\right] = \frac{d}{dt}\mathcal{L}(u(t)).
\end{aligned}$$

Next, by (6.4.10) and the assumption (6.4.5), integrating (6.4.16) over the interval $[0, T]$, we obtain

(6.4.17)
$$\int_0^T \|u_t(s)\|^2_{H^{-1}(\Omega)} ds = \mathcal{L}(u_0) - \mathcal{L}(u(T))$$
$$= \mathcal{L}(u_0) - \frac{1}{2}\int_\Omega |\nabla u|^2 dx - \int_\Omega F(u(T))dx \le \mathcal{L}(u_0) + M|\Omega|.$$

Since the right hand side of the last inequality is independent of T, we infer that

$$\int_0^\infty \|u_t(s)\|^2_{H^{-1}(\Omega)} ds \le \mathcal{L}(u_0) + M|\Omega|, \tag{6.4.18}$$

which concludes the first step of the estimate.

Step 2. We start with the simple observation that, for any function $\phi \in H^1(\Omega)$ with zero average $\overline{\phi}$,

$$\|\phi\|^2_{L^2(\Omega)} \le c\|\nabla\phi\|_{L^2(\Omega)}\|\phi\|_{H^{-1}(\Omega)}. \tag{6.4.19}$$

Multiplying (6.4.1) by $[k(u)]_t$, we obtain

$$\int_\Omega u_t\,[k(u)]_t dx = \int_\Omega \Delta[k(u)]\,[k(u)]_t dx \tag{6.4.20}$$
$$= -\frac{1}{2}\frac{d}{dt}\int_\Omega |\nabla[k(u)]|^2 dx$$

and further

$$\int_\Omega u_t\,[k(u)]_t dx = \int_\Omega u_t[-\Delta u + f(u)]_t dx = \int_\Omega |\nabla u_t|^2 dx + \int_\Omega f'(u)\,u_t^2 dx.$$

Hence, by (6.4.20), (6.4.6) it follows that

(6.4.21)
$$\frac{1}{2}\frac{d}{dt}\int_\Omega |\nabla[k(u)]|^2 dx \le -\int_\Omega |\nabla u_t|^2 dx + \lambda\int_\Omega u_t^2 dx$$
$$\le -\int_\Omega |\nabla u_t|^2 dx + \lambda c\Big(\int_\Omega |\nabla u_t|^2 dx\ \|u_t\|^2_{H^{-1}(\Omega)}\Big)^{\frac{1}{2}}$$
$$\le \frac{\lambda^2 c^2}{4}\|u_t\|^2_{H^{-1}(\Omega)},$$

where the condition (6.4.19) and the Cauchy inequality were used.

Step 3. By definition (6.4.4), $k(u) = -\Delta u + f(u)$, so that

(6.4.22)
$$\int_\Omega |\Delta u|^2 dx = -\int_\Omega k(u)\Delta u dx + \int_\Omega f(u)\Delta u dx$$
$$= \int_\Omega \nabla[k(u)]\nabla u dx - \int_\Omega f'(u)|\nabla u|^2 dx$$
$$\le \frac{1}{2}\int_\Omega |\nabla[k(u)]|^2 dx + \Big(\frac{1}{2} + \lambda\Big)\int_\Omega |\nabla u|^2 dx,$$

as a consequence of the Cauchy inequality and the assumption (6.4.6).

Step 4. For the X^α solutions, $\alpha \in [\frac{3}{4}, 1)$, of (6.4.1)-(6.4.3) we are now able to give its X^α a priori estimate global in time. Note that for such X^α solutions the quantity $\nabla[k(u)](t)$ is well defined and belongs to $L^2(\Omega)$ as long as the solution exists. Integrating (6.4.21) over $(0,t)$, we get

$$\int_\Omega |\nabla[k(u(t))]|^2 dx - \int_\Omega |\nabla[k(u_0)]|^2 dx$$
$$\leq \frac{\lambda^2 c^2}{2} \int_0^t \|u_t(s)\|^2_{H^{-1}(\Omega)} ds \leq \frac{\lambda^2 c^2}{2} (\mathcal{L}(u_0) + M|\Omega|),$$

as a result of (6.4.18). For $u_0 \in X^\alpha$, $\alpha \in [\frac{3}{4}, 1)$ we thus obtain an estimate global in time

$$\int_\Omega |\nabla[k(u(t))]|^2 dx \leq \int_\Omega |\nabla[k(u_0)]|^2 dx + \frac{\lambda^2 c^2}{2} (\mathcal{L}(u_0) + M|\Omega|), \tag{6.4.23}$$

which, together with (6.4.22) and (6.4.12), implies that

(6.4.24)
$$\int_\Omega |\Delta u|^2 dx + |\overline{u}_0|^2 \leq \frac{1}{2} \left(\int_\Omega |\nabla[k(u_0)]|^2 dx + \frac{\lambda^2 c^2}{2} (\mathcal{L}(u_0) + M|\Omega|) \right)$$
$$+ (1 + 2\lambda) (\mathcal{L}(u_0) + M|\Omega|) + |\overline{u}_0|^2.$$

Since the quantity

$$\left(\|\Delta u\|^2_{L^2(\Omega)} + |\overline{u}|^2 \right)^{\frac{1}{2}} \tag{6.4.25}$$

defines a norm equivalent to $H^2(\Omega)$ norm on

$$X^{\frac{1}{2}} = \left\{ \phi \in H^2(\Omega) : \frac{\partial \phi}{\partial N} = 0 \text{ on } \partial\Omega \right\}$$

(see (1.3.76)), condition (6.4.24) is the required $X^{\frac{1}{2}}$ a priori estimate of the solution.

Formally, to fulfill the assumptions of Theorem 3.1.1 we need to rewrite (6.4.1) in a different form:

$$u_t = -(\Delta^2 u + \rho u) + \Delta (f(u)) + \rho u, \; x \in \Omega \subset R^n, \; t > 0,$$

with some fixed $\rho > 0$. Then the condition

$$0 < \rho \leq Re\; \sigma(A) = Re\; \sigma(\Delta^2 + \rho I)$$

holds. The estimate (6.4.24) and the inclusions (6.4.14), $H^2(\Omega) \subset L^\infty(\Omega)$, $H^2(\Omega) \subset W^{1,4}(\Omega)$, $n \leq 3$, are all that we need to justify global solvability of (6.4.1)-(6.4.3) in X^α, $\alpha \in [\frac{3}{4}, 1)$, based on our main abstract Theorem 3.1.1. With

(6.4.24) giving an a priori estimate in $Y = H^2(\Omega)$ the subordination condition (3.1.4) now has the form

(6.4.26)
$$\begin{aligned}
&\|\Delta f(u(t)) + \rho u\|_{L^2(\Omega)} \\
&\leq \|f'(u(t))\|_{L^\infty(\Omega)}\|\Delta u\|_{L^2(\Omega)} + \|f''(u(t))\|_{L^\infty(\Omega)}\|\nabla u\|^2_{L^4(\Omega)} + \|\rho u\|_{L^2(\Omega)} \\
&\leq \|f'(u(t))\|_{L^\infty(\Omega)}\|u\|_{H^2(\Omega)} + c\|f''(u(t))\|_{L^\infty(\Omega)}\|u\|^2_{H^2(\Omega)} + \|\rho u\|_{L^2(\Omega)} \\
&\leq \sup_{|s|\leq C\|u(t)\|_{H^2(\Omega)}} (|f'(s)| + c|f''(s)| + \rho)\,(1 + \|u(t)\|^2_{H^2(\Omega)}) \\
&=: g(\|u(t)\|_{H^2(\Omega)}),
\end{aligned}$$

with an increasing function $g : [0, +\infty) \to [0, +\infty)$. Global solvability of (6.4.1)-(6.4.3) follows from Theorem 3.1.1 □

Now we extend the result of Proposition 6.4.1 to cover the case $\alpha \in [\frac{1}{2}, \frac{3}{4}]$. For such values of α the quantity $\nabla k(u(0))$ is not properly defined and we need to modify the previous considerations using the smoothing property of the semigroup. In fact to justify global solvability of (6.4.1)-(6.4.3) also for $\alpha \in [\frac{1}{2}, \frac{3}{4}]$, we need to complete the above considerations showing that images of bounded sets in $X^{\frac{1}{2}}$ are bounded in $X^{\frac{3}{4}}$ for small $t > 0$. But such a property is true in general for solutions of (2.1.1) local in time under Assumption 2.1.1.

From Proposition 3.2.1 with $\alpha = \frac{1}{2}$, $\varepsilon = \frac{1}{4}$, we have

Proposition 6.4.2. *For any bounded set $B \subset X^{\frac{1}{2}}$ there is a time $\tau_B > 0$ such that solutions of* (6.4.1)-(6.4.3) *originating in B exist and are bounded in $X^{\frac{3}{4}}$ (uniformly with respect to $u_0 \in B$) for any t taken from* $(0, \tau_B)$.

As a consequence of Proposition 6.4.2 the subordination condition (6.4.26) is effective also for $\alpha \in [\frac{1}{2}, \frac{3}{4}]$, $t \geq \frac{\tau_B}{2}$ and arbitrary u_0 from bounded set $B \subset X^{\frac{1}{2}}$. Global solvability of (6.4.1)-(6.4.3) follows thus from Theorem 3.1.1.

Stationary solutions of the problem (6.4.1)-(6.4.3). The set of stationary solutions of (6.4.1)-(6.4.3) consists of all $X^{\frac{1}{2}}$ solutions of second order elliptic problems

$$\begin{cases} -\Delta v + f(v) = a, \\ \frac{\partial v}{\partial N} = 0 \text{ on } \partial\Omega, \\ \overline{v} = \overline{u}_0 = const. \end{cases} \tag{6.4.27}$$

where $a = |\Omega|^{-1} \int_\Omega f(v)dx$. To construct the global attractor for (6.4.1)-(6.4.3) we need to estimate the $D(A)$ norm of solutions to (6.4.27) in terms of $\overline{u}_0$, Ω and the constants appearing in (6.4.5)-(6.4.7).

First a priori estimate. Multiplying (6.4.27) by v, integrating over Ω, we obtain

$$\int_\Omega |\nabla v|^2 dx + \int_\Omega f(v)v dx = a \int_\Omega v dx,$$

or, using (6.4.7),

$$\int_\Omega |\nabla v|^2 dx = |\Omega|^{-1} \int_\Omega f(v)dx \int_\Omega v dx - \int_\Omega f(v)v dx = - \int_\Omega f(v)(v-\overline{v})dx \leq \beta|\Omega|. \tag{6.4.28}$$

Second a priori estimate. Multiplying (6.4.27) by Δv and integrating over Ω, we find

$$-\int_\Omega (\Delta v)^2 dx + \int_\Omega \Delta v f(v) dx = a \int_\Omega \Delta v dx = 0,$$

hence, by (6.4.6),

$$\int_\Omega (\Delta v)^2 dx = \int_\Omega \Delta v f(v) dx = -\int_\Omega f'(v)|\nabla v|^2 dx \leq \lambda \int_\Omega |\nabla v|^2 dx \leq \lambda\beta|\Omega| \tag{6.4.29}$$

as a consequence of (6.4.28).

Higher order estimates. As a consequence of the boundary condition in (6.4.27), for sufficiently smooth solutions

$$\frac{\partial(\Delta v)}{\partial N} = \frac{\partial(f(v))}{\partial N} = f'(v)\frac{\partial v}{\partial N} = 0 \text{ on } \partial\Omega.$$

Therefore, multiplying (6.4.27) by $\Delta^2 v$, integrating, we get

$$\int_\Omega (\nabla\Delta v)^2 dx - \int_\Omega f'(v)\nabla v \nabla\Delta v dx = 0,$$

or further

$$\begin{aligned}\int_\Omega (\nabla\Delta v)^2 dx &\leq |\int_\Omega f'(v)\nabla v\nabla\Delta v dx| \\ &\leq \|f'(v)\|_{L^\infty(\Omega)}\|\nabla v\|_{L^2(\Omega)}\|\nabla\Delta v\|_{L^2(\Omega)} \\ &\leq \frac{1}{2}\|f'(v)\|^2_{L^\infty(\Omega)}\|\nabla v\|^2_{L^2(\Omega)} + \frac{1}{2}\|\nabla\Delta v\|^2_{L^2(\Omega)},\end{aligned}$$

which gives the $L^2(\Omega)$ estimate of $|\nabla\Delta v|$ thanks to the estimates (6.4.28), (6.4.29) and the inclusion $H^2(\Omega) \subset L^\infty(\Omega)$. The $D(A)$ norm estimate will be obtained by imposing (6.4.27) with the Laplacian and multiplying the result by $\Delta^2 v$.

The semigroup, further properties of stationary solutions. Recall that in previous considerations a priori estimates in $X^{\frac{1}{2}}$ norm of the X^α solutions, $\alpha \in [\frac{1}{2}, 1)$, have been given. Together with the abstract Theorem 3.1.1 this guarantees global solvability of (6.4.1)-(6.4.3) in X^α, $\alpha \in [\frac{1}{2}, 1)$, and allows us to define a semigroup $T(t) : X^\alpha \to X^\alpha$ by the formula (3.1.1). There is no need to mark in the notation of $T(t)$ the dependence on $\alpha \in [\frac{1}{2}, 1)$ since, at least for $t > 0$, these semigroups coincide as a result of the smoothing action of $T(t)$.

Now we study properties of stationary solutions of (6.4.1)-(6.4.3). We have

Lemma 6.4.1. *If for an X^α solution w ($\alpha \in [\frac{1}{2}, 1)$) of* (6.4.1)-(6.4.3)

$$\mathcal{L}(w(t)) = const., \ t > 0,$$

then w is a stationary solution of (6.4.1)-(6.4.3).

Proof. The solution $w(t)$ belongs to $D(A)$ for each $t > 0$. As a consequence of (6.4.16) and (6.4.1),

$$\forall_{t>0}\ 0 = \frac{d}{dt}\mathcal{L}(w(t)) = -\int_\Omega |\nabla k(u(t))|^2 dx;$$

therefore

$$\forall_{t>0}\ |\nabla(-\Delta w(t) + f(w(t))| = 0 \ \text{ a. e. in } \ \Omega. \tag{6.4.30}$$

From the well known properties of the distributional derivatives (see [SZ, p. 92]), the function under the gradient in (6.4.30) is independent of x. In the presence of the inclusion $D(A) \subset C^2(\overline{\Omega})$, $n \le 3$, condition (6.4.30) gives

$$\forall_{t>0}\ -\Delta w(t) + f(w(t)) = c(t) \ \text{ for every } \ x \in \Omega. \tag{6.4.31}$$

Taking Laplacian of both sides of (6.4.31), we find

$$-\Delta[\Delta w(t) + f(w(t))] = 0, \ t > 0, \ x \in \Omega. \tag{6.4.32}$$

Since $w(t)$ is a solution to (6.4.1)-(6.4.3), this shows that

$$\frac{dw(t)}{dt} = 0 \ \text{ for } \ t > 0,$$

hence w is time independent and $w(t) \equiv w_0$ for $t \ge 0$. The proof is complete. □

Lemma 6.4.2. *An element $w \in D(A)$ is a stationary solution of* (6.4.1)-(6.4.3) *if and only if w is an $H^2(\Omega)$ solution of the problem* (6.4.27).

Proof. The 'if' part follows since any $H^2(\Omega)$ solution of (6.4.27) belongs to $D(A)$ (which is a consequence of the elliptic regularity theory when $\partial\Omega \in C^4$), so that w is then a time independent solution of (6.4.1)-(6.4.3).

The 'only if' part follows from the considerations leading to (6.4.31) in the proof of Lemma 6.4.1. □

Point dissipativeness of the semigroup $\{T(t)\}$. Since an arbitrary constant $w \equiv const.$ in Ω is a stationary solution to (6.4.1)-(6.4.3), there is no hope of finding a compact global attractor in the whole space X^α. To overcome this difficulty, for fixed $\delta > 0$, consider the complete metric space

$$H^\alpha_\gamma := \{\phi \in X^\alpha : |\overline{\phi}| \le \gamma\}. \tag{6.4.33}$$

Due to (6.4.9) the spaces H^α_γ are positively invariant with respect to $T(t)$. Therefore, we will restrict all further considerations to the space H^α_γ.

Denote by $\mathcal{D}$ the set of all fixed points of the semigroup $\{T(t)\}$ on $X^{\frac{1}{2}}$. By the elliptic regularity theory $\mathcal{D} \subset D(A)$, hence $\mathcal{D}$ is the common set of fixed points of all semigroups generated by (6.4.1)-(6.4.3) on X^α, $\alpha \in [\frac{1}{2}, 1)$. For any $\gamma > 0$ write

$$\mathcal{D}_\gamma := \mathcal{D} \cap H^\alpha_\gamma; \tag{6.4.34}$$

hence $\mathcal{D}_\gamma$ is a set of stationary solutions to (6.4.1)-(6.4.3) with space averages satisfying $|\overline{v}| \leq \gamma$.

The following *point dissipativeness* property of the semigroup $\{T(t)\}$ is a consequence of the properties of the Lyapunov function (6.4.11).

Lemma 6.4.3. *There is a bounded set $\mathcal{B} \subset H^\alpha_\gamma$ attracting each point of H^α_γ.*

Proof. Define $\mathcal{B}$ as

$$\mathcal{B} := \bigcup_{v \in H^\alpha_\gamma} \omega(v),$$

where $\omega(v)$ is the ω-limit set of v. Since $T(t)$ (considered on H^α_γ) was shown to take bounded sets into bounded sets and the operator A has compact resolvent it follows that $T(t)$ is compact for $t > 0$. This shows that $\mathcal{B}$ attracts each point of H^α_γ (see [HA 2, Lemma 3.2.1]). It suffices to prove that $\mathcal{B}$ is bounded.

For every $v_0 \in \mathcal{B}$ there is $u_0 \in H^\alpha_\gamma$ such that $v_0 \in \omega(u_0)$. Setting $u = T(t)u_0$ as an argument of the Lyapunov function (6.4.11) we obtain

$$\exists_{l \in R} \lim_{t \to +\infty} \mathcal{L}(T(t)u_0) = l,$$

since $\mathcal{L}(T(t)u_0)$ is nonincreasing and bounded below. Also, the Lyapunov function must be identically equal to l along the trajectory of any element of $\omega(u_0)$. Thus

$$\mathcal{L}(T(t)v_0) \equiv l \ \text{ for } \ t \geq 0,$$

so that, according to Lemma 6.4.1, v_0 must be a stationary solution of (6.4.1)-(6.4.3). But the set $\mathcal{D}_\gamma$ of stationary solutions was bounded in $D(A)$. Hence $\mathcal{D}_\gamma$ is also bounded in X^α with any $\alpha \in [\frac{1}{2}, 1)$. This shows that $\mathcal{B}$ is bounded and completes the proof of Lemma 6.4.3. □

As a direct consequence of Theorem 4.2.3 we claim:

Proposition 6.4.3. *The semigroup $\{T(t)\}$ generated by (6.4.1)-(6.4.3) on a complete metric space H^α_γ possesses a global attractor. Moreover, this attractor is independent of $\alpha \in [\frac{1}{2}, 1)$.*

6.5. Burgers equation

A simple physical model for viscous fluid motion is the Burgers equation [DL 2]:

$$\tag{6.5.1} \begin{cases} v_t = \nu v_{xx} + Uv - \frac{1}{2}(v^2)_x, \ t > 0, \ x \in (0, \pi), \\ U_t = P - \nu U - \int_0^\pi v^2(t,x)dx, \ t > 0, \\ v(t,0) = v(t,\pi) = 0, \ t > 0, \\ v(0,x) = v_0(x), \ x \in (0,\pi), \\ U(0) = U_0, \end{cases}$$

where $\nu > 0$ denotes viscosity and $P > 0$ stands for a constant pressure drop.

Writing $A_1 = -\nu \frac{\partial^2}{\partial x^2}$ we observe that A_1 is a sectorial operator in $L^p(0,\pi)$. Moreover $A_2 := \nu I$ with $D(A_2) = R^1$, as bounded and linear (see Example 1.3.1), is also sectorial and hence the problem (6.5.1) admits an abstract formulation (2.1.1) with the product operator $A_p := A_1 \times A_2$ considered on the domain $D(A_p) = (W_0^{1,p}(0,\pi) \cap W^{2,p}(0,\pi)) \times R^1$ being sectorial in $L^p(0,\pi) \times R^1$ (see Example 1.3.2). In this example the substitution operator F_p is defined as

$$F_p(U,v) := \left(Uv - \frac{1}{2}(v^2)_x, P - \int_0^\pi v^2(t,x)dx \right).$$

Whenever $p > 1$ and $2\alpha > \max\{1, \frac{1}{2} + \frac{1}{p}\}$ it may be seen that F_p acting from $X_p^\alpha = D(A_{1p}^\alpha) \times R^1$ into $L^p(0,\pi) \times R^1$ is Lipschitz continuous on bounded sets and hence (6.5.1) is shown to generate a local semigroup on X_p^α.

In order to derive a suitable a priori estimate for (6.5.1) let us multiply the first equation by v, integrate it over $(0,\pi)$ under the homogeneous boundary conditions, multiply the second equation by U and next add the results. We obtain

$$\tag{6.5.2} \frac{1}{2}\frac{d}{dt}\left(\int_0^\pi v^2 dx + U^2 \right) = -\nu \left(\int_0^\pi v_x^2 dx + U^2 \right) + PU$$

and further, using the Poincaré and Cauchy inequalities,

$$\tag{6.5.3} \frac{1}{2}\frac{d}{dt}\left(\int_0^\pi v^2 dx + U^2 \right) \le -\nu \left(\int_0^\pi v^2 dx + U^2 \right) + \frac{P^2}{2\varepsilon} + \frac{\varepsilon U^2}{2}.$$

When $\varepsilon = \nu$ is inserted in (6.5.3), we come to the differential inequality

$$\tag{6.5.4} \frac{d}{dt}\left(\int_0^\pi v^2 dx + U^2 \right) \le -\nu \left(\int_0^\pi v^2 dx + U^2 \right) + \frac{P^2}{\nu},$$

which leads to the estimate

$$\tag{6.5.5} \int_0^\pi v^2 dx + U^2 \le \left(\int_0^\pi v_0^2 dx + U_0^2 \right) e^{-\nu t} + \frac{P^2}{\nu}\frac{1 - e^{-\nu t}}{\nu},$$

so that asymptotically we get

$$\tag{6.5.6} \limsup_{t \to +\infty} \left(\int_0^\pi v^2 dx + U^2 \right) \le \frac{P^2}{\nu^2} \quad \text{for all } (v_0, U_0) \in X_p^\alpha.$$

Consider now the first equation in (6.5.1) separately where $U = U(t)$ will be treated as a given coefficient bounded in $[0, \infty)$ and asymptotically estimated independently of U_0 (according to (6.5.5)-(6.5.6)). The iterative estimates of Lemma 9.3.1 will be used on this equation.

Multiplying the equation by v^{2^k-1}, $k = 2, 3, \dots$, integrating over $(0, \pi)$, we get the estimate

$$\int_0^\pi v_t v^{2^k-1} dx = \nu \int_0^\pi v_{xx} v^{2^k-1} dx + U \int_0^\pi v^{2^k} dx;$$

since the last term vanishes

$$-\frac{1}{2}\int_0^\pi (v^2)_x v^{2^k-1} dx = -\frac{1}{2^k+1}\int_0^\pi (v^{2^k+1})_x dx = 0, \ k \in N.$$

The last observation allows us to estimate the $L^\infty(0, \pi)$ norm of v through its $L^2(0, \pi)$ norm precisely as in the proof of Lemma 9.3.1; note also, that the remaining part Uv of the nonlinearity satisfies

$$v\, Uv \le Cv^2, \tag{6.5.7}$$

where C is the bound of $|U|$ resulting from (6.5.5). We thus claim an estimate

$$\sup_{t\ge 0} \|v(t, \cdot)\|_{L^\infty(0,\pi)} \le const. \max\left\{\sup_{t\ge 0} \|v(t, \cdot)\|_{L^2(0,\pi)}, 1\right\}. \tag{6.5.8}$$

As follows from (6.5.6) the constant C appearing in (6.5.7) is asymptotically independent of U_0:

$$\forall_{U_0} \forall_{\varepsilon>0} \exists_{t_0>0} \forall_{t\ge t_0}\ vUv \le \left(\frac{P}{\nu} + \varepsilon\right) v^2.$$

Estimating as in the proof of Lemma 9.3.1 below (with $D = 0$, $n = 1$) for $t \ge t_0$, we find that

$$\frac{d}{dt}\int_0^\pi v^{2^k} dx \le -2^k \int_0^\pi v^{2^k} dx + const.(2^k)^{\frac{3}{2}} \left(\int_0^\pi v^{2^{k-1}} dx\right)^2, \tag{6.5.9}$$

$k = 2, 3, \dots$, with *const.* independent of U_0. Set $k = 2$; then from (6.5.6), as a consequence of Lemma 1.2.5, with $f(t) = const.4^{\frac{3}{2}} \left(\int_0^\pi v^2(t, x) dx\right)^2$, we obtain

(6.5.10)

$$\limsup_{t\to+\infty} \int_0^\pi v^4(t, x) dx \le 2const. \limsup_{t\to+\infty} \left(\int_0^\pi v^2(t, x) dx\right)^2 \le 2const. \frac{P^4}{\nu^4},$$

or equivalently

$$\limsup_{t\to+\infty} \|v(t)\|_{L^4(0,\pi)} \le (2const.)^{\frac{1}{4}} \frac{P}{\nu}. \tag{6.5.11}$$

Applying successively Lemma 1.2.5 to (6.5.9) we get simultaneously the bounds

$$\limsup_{t\to+\infty} \|v(t)\|_{L^{2^k}(0,\pi)} \le const._k \frac{P}{\nu},\ k = 2,3,\dots, \tag{6.5.12}$$

which give the required asymptotic estimate of v in $L^r(0,\pi)$ with r arbitrarily large.

Note that by Proposition 1.2.3 the resolvent of A_1 is compact. Fixing arbitrary α satisfying the condition $2\alpha > \max\{1, \frac{1}{2}+\frac{1}{p}\}$ the assumption (A_2) will be satisfied with $Y = L^r(0,\pi)\times R^1$, r sufficiently large, since

(6.5.13)

$$\begin{aligned}\|F_p(U,v)\|_{L^p(0,\pi)\times R} &= \|Uv - vv_x\|_{L^p(0,\pi)} + \left|P - \int_0^\pi v^2(t,x)dx\right| \\ &\le |U|\|v\|_{L^p(0,\pi)} + \|vv_x\|_{L^p(0,\pi)} + P + \|v\|^2_{L^2(0,\pi)}\end{aligned}$$

and in the light of the estimates (6.5.5), (6.5.6), (6.5.8), (6.5.12) we need only estimate the term $\|vv_x\|_{L^p(0,\pi)}$.

First by the Young inequality (with $m = \frac{p+\delta}{p}$, $\varepsilon = 1$) we get

$$|vv_x| \le \frac{p}{p+\delta}|v_x|^{\frac{p+\delta}{p}} + \frac{\delta}{p+\delta}|v|^{\frac{p+\delta}{\delta}} \tag{6.5.14}$$

with $\delta > 0$ arbitrary. Next, the Nirenberg-Gagliardo inequality (see Remark 1.2.1) shows that

$$\begin{aligned}\left\||v_x|^{\frac{p+\delta}{p}}\right\|_{L^p(0,\pi)} &= \|v_x\|^{\frac{p+\delta}{p}}_{L^{p+\delta}(0,\pi)} \\ &\le \left(const.\|v\|^\theta_{W^{2\alpha,p}(0,\pi)}\|v\|^{1-\theta}_{L^p(0,\pi)}\right)^{\frac{p+\delta}{p}} \\ &= \left(const.\|v\|^{1-\theta}_{L^p(0,\pi)}\right)^{\frac{p+\delta}{p}} \|v\|^{\theta\frac{p+\delta}{p}}_{X^\alpha_p},\end{aligned} \tag{6.5.15}$$

where $1+\frac{1}{p} < 2\alpha\theta + \frac{1}{p+\delta}$ (note that $2\alpha > \max\left\{1, \frac{1}{2}+\frac{1}{p}\right\}$); we need also the restriction $\theta\frac{p+\delta}{p} < 1$ to be satisfied. For fixed α and p we can choose $\theta \in [0,1)$ and (small) $\delta > 0$, such that the above restrictions are satisfied. The estimates (6.5.5), (6.5.8) and (6.5.15) are sufficient to verify condition (A_2) since, thanks to (6.5.14),

$$\|vv_x\|_{L^p(0,\pi)} \le g\left(\|v\|_{L^{\frac{(p+\delta)p}{\delta}}(0,\pi)}\right)\left(1 + \|v\|^{\theta\frac{p+\delta}{p}}_{X^\alpha_p}\right)$$

and we can take $Y = L^{\frac{(p+\delta)p}{\delta}}(0,\pi)\times R^1$ in (A_2).

6.6. Navier-Stokes equations in low dimension ($n \le 2$)

The huge number of publications concerning the Navier-Stokes system makes it impossible to mention all of them. We shall refer here only to [HE 1, pp. 79-81], [TE 1, pp. 102-113], [WA 5] [GI-MI] for basic concepts and techniques. Although further on, in Chapter 8 the Navier-Stokes system will be reconsidered in higher space dimension, some introductory facts are - for convenience - repeated separately in each of these cases.

Consider

$$(6.6.1)\qquad \begin{cases} u_t = \nu\Delta u - (u,\nabla)u + h - \nabla p \ \text{ in } \ (0,T)\times\Omega, \\ div\ u = 0, \ \text{ in } \ (0,T)\times\Omega, \\ u = 0 \ \text{ on } \ \partial\Omega\times(0,T), \\ u(0,x) = u_0(x) \ \text{ in } \ \Omega, \end{cases}$$

where Ω is a bounded domain in R^2 with $C^{2+\rho}$ boundary $\partial\Omega$, $0<\rho<1$, $u = (u^1(t,x),u^2(t,x))$ denotes velocity, ν constant viscosity, $p(t,x)$ pressure and $h = (h^1(x),h^2(x))$ the external forces. Projecting this problem (P_r) onto a subspace $X_r \subset [L^r(\Omega)]^2$, $1<r<\infty$, of divergence-free functions

$$X_r = cl_{[L^r(\Omega)]^2}\{\phi \in [C_0^\infty(\Omega)]^2 : div\,\phi = 0\}$$

we get (6.6.1) in an abstract form

$$(6.6.2)\qquad u_t + A_r u = F_r u + P_r h, \ t>0, \ u(0) = u_0,$$

with sectorial product operator

$$A_r = -\nu P_r(\Delta,\Delta)$$

(see [GI-MI, Lemma 1.1], [HE 1, pp. 80-81]) defined on the domain $D(A_r) = X_r \cap \{\phi \in [W^{2,r}(D)]^2 : \phi|_{\partial\Omega} = 0\}$ and nonlinear term $F_r u = -P_r(u,\nabla)u$. Compactness of the resolvent $(A_r + \lambda I)^{-1} : X_r \longrightarrow X_r$, $\lambda \in \sigma(A_r)$ follows from the estimate in [GI-MI, Lemma 3.1] with $m=0$ and compactness of the embedding $[W^{2,r}(\Omega)]^2 \subset [L^r(\Omega)]^2$.

We choose $r \in (2,\infty)$ and $\alpha_0 \in (0,1)$ satisfying

$$(6.6.3)\qquad 1 < 2\left(\alpha_0 - \frac{1}{r}\right).$$

From [GI-MI, p. 269] for $0 \leq \alpha \leq 1$, it is known that

$$X_r^\alpha = D(A_r^\alpha) = [X_r, D(A_r)]_\alpha$$

(the symbol $[\cdot,\cdot]_\alpha$ denotes here a complex interpolation space; see [TR]) and, whenever $\alpha \geq 0$, X_r^α is continuously embedded in the space of Bessel potentials $[H_r^{2\alpha}(\Omega)]^2$. In particular, for $\alpha \in [\alpha_0, 1)$, X_r^α is a subset of $[L^\infty(\Omega)]^2 \cap [H_r^1(\Omega)]^2$ and the nonlinear term $F_r : X_r^\alpha \longrightarrow X_r$ is Lipschitz continuous on bounded sets. Indeed, for $\phi,\psi \in X_r^\alpha$ we have

$$\begin{aligned} \|F_r\phi - F_r\psi\|_{X^r} &\leq C_r\|\phi\|_{[W^{1,r}(\Omega)]^2}\|\phi-\psi\|_{[W^{1,r}(\Omega)]^2} \\ &\quad + C_r\|\phi-\psi\|_{[W^{1,r}(\Omega)]^2}\|\psi\|_{[W^{1,r}(\Omega)]^2} \\ &\leq C_r \max\left\{\|\phi\|_{[W^{1,r}(\Omega)]^2}, \|\psi\|_{[W^{1,r}(\Omega)]^2}\right\}\|\phi-\psi\|_{[W^{1,r}(\Omega)]^2} \end{aligned}$$

(see [GI-MI, Lemma 3.2]), which justifies local solvability of (6.6.2).

Following [TE 1, pp. 107-109] we shall estimate the solutions of (6.6.1) in the $[H_0^1(\Omega)]^2$ norm.

First a priori estimate. Multiplying the first equation in (6.6.1) in $[L^2(\Omega)]^2$ by u, we get

$$\frac{1}{2}\frac{d}{dt}\int_\Omega \sum_{i=1}^2 (u^i)^2 dx + \nu \int_\Omega \sum_{i,k=1}^2 \left(\frac{\partial u^i}{\partial x_k}\right)^2 dx = \int_\Omega \sum_{i=1}^2 h^i u^i dx,$$

or equivalently

$$\frac{1}{2}\frac{d}{dt}\|u\|^2_{[L^2(\Omega)]^2} + \nu\|\nabla u\|^2_{[L^2(\Omega)]^2} = \langle h, u\rangle_{[L^2(\Omega)]^2}.$$

By the Poincaré inequality

$$\|\phi\|^2_{[L^2(\Omega)]^2} \leq C_\Omega \|\nabla\phi\|^2_{[L^2(\Omega)]^2},\ \phi \in [H^1_0(\Omega)]^2,$$

and by the Cauchy inequality we thus infer that

(6.6.4)

$$\begin{aligned}\frac{1}{2}\frac{d}{dt}\|u\|^2_{[L^2(\Omega)]^2} &\leq -\nu\|\nabla u\|^2_{[L^2(\Omega)]^2} + \frac{1}{2\varepsilon}\|u\|^2_{[L^2(\Omega)]^2} + \frac{\varepsilon}{2}\|h\|^2_{[L^2(\Omega)]^2} \\ &\leq \left(-\frac{\nu}{C_\Omega} + \frac{1}{2\varepsilon}\right)\|u\|^2_{[L^2(\Omega)]^2} + \frac{\varepsilon}{2}\|h\|^2_{[L^2(\Omega)]^2}.\end{aligned}$$

For $\varepsilon = \frac{C_\Omega}{\nu}$ the condition (6.6.4) leads through Lemma 1.2.4 to the estimate

$$\|u(t)\|^2_{[L^2(\Omega)]^2} \leq \max\left\{\|u_0\|^2_{[L^2(\Omega)]^2},\ \left(\frac{C_\Omega}{\nu}\right)^2 \|h\|^2_{[L^2(\Omega)]^2}\right\},\ t \geq 0, \tag{6.6.5}$$

and an asymptotic estimate

$$\limsup_{t\to\infty} \|u(t)\|^2_{[L^2(\Omega)]^2} \leq \left(\frac{C_\Omega}{\nu}\right)^2 \|h\|^2_{[L^2(\Omega)]^2} \tag{6.6.6}$$

which is independent of u_0. Returning to (6.6.4), for fixed $r > 0$ and $\varepsilon = 1$, we find that

$$\begin{aligned}\int_t^{t+r} \|\nabla u(s,\cdot)\|^2_{[L^2(\Omega)]^2} ds &\leq \frac{1}{2}\|u(t,\cdot)\|^2_{[L^2(\Omega)]^2} - \frac{1}{2}\|u(t+r,\cdot)\|^2_{[L^2(\Omega)]^2} \\ &+ \frac{1}{2}\int_t^{t+r} \left(\|u(s,\cdot)\|^2_{[L^2(\Omega)]^2} + \|h\|^2_{[L^2(\Omega)]^2}\right) ds.\end{aligned}$$

Choosing, according to (6.6.6), the value $t(u_0)$ in such a way that

$$\|u(t)\|^2_{[L^2(\Omega)]^2} \leq \left(\frac{C_\Omega}{\nu} + 1\right)^2 \|h\|^2_{[L^2(\Omega)]^2},\ t \geq t(u_0), \tag{6.6.7}$$

we find independently of u_0 the estimate

(6.6.8)

$$\begin{aligned}&\int_t^{t+r} \|\nabla u(s,\cdot)\|^2_{[L^2(\Omega)]^2} ds \\ &\leq \left[\left(\frac{C_\Omega}{\nu} + 1\right)^2 \left(1 + \frac{1}{2}r\right) + \frac{1}{2}r\right] \|h\|^2_{[L^2(\Omega)]^2},\quad \text{for } t \geq t(u_0).\end{aligned}$$

Second a priori estimate. Multiplying (6.6.2) in $[L^2(\Omega)]^2$ by $A_r u$, we find that

$$\begin{aligned}(6.6.9)\qquad &\frac{\nu}{2}\frac{d}{dt}\|\nabla u\|^2_{[L^2(\Omega)]^2} + \|A_r u\|^2_{[L^2(\Omega)]^2} \\ &= \langle -P_r(u,\nabla)u, A_r u\rangle_{[L^2(\Omega)]^2} + \langle P_r h, A_r u\rangle_{[L^2(\Omega)]^2}.\end{aligned}$$

Using the estimate of [GI-MI, Lemma 3.3 (i)] with $c = C_{0,2} > 0$,

$$(6.6.10)\qquad \|P_r h\|_{[L^2(\Omega)]^2} \le c\|h\|_{[L^2(\Omega)]^2},$$

the nonlinear term will be estimated as

$$\begin{aligned}&\|P_r(u,\nabla)u\|_{[L^2(\Omega)]^2} \le c\|(u,\nabla)u\|_{[L^2(\Omega)]^2} \\ &= c\left(\sum_{i=1}^{2}\int_\Omega\left(\sum_{j=1}^{2} u^j\frac{\partial u^i}{\partial x_j}\right)^2 dx\right)^{\frac12} \le c\left(2\sum_{i,j=1}^{2}\|u^j\|^2_{L^\infty(\Omega)}\|\nabla u^i\|^2_{L^2(\Omega)}\right)^{\frac12} \\ &\le const._1\|u\|^{\frac12}_{[L^2(\Omega)]^2}\|u\|^{\frac12}_{[H^2(\Omega)]^2}\|\nabla u\|_{[L^2(\Omega)]^2} \\ &\le const.\|u\|^{\frac12}_{[L^2(\Omega)]^2}\|A_r u\|^{\frac12}_{[L^2(\Omega)]^2}\|\nabla u\|_{[L^2(\Omega)]^2},\end{aligned}$$

where the Agmon inequality (1.2.44) and the Calderón-Zygmund type estimate for the Stokes operator A_r (see [GI-MI, Lemma 3.1], [TE 2, Proposition 2.3, Chapter I])

$$\exists_{C_{0,2}>0}\ \forall_{v\in D(A_2)}\ \|v\|_{[H^2(\Omega)]^2} \le C_{0,2}\|A_r v\|_{[L^2(\Omega)]^2}$$

have been used. Next, the Cauchy inequality, the Young inequality (with $m = \frac43$) and (6.6.10) lead to

$$\begin{aligned}(6.6.11)\qquad &\frac{\nu}{2}\frac{d}{dt}\|\nabla u\|^2_{[L^2(\Omega)]^2} + \|A_r u\|^2_{[L^2(\Omega)]^2} \\ &\le const.\|u\|^{\frac12}_{[L^2(\Omega)]^2}\|A_r u\|^{\frac32}_{[L^2(\Omega)]^2}\|\nabla u\|_{[L^2(\Omega)]^2} \\ &\quad + \frac14\|A_r u\|^2_{[L^2(\Omega)]^2} + \|P_r h\|^2_{[L^2(\Omega)]^2} \\ &\le \frac12\|A_r u\|^2_{[L^2(\Omega)]^2} + const.'\|u\|^2_{[L^2(\Omega)]^2}\|\nabla u\|^4_{[L^2(\Omega)]^2} + c\|h\|^2_{[L^2(\Omega)]^2}.\end{aligned}$$

Setting $y(t) = \|\nabla u(t)\|^2_{[L^2(\Omega)]^2}$ and using the Calderón-Zygmund type estimate $y(t) \le C^2_{0,2}\|A_r u(t)\|^2_{[L^2(\Omega)]^2}$, we arrive at the differential inequality

$$\begin{aligned}(6.6.12)\qquad &y'(t) \le \frac{2c}{\nu}\|h\|^2_{[L^2(\Omega)]^2} \\ &\qquad + y(t)\frac{2}{\nu}\left(const.'\|u\|^2_{[L^2(\Omega)]^2}\|\nabla u\|^2_{[L^2(\Omega)]^2} - \frac{1}{2C^2_{0,2}}\right).\end{aligned}$$

Given the earlier a priori estimates (6.6.7) and (6.6.8) we can use on (6.6.12) Lemma 1.2.8 with

$$g := \frac{2}{\nu}\left(const.'\|u\|^2_{[L^2(\Omega)]^2}\|\nabla u\|^2_{[L^2(\Omega)]^2} - \frac{1}{2C^2_{0,2}}\right), \quad h := \frac{2c}{\nu}\|h\|^2_{[L^2(\Omega)]^2}$$

and get the independent of u_0 estimate

$$\|\nabla u(t)\|_{[L^2(\Omega)]^2} = \|u(t)\|_{[H^1_0(\Omega)]^2} \leq const.'', \text{ for } t \geq t(u_0), \tag{6.6.13}$$

where $t(u_0)$ is the same as in (6.6.8) and $const.''$ is independent of u_0.

Based on [GI-MI, Lemma 3.3] we can now estimate the right side of (6.6.2) as follows:

(6.6.14)

$$\begin{aligned}\|F_r u + P_r h\|_{[L^r(\Omega)]^2} &\leq \|P_r(u,\nabla)u\|_{[L^r(\Omega)]^2} + C_{0,r}\|h\|_{[L^r(\Omega)]^2}\\ &\leq C_r\|u\|^2_{[W^{1,r}(\Omega)]^2} + C_{0,r}\|h\|_{[L^r(\Omega)]^2}.\end{aligned}$$

A simple application of the Nirenberg-Gagliardo inequality (1.2.53) gives next

$$\|u\|_{[W^{1,r}(\Omega)]^2} \leq c\|u\|^{1-\theta}_{[H^1_0(\Omega)]^2}\|u\|^{\theta}_{X_r^{1-\varepsilon}}$$

($\theta > \frac{1-\frac{2}{r}}{2-2\varepsilon-\frac{2}{r}}$ and $\varepsilon \in (0,\frac{1}{r})$ is arbitrary, so that $2\theta < 1$). Hence, (6.6.14) may be extended to a suitable form of the condition (3.1.4), and we obtain

$$\|F_r u + P_r h\|_{[L^r(\Omega)]^2} \leq C^2\|u\|^{2-2\theta}_{[H^1_0(\Omega)]^2}\|u\|^{2\theta}_{X_r^{1-\varepsilon}} + C_{0,r}\|h\|_{[L^r(\Omega)]^2}, \tag{6.6.15}$$

which together with (6.6.13) justifies the applicability of the general theory of Chapter 3. Existence of a global attractor for the semigroup generated by (6.6.1) on X^α_r with $r > 2$ and $\alpha \in (\max\{1-\frac{1}{r}, \frac{1}{2}+\frac{1}{r}\}, 1)$ now follows from Theorem 4.2.1.

6.7. Cauchy problems in the half-space $R^+ \times R^n$

In this section two examples of the Cauchy problem in $R^+ \times R^n$ will be discussed. The proof of X^α solvability global in time is, for such problems, essentially the same as in the case of bounded Ω. However, due to the noncompactness of the resolvent, the existence of a global attractor is more difficult to obtain. For the examples below we shall show that any X^α solution corresponding to the initial data from a weighted $L^p(R^n;\rho)$ space will stay in this space for all $t > 0$. Then the existence of a global attractor will be a consequence of Theorem 4.2.4.

Let us introduce an auxiliary Banach space Z for which the inclusion $X^{\frac{1}{2}} \cap Z \subset X$ (where $X = L^p(R^n)$, $X^{\frac{1}{2}} = W^{1,p}(R^n) = H^1_p(R^n)$ and $p > n$) will be compact:

(6.7.1)

$$\begin{aligned}Z &:= L^p(R^n;(1+|x|^2)^\nu)\\ &= \left\{\psi \in L^p(R^n) : \|\psi\|^p_{L^p(R^n;\rho)} = \int_{R^n}|\psi(x)|^p(1+|x|^2)^\nu dx < +\infty\right\}.\end{aligned}$$

Lemma 6.7.1. *$W^{1,p}(R^n) \cap L^p(R^n;\rho)$ is compactly embedded in $L^p(R^n)$ for all $p \in [1,\infty)$.*

Proof. We need to show that any set B bounded in $W^{1,p}(R^n) \cap L^p(R^n;\rho)$ (normed by the sum of the norms) is precompact in $L^p(R^n)$.

For a given $\varepsilon > 0$ we shall construct an ε-covering in $L^p(R^n)$ of the set B (see [B-V 1, Lemma 2.16]). Since for arbitrary $\psi \in B$

$$(6.7.2) \qquad \begin{aligned}\int_{|x|\geq r} |\psi(x)|^p dx &= \int_{|x|\geq r} |\psi(x)|^p (1+|x|^2)^{\nu}(1+|x|^2)^{-\nu} dx \\ &\leq \|\psi\|^p_{L^p(R^n;\rho)}(1+r^2)^{-\nu},\end{aligned}$$

we will choose r so large that $\sup_{\psi\in B} \|\psi\|_{L^p(R^n;\rho)}(1+r^2)^{-\frac{\nu}{p}} < \frac{\varepsilon}{2}$. Next, as a consequence of the Sobolev embeddings, the inclusion $W^{1,p}(\Omega_r) \subset L^p(\Omega_r)$, $\Omega_r = \{x \in R^n : |x| < r\}$, is compact. Therefore there exists a finite set of elements $e_1, e_2, \dots, e_m \in L^p(\Omega_r)$ such that balls centered at $e_1, e_2, \dots, e_m$ with radii $\frac{\varepsilon}{2}$ cover B' - the set of restrictions of the elements of B to Ω_r. Extending the functions $e_1, e_2, \dots, e_m$ as zero outside Ω_r we thus obtain a finite ε-covering of B in $L^p(R^n)$. The proof is complete. □

Remark 6.7.1. It is evident from the above proof that the exact form of a weight function ρ is inessential as long as ρ is strictly positive and increasing to $+\infty$ when $|x| \to \infty$. Let us also refer the reader to the results concerning compact embedding of $W^{1,p}(R^n)$ contained in [CHA].

Remark 6.7.2. The reason for choosing Z as a weighted $L^p(R^n;\rho)$ space with increasing ρ is that we need to exclude from the space $W^{1,p}(R^n) \cap Z$ the *travelling waves*, that is families $\{v_t\}_{t\geq 0}$ given by

$$(6.7.3) \quad v_t(x) = v_0(x+at),\ t>0,\ v_0 \in W^{1,p}(R^n),\ v_0 \neq 0,\ 0 \neq a \in R^n.$$

Such a family, bounded in $W^{1,p}(R^n)$, is not compact in $L^p(R^n)$ (see [D-S]).

We are now ready to present particular examples.

Example 6.7.1. Our first example is the second order parabolic initial value problem studied in [B-V 1]:

$$(6.7.4) \qquad \begin{cases} u_t = \mu\Delta u - \lambda_0 u - f(u) - g(x),\ (t,x) \in R^+ \times R^n, \\ u(0,x) = u_0(x),\ x \in R^n. \end{cases}$$

We shall assume that λ_0 in (6.7.4) is positive, $f : R \to R$ is locally Lipschitz continuous, satisfies equality $f(0) = 0$ and the one side growth restriction:

$$(6.7.5) \qquad -f(u)u \leq \frac{\lambda_0}{2}u^2.$$

We shall also assume that $g \in L^\infty(R^n) \cap L^p(R^n;(1+|x|^2)^\nu)$ with some even integer $p > n$ and consider the operator $-A := \mu\Delta - \lambda_0 I$ on $L^p(R^n)$ with $D(A) = W^{2,p}(R^n)$ and $X^{\frac{1}{2}} = W^{1,p}(R^n)$ (see [TR, Theorem 2.5.3]).

Evidently, Assumption 2.1.1 is satisfied (see [TR, Lemma 2.5.2]) so that local $X^{\frac{1}{2}}$ solvability of (6.7.4) follows via Theorem 2.1.1. The following lemma provides estimates uniform in time of solutions, which will be used to show that the requirements of (A_2) are satisfied.

Lemma 6.7.2. *For all $u_0 \in X^{\frac{1}{2}} \cap L^p(R^n;(1+\varepsilon_0|x|^2)^\nu)$ the existing local $X^{\frac{1}{2}}$ solutions $u(t,u_0)$ of* (6.7.4) *are estimated uniformly in time in the space $L^\infty(R^n) \cap L^p(R^n;(1+\varepsilon_0|x|^2)^\nu)$ as stated in* (6.7.8), (6.7.15) *below. Furthermore an estimate in this space, asymptotically independent of $u_0 \in X^{\frac{1}{2}}$, is given in* (6.7.9), (6.7.16) *below.*

Proof. Since $p > n$ we have $X^{\frac{1}{2}} \subset L^\infty(R^n)$, so that $u(t,u_0)$ belongs to $L^\infty(R^n)$ for each $t \geq 0$ while $u_0 \in X^{\frac{1}{2}}$. Estimating (6.7.4) in a standard way, for $k \in \{p, p+1, \dots\}$ we obtain

(6.7.6)
$$\begin{aligned}\frac{1}{2k}\frac{d}{dt}\int_{R^n} u^{2k}dx &= -\mu\frac{2k-1}{k^2}\int_{R^n}|\nabla(u^k)|^2dx - \lambda_0\int_{R^n}u^{2k}dx \\ &\quad -\int_{R^n} f(u)u\,u^{2k-2}dx - \int_{R^n} g(x)u^{2k-1}dx \\ &\leq -\frac{\lambda_0}{2}\int_{R^n}u^{2k}dx + \left[\int_{R^n}u^{2k}dx\right]^{\frac{2k-1}{2k}}\left[\int_{R^n}g^{2k}(x)dx\right]^{\frac{1}{2k}} \\ &\leq -\frac{\lambda_0}{4}\int_{R^n}u^{2k}dx + \frac{1}{2k}\left[\frac{k\lambda_0}{2(2k-1)}\right]^{-(2k-1)}\int_{R^n}g^{2k}(x)dx.\end{aligned}$$

Solving the differential inequality (6.7.6) we find that

$$\begin{aligned}\|u\|_{L^{2k}(R^n)} &\leq \|u_0\|_{L^{2k}(R^n)}e^{-\frac{\lambda_0 t}{4}} \\ &\quad + \left[\frac{2}{\lambda_0 k}\right]^{\frac{1}{2k}}\left[\frac{2(2k-1)}{\lambda_0 k}\right]^{\frac{2k-1}{2k}}\|g\|_{L^{2k}(R^n)},\end{aligned} \tag{6.7.7}$$

which implies the required estimates

$$\|u\|_{L^\infty(R^n)} \leq \max\left\{\|u_0\|_{L^\infty(R^n)}, \frac{4}{\lambda_0}\|g\|_{L^\infty(R^n)}\right\} \tag{6.7.8}$$

and

$$\limsup_{t\to+\infty}\|u\|_{L^\infty(R^n)} \leq \frac{4}{\lambda_0}\|g\|_{L^\infty(R^n)}. \tag{6.7.9}$$

Below, the counterparts of the estimates (6.7.8) and (6.7.9) in the $L^p(R^n;(1+\varepsilon|x|^2)^\nu)$ norm will be shown. Define a sequence $\{\phi_N\}$ of smooth real valued functions:

$$\phi_N(x) = \begin{cases}(1+\varepsilon|x|^2)^\nu, & \text{for } |x| \leq N, \\ (1+\varepsilon N^2)^\nu, & \text{otherwise.}\end{cases} \tag{6.7.10}$$

Such cut-off functions belong to $W^{1,p}_{loc.}(R^n)$ (see [GI-TR, Theorem 7.8]), the integration by parts formula used below is valid and the estimate (see [B-V 1, formula (1.9)])

$$|\nabla\phi_N| \le \nu n\sqrt{\varepsilon}\phi_N, \; N = 1,2,\dots, \tag{6.7.11}$$

holds for them. Multiplication of (6.7.4) by $u^{p-1}\phi_N$ (note that p is an even integer) and the Young inequality (Lemma 1.2.2) give

(6.7.12)
$$\begin{aligned}
\frac{1}{p}\frac{d}{dt}\int_{R^n} u^p\phi_N(x)dx &= -\mu\int_{R^n}\nabla u\nabla(u^{p-1}\phi_N(x))dx - \lambda_0\int_{R^n} u^p\phi_N(x)dx \\
&\quad - \int_{R^n} f(u)u\, u^{p-2}\phi_N(x)dx - \int_{R^n} g(x)u^{p-1}\phi_N(x)dx \\
&\le -\frac{\mu}{(\frac{p}{2})^2}\left(p-1-\frac{\delta}{2}\right)\int_{R^n}|\nabla(u^{\frac{p}{2}})|^2\phi_N(x)dx \\
&\quad + \left(\mu\frac{\nu^2n^2\varepsilon}{2\delta} - \frac{\lambda_0}{2} + \gamma\right)\int_{R^n}|u|^p\phi_N(x)dx + C_\gamma\int_{R^n}|g(x)|^p\phi_N(x)dx,
\end{aligned}$$

for any positive δ and γ. Choosing next $\delta = \delta_0$, $\gamma = \gamma_0$ and $\varepsilon = \varepsilon_0$ satisfying

$$\begin{cases} p-1-\frac{\delta_0}{2} = 0, \\ \mu\frac{\nu^2n^2\varepsilon_0}{2\delta_0} + \gamma_0 = \frac{\lambda_0}{4}, \end{cases} \tag{6.7.13}$$

and inserting these values of parameters into (6.7.12) we obtain that

$$\begin{aligned}
&\frac{1}{p}\frac{d}{dt}\int_{R^n}|u|^p\phi_N(x)dx \\
&\quad\le -\frac{\lambda_0}{4}\int_{R^n}|u|^p\phi_N(x)dx + C_{\gamma_0}\int_{R^n}|g(x)|^p\phi_N(x)dx.
\end{aligned} \tag{6.7.14}$$

Using Lemma 1.2.4 and letting N tend to infinity we verify that both

(6.7.15)
$$\begin{aligned}
&\|u(t,u_0)\|_{L^p(R^n;(1+\varepsilon_0|x|^2)^\nu)} \\
&\le \max\left\{\|u_0\|_{L^p(R^n;(1+\varepsilon_0|x|^2)^\nu)}, \left[\frac{4C_{\gamma_0}}{\lambda_0}\right]^{\frac{1}{p}}\|g\|_{L^p(R^n;(1+\varepsilon_0|x|^2)^\nu)}\right\}
\end{aligned}$$

and

$$\limsup_{t\to+\infty}\|u(t,u_0)\|_{L^p(R^n;(1+\varepsilon_0|x|^2)^\nu)} \le \left[\frac{4C_{\gamma_0}}{\lambda_0}\right]^{\frac{1}{p}}\|g\|_{L^p(R^n;(1+\varepsilon_0|x|^2)^\nu)} \tag{6.7.16}$$

hold. Lemma 6.7.2 is proved. □

Theorem 6.7.1. *The Cauchy problem* (6.7.4) *generates on the phase space* $X^{\frac{1}{2}}$ *a* C^0 *semigroup of global* $X^{\frac{1}{2}}$ *solutions which is bounded dissipative and keeps orbits of bounded sets bounded.*

Proof. Referring to Theorem 3.1.1 we shall check that all the requirements of (A_2) are satisfied with $Y = L^p(R^n) \cap L^\infty(R^n)$ ($\|\cdot\|_Y := \|\cdot\|_{L^p(R^n)} + \|\cdot\|_{L^\infty(R^n)}$). Subordination of the nonlinear term to $A^{\frac{1}{2}}$ required in the condition (3.1.4) follows from (6.7.7), (6.7.8) and local Lipschitz continuity of f. Indeed, since $p = 2k$ for some $k \geq 1$ and $f(0) = 0$, we have

(6.7.17)
$$\|f(u(t,u_0)) + g\|_{L^p(R^n)} \leq L_{\|u(t,u_0)\|_{L^p(R^n)} + \|u(t,u_0)\|_{L^\infty(R^n)}} \times \left(\|u(t,u_0)\|_{L^p(R^n)} + \|u(t,u_0)\|_{L^\infty(R^n)}\right) + \|g\|_{L^p(R^n)}$$

where

$$L_s := \inf\{L \geq 0 : \ \forall_{|y| \leq s} \ |f(y)| \leq L|y|\}.$$

Note that the right side of (6.7.17) is a nondecreasing function

$$[0, +\infty) \ni s \to sL_s + \|g\|_{L^p(R^n)} \in [0, +\infty).$$

Inequalities (6.7.7), (6.7.8) thus give a variant of the estimate (3.1.3) (with $\alpha = \frac{1}{2}$) required in Theorem 3.1.1 for the global solvability of (6.7.4) in $X^{\frac{1}{2}}$. Moreover, (6.7.7) and (6.7.8) are asymptotically independent of the initial condition so that the semigroup generated by (6.7.4) on $X^{\frac{1}{2}}$ is bounded dissipative according to Corollary 4.1.3. □

Let $B_{L^p(R^n;(1+\varepsilon_0|x|^2)^\nu)}(r)$ be a ball in $L^p(R^n; (1+\varepsilon_0|x|^2)^\nu)$ with radius r centered at zero. Write

(6.7.18)
$$V_r^{\frac{1}{2}} := B_{L^p(R^n;(1+\varepsilon_0|x|^2)^\nu)}(r) \cap X^{\frac{1}{2}}.$$

We then have

Theorem 6.7.2. *For each* $r \geq \left[\frac{4C_{\gamma_0}}{\lambda_0}\right]^{\frac{1}{p}} \|g\|_{L^p(R^n;(1+\varepsilon_0|x|^2)^\nu)}$ *the semigroup generated by the problem* (6.7.4) *on* $X^{\frac{1}{2}}$ *has a global attractor* $\mathcal{A}_r$ *in a complete metric space* $cl_{X^{\frac{1}{2}}} V_r^{\frac{1}{2}}$. *Moreover, the sets* $\mathcal{A}_r$ *are independent of* $r \geq \left[\frac{4C_{\gamma_0}}{\lambda_0}\right]^{\frac{1}{p}} \|g\|_{L^p(R^n;(1+\varepsilon_0|x|^2)^\nu)}$ *and* $T(t) : cl_{X^{\frac{1}{2}}} V_r^{\frac{1}{2}} \to cl_{X^{\frac{1}{2}}} V_r^{\frac{1}{2}}$ *is compact for* $t > 0$.

Proof. Given Lemma 6.7.2, solutions starting from $u_0 \in L^p(R^n; (1 + \varepsilon_0|x|^2)^\nu) \cap X^{\frac{1}{2}}$ will stay in this space for all $t \geq 0$. Furthermore, for each $r \geq \left[\frac{4C_{\gamma_0}}{\lambda_0}\right]^{\frac{1}{p}} \|g\|_{L^p(R^n;(1+\varepsilon_0|x|^2)^\nu)}$ the set $V_r^{\frac{1}{2}}$ is positively invariant, whereas the inclusion $L^p(R^n; (1+\varepsilon_0|x|^2)^\nu) \cap X^{\frac{1}{2}} \subset L^p(R^n)$ is compact (see Lemma 6.7.1). Hence the assumptions of Theorem 4.2.4 are satisfied so that the semigroup $\{T(t)\}$ generated by (6.7.4) on $X^{\frac{1}{2}}$ can be restricted to a complete metric space $cl_{X^{\frac{1}{2}}} V_r^{\frac{1}{2}}$ and the operators $T(t) : cl_{X^{\frac{1}{2}}} V_r^{\frac{1}{2}} \to cl_{X^{\frac{1}{2}}} V_r^{\frac{1}{2}}$ are compact for $t > 0$. Therefore, the existence of the global attractor for $\{T(t)\}$ restricted to $cl_{X^{\frac{1}{2}}} V_r^{\frac{1}{2}}$ follows via Theorem 4.2.4. The proof is complete. □

Example 6.7.2. Our second example will be the Cauchy problem with hydrodynamic type nonlinearity studied earlier in [AB, p. 94]:

$$\begin{cases} u_t = \mu\Delta u - \lambda u - u\langle l, \nabla u\rangle + f(x), \ (t,x) \in R^+ \times R^n, \\ u(0,x) = u_0(x), \ x \in R^n. \end{cases} \tag{6.7.19}$$

The data l, f in (6.7.19) are required to satisfy the following conditions:

$$l \in [W^{1,\infty}(R^n)]^n, \ \langle l, \nabla\rangle = \sum_{j=1}^n \frac{\partial l_j}{\partial x_j} = 0, \tag{6.7.20}$$

$$f \in L^\infty(R^n) \cap L^p(R^n; (1+|x|^2)^\nu), \quad \text{with some even } p > n, \tag{6.7.21}$$

$$\mu \text{ and } \lambda \text{ are positive constants.} \tag{6.7.22}$$

The abstract setting of (6.7.19) in the form (2.1.1) is generically the same as in the previous example. Therefore, to obtain for (6.7.19) results similar to Theorems 6.7.1, 6.7.2 it suffices to show the counterpart of the estimates (6.7.8), (6.7.9) and (6.7.15), (6.7.16).

From (6.7.19), for $k \in \{p, p+1, \dots\}$, we have

$$\begin{aligned} \int_{R^n} \langle l, \nabla u\rangle u^{2k} dx &= \frac{1}{2k+1}\int_{R^n} \langle l, \nabla[u^{2k+1}]\rangle dx \\ &= -\frac{1}{2k+1}\int_{R^n} \langle l, \nabla\rangle u^{2k+1} dx = 0, \end{aligned} \tag{6.7.23}$$

which is typical for hydrodynamic type nonlinearities. As a consequence of (6.7.23) we obtain from (6.7.19) the estimate

$$\begin{aligned} \frac{1}{2k}\frac{d}{dt}\int_{R^n} u^{2k} dx &= -\mu\frac{2k-1}{k^2}\int_{R^n} |\nabla(u^k)|^2 dx \\ &\quad - \lambda\int_{R^n} u^{2k} dx - \int_{R^n} f(x) u^{2k-1} dx \\ &\le -\lambda\int_{R^n} u^{2k} dx + \left[\int_{R^n} u^{2k} dx\right]^{\frac{2k-1}{2k}} \left[\int_{R^n} f^{2k}(x) dx\right]^{\frac{1}{2k}}, \end{aligned} \tag{6.7.24}$$

and hence the $L^\infty(R^n)$ bound for the solutions of (6.7.19) follows after the same calculations as in (6.7.6)-(6.7.9).

To find the bound in weighted space consider the equality

$$\begin{aligned} \frac{1}{p}\frac{d}{dt}\int_{R^n} u^p \phi_N(x) dx &= -\mu\int_{R^n} \nabla u \nabla[u^{p-1}\phi_N(x)] dx - \lambda\int_{R^n} u^p \phi_N(x) dx \\ &\quad - \int_{R^n} \langle l, \nabla u\rangle u^p \phi_N(x) dx + \int_{R^n} f(x) u^{p-1}\phi_N(x) dx. \end{aligned} \tag{6.7.25}$$

Comparing (6.7.25) with (6.7.12) it is seen that $\int_{R^n}\langle l, \nabla u\rangle u^p \phi_N(x)dx$ is the only term that did not appear previously in (6.7.12) and must be estimated separately. Given (6.7.11), (6.7.20) and the $\|u(t,u_0)\|_{L^\infty(R^n)}$ estimate justified above we have

$$\begin{aligned}|\nabla\phi_N| &\leq \nu n\sqrt{\varepsilon}\phi_N,\\ |l| &\leq C_l,\\ \|u(t,u_0)\|_{L^\infty(R^n)} &\leq C_{\|u_0\|_{L^\infty(R^n)},\|f\|_{L^\infty(R^n)}},\end{aligned}$$

from which it follows that

(6.7.26)

$$\begin{aligned}-\int_{R^n}\langle l,\nabla u\rangle u^p\phi_N dx &= -\frac{1}{p+1}\int_{R^n}\langle l,\nabla[u^{p+1}]\rangle\phi_N dx\\ &= \frac{1}{p+1}\int_{R^n}\langle l,\nabla\phi_N\rangle u^{p+1}dx \leq const.\ \sqrt{\varepsilon}\int_{R^n}u^p\phi_N dx.\end{aligned}$$

The required estimate of solutions of (6.7.19) in $L^p(R^n;(1+\varepsilon|x|^2)^\nu)$ and the existence of the attractor for (6.7.19) may be thus inferred analogously as in the previous example. Note only that the nonlinearity is now estimated as follows:

$$\begin{aligned}\|u\langle l,\nabla u\rangle + f\|_{L^p(R^n)} &\leq \sum_{j=1}^{n}\left(\int_\Omega |u|^p|l_j|^p\left|\frac{\partial u}{\partial x_j}\right|^p dx\right)^{\frac{1}{p}} + \|f\|_{L^p(R^n)}\\ &\leq const._1\|u\|_{L^\infty(R^n)}\|u\|_{X^{\frac{1}{2}}} + \|f\|_{L^p(R^n)}\\ &\leq const._2\|u\|_{L^\infty(R^n)}\|u\|_{L^p(R^n)}^{1-\frac{1}{2\alpha}}\|u\|_{X^\alpha}^{\frac{1}{2\alpha}} + \|f\|_{L^p(R^n)}\\ &\leq \left(const._2(\|u\|_{L^p(R^n)} + \|u\|_{L^\infty(R^n)})^{2-\frac{1}{2\alpha}} + \|f\|_{L^p(R^n)}\right)\left(1+\|u\|_{X^\alpha}^{\frac{1}{2\alpha}}\right),\end{aligned}$$

for $\alpha \in \left(\frac{1}{2},1\right)$.

Bibliographical notes.

Since it is impossible to mention all important references which relate to the examples presented only very general comments will be given. We expect that a number of new examples connected with equations having fractional powers of a sectorial operator in the main part (see [B-K-W], [C-C]) will be shown to fall in the abstract setting of this monograph.

- Note that more examples of global attractors for parabolic problems can be found in [TE 1], [HA 2], [B-V 2], [C-C-D 1].
- Similar considerations of the Cahn-Hilliard problem were given earlier in our papers [DL 1], [C-D 2]; however, the present formulation is based on [L-Z].
- Sections 6.3, 6.5, 6.6 are presented here in a new way. Finally, Section 6.7 follows the presentation of [C-D 4].
- Among a great number of publications devoted to asymptotics of the Navier-Stokes equation we recall here only the, today classical, result of C. Foias and G. Prodi [F-P] and the recent result by G. R. Sell [SEL].

CHAPTER 7

Backward uniqueness and regularity of solutions

This chapter provides additional description of the solutions to (2.1.1). Although we go here beyond the global stability (and even global solvability) of (2.1.1) in the large we shall justify, nevertheless, the important properties of the semigroups corresponding to examples previously studied: invertibility of the flow on a global attractor and its regularity.

7.1. Invertible processes

Processes described by parabolic equations are not, in general, invertible in time. Therefore one of the most exciting properties of the flow on global attractors related to certain such equations is the possible invertibility of the flow.

If $\mathcal{A}$ is an invariant set, although to each point $v \in \mathcal{A}$ corresponds a complete orbit through v lying in $\mathcal{A}$, we can hardly be sure that such an orbit is unique unless the semigroup $\{T(t)\}$ we deal with is one-to-one; i.e. $T(t)$ is a one-to-one map for each $t \geq 0$. In the latter case the phase space V is filled by the nonintersecting curves such that through each point of V passes one and only one integral curve of (2.1.1); obviously in this case $\{T(t)\}$ on $\mathcal{A}$ can be extended to a group.

Invertibility in time of the systems governed by evolutionary equations is related to the notion of *backward uniqueness* (see [HE 1, p. 208], [FR 1, p. 181]). The following is the formulation of [TE 1].

Backward uniqueness. *The general evolutionary equation*

$$\frac{du}{dt}(t) + \mathcal{N}\big(u(t)\big) = 0 \tag{7.1.1}$$

has the backward uniqueness property if and only if for any $u(\cdot)$, $\tilde{u}(\cdot)$ *satisfying* (7.1.1) *on some interval* $(\tau - \varepsilon, \tau]$ *the equality*

$$u(\tau) = \tilde{u}(\tau),$$

implies that $u(t) = \tilde{u}(t)$ *for all* $t \leq \tau$ *for which both* u *and* $\tilde{u}$ *are defined.*

Remark 7.1.1. There are many simple examples of parabolic problems for which the above condition is violated. In particular, for the positive solutions of the problem

$$u_t = \Delta u - \lambda u^\theta, \ \lambda > 0, \ \theta \in (0,1), \tag{7.1.2}$$

considered with Dirichlet data,

$$u = 0 \ \text{ on } \ \partial\Omega, \ u(0,x) = u_0(x) \ \text{ in } \Omega, \tag{7.1.3}$$

the *extinction phenomenon* has been observed; *nonnegative solutions to* (7.1.2)-(7.1.3) *decay to zero in a finite time* (see [F-H], [KAL], [DL 6]). More precisely, for every smooth, nonnegative u_0 there exists a positive time t_{u_0} such that $u(t,x) \equiv 0$ for $t \geq t_{u_0}$. This type of behavior will not be studied here. We have only pointed out problems for which backward uniqueness fails.
To study the existence of solutions to (7.1.2), (7.1.3) we need to go beyond the theory introduced earlier since the nonlinear term in that problem is not Lipschitz continuous. Remaining, however, within the framework of ideas of the present book the validity of Theorems 1, 2 of [L-M] (with $D = X = L^p(\Omega)$) for the abstractly formulated (7.1.2), (7.1.3) will be shown in Section 8.1. This will justify the existence of a solution to (7.1.2), (7.1.3) global in time (see Example 8.1.1 and Remark 8.1.1).

Various theorems concerning backward uniqueness can be found in [FR 1], [FR 2], [HE 1], [TE 1]. The considerations below are based on the result of [TE 1, Chapter III]. Our goal here is to show that the equations belonging to the large, although specific, class of problems introduced in Section 5.1 possess the backward uniqueness property. This, in particular, constitutes an additional motivation for the separate study of these problems. Note that all we require of the right hand side term f is to satisfy a local Lipschitz condition. The latter assumption on f is essential for the invertibility of a flow. Indeed, if f violates the Lipschitz condition even in one point, then - recalling the extinction phenomenon - the flow may not possess this property.

Processes in Hilbert spaces. Let H and $D(G)$ be two Hilbert spaces such that $D(G) \subset H$ is the domain of a linear, positive, self-adjoint operator $G : D(G) \to H$. As shown in Proposition 1.3.3, such an operator G is sectorial. Note further that $H^{\frac{1}{2}} = D(G^{\frac{1}{2}}) = [H, D(G)]_{\frac{1}{2}}$ is a Hilbert space (see [TR, Section 1.18.10]), $G^{\frac{1}{2}}$ is again a linear, positive, self-adjoint operator (see [HE 1, Section 1.4]) and, consequently,

$$\|v\|^2_{H^{\frac{1}{2}}} = \|G^{\frac{1}{2}}v\|^2_H = (G^{\frac{1}{2}}v, G^{\frac{1}{2}}v)_H = (Gv, v)_H, \ v \in D(G).$$

Proposition 7.1.1. *Let $\mathcal{W}$ be a Banach space continuously embedded in $H^{\frac{1}{2}}$ and let*

$$w \in C([0,\tau], \mathcal{W}) \cap C^1((0,\tau); H) \cap C((0,\tau); D(G)). \tag{7.1.4}$$

Consider further a function $h : (0,\tau) \times R(w) \to H$ ($R(w) \subset \mathcal{W}$ being the range of w) such that, for some $k \in L^2(0,\tau)$,

$$\|h(t, w(t))\|_H \leq k(t)\|w(t)\|_{H^{\frac{1}{2}}}, \ t \in (0,\tau). \tag{7.1.5}$$

If w fulfills in H (in the sense of [TE 1, Chapter II] *or* [LI]*) an equation*

$$\dot{w}(t) + Gw(t) = h(t,w),\ t \in (0,\tau), \tag{7.1.6}$$

and, for some $\tau_0 \in (0,\tau)$,

$$w(\tau_0) = 0, \tag{7.1.7}$$

then

$$w(t) = 0 \ \textit{for all} \ t \in [0,\tau_0].$$

Remark 7.1.2. Essentially, this result mirrors [TE 1, Lemmas 6.1, 6.2] and is based on estimates of $\Lambda(t) := \frac{\|w(t)\|_{H^{\frac{1}{2}}}}{\|w(t)\|_H}$ and $\Lambda_1(t) := \log \frac{1}{\|w(t)\|_H}$. It is true that in (7.1.4) we require of w more regularity than necessary from the point of view of the theory in [TE 1]. However, the condition (7.1.4) corresponds to the smoothness of the solutions that are considered in this book. In particular, the functions Λ and Λ_1 are of class C^1 in a neighborhood of any point $t > 0$ for which $\|w(t)\|_H > 0$.

Proof. Let us sketch the proof of Proposition 7.1.1. Suppose that (7.1.7) holds and, for some $t_0 \in (0,\tau_0)$, $\|w(t_0)\|_H \neq 0$. Based on the continuity of $\|w(\cdot)\|_H$ take further $c \in (0,\tau_0 - t_0)$ such that $\|w(t_0+c)\|_H = 0$ and $\|w(t)\|_H > 0$ for $t \in [t_0, t_0+c)$. Differentiating the functions Λ, Λ_1 on (t_0,t_0+c) and using (7.1.5) we obtain the estimates

$$\begin{cases} \Lambda'(t) \le 2k^2(t)\Lambda(t), & t \in (t_0,t_0+c), \\ \Lambda_1'(t) \le 2\Lambda(t) + k^2(t), & t \in (t_0,t_0+c) \end{cases} \tag{7.1.8}$$

(see [TE 1, pp. 169-170] for detailed calculations). However, if the inequalities (7.1.8) hold, then $\Lambda_1(t)$ remains bounded when $t \to t_0 + c^-$, which produces the contradiction. □

Remark 7.1.3. The space $\mathcal{W}$ in Proposition 7.1.1 plays only an auxiliary role. If $\mathcal{W} = H^{\frac{1}{2}}$, then Proposition 7.1.1 coincides with Lemma 6.1 of [TE 1]. The choice of $\mathcal{W} = H^{\frac{1}{2}}$ is satisfactory from the point of view of many applications. For instance, this choice may be made in the case of (2.1.1) with $X := H$ being a Hilbert space and nonlinear term F satisfying the Lipschitz condition between $H^{\frac{1}{2}}$ and H (i.e. when $H^{\frac{1}{2}}$ solutions are available). Indeed we then have

$$\dot{u} + Au = F(u),\ u(0) = u_0,$$

with local H^α solutions satisfying (7.1.4) for any $\tau \in (0,\tau_{u_0})$. Setting $w := u - v$, condition (7.1.5) will be satisfied provided that we justify the estimate

$$\|F(u(t)) - F(\tilde{u}(t))\|_H \le k(t)\|u(t) - \tilde{u}(t)\|_{H^{\frac{1}{2}}},$$

with a certain $k \in L^2(0,\tau)$.

Let us develop the latter remark in more detail.

Example 7.1.1. For $t \geq 0$ consider $H^{\frac{1}{2}}$ solutions of the Cahn-Hilliard problem (6.4.1)-(6.4.3) in space dimension $n \leq 3$, where

$$H = L^2(\Omega),\ D(A) = H^4_{\{\frac{\partial}{\partial N}, \frac{\partial \Delta}{\partial N}\}}(\Omega),\ H^{\frac{1}{2}} = H^2_{\{\frac{\partial}{\partial N}\}}(\Omega).$$

It has already been shown in (6.4.15) that for each bounded subset $\mathcal{U}$ of $H^{\frac{1}{2}}$

$$\|\Delta f(\phi) - \Delta f(\psi)\|_H \leq L_{\mathcal{U}} \|\phi - \psi\|_{H^{\frac{1}{2}}},\ \phi, \psi \in \mathcal{U}. \tag{7.1.9}$$

Let u, $\tilde{u}$ be a pair of $H^{\frac{1}{2}}$ solutions and suppose that $u(\tau_0) = \tilde{u}(\tau_0)$ for some $\tau_0 > 0$. Take any $t_0 > \tau_0$ and define $w = u - \tilde{u}$ on $(0, t_0)$. Consider further $(\Delta f(u) - \Delta f(\tilde{u}))$ as a counterpart of $h(\cdot, w)$. Using (7.1.9) with $\mathcal{U} := B_{H^{\frac{1}{2}}}(r_0)$ being a ball in $H^{\frac{1}{2}}$ centered at zero with radius $r_0 := \sup_{t \in [0,t_0]} \|u(t)\|_{X^{\frac{1}{2}}} + \sup_{t \in [0,t_0]} \|\tilde{u}(t)\|_{X^{\frac{1}{2}}}$, we obtain immediately the estimate

$$\|h(t, w)\|_{H^{\frac{1}{2}}} = \|\Delta f(u) - \Delta f(\tilde{u})\|_H \leq L_{r_0} \|w\|_{H^{\frac{1}{2}}},\ t \in (0, t_0).$$

Next we use Proposition 7.1.1 with $\mathcal{W} = H^{\frac{1}{2}}$ to conclude that $u \equiv \tilde{u}$ on $[0, \tau_0]$. Therefore, both $H^{\frac{1}{2}}$ solutions coincide as long as they are defined. This justifies the backward uniqueness property for the Cahn-Hilliard equation. The same reasoning is true for H^α solutions of this problem with any $\alpha \in (\frac{1}{2}, 1)$ since they are contained in the set of solutions considered above.

Example 7.1.2. Consider the Burgers equation (6.5.1) in a Hilbert space $H = L^2(0, \pi) \times R$ equipped with a scalar product

$$\langle(\phi, U), (\psi, \tilde{U})\rangle_H = \langle\phi, \psi\rangle_{L^2(\Omega)} + U\tilde{U},\ (\phi, U), (\psi, \tilde{U}) \in H.$$

It is clear that a product operator $A = (-\nu\frac{\partial^2}{\partial x^2}, \nu I)$ is self-adjoint and positive definite in H, whereas $H^{\frac{1}{2}} = H^1_0(0, \pi) \times R$ ($D(A) = (H^2(0, \pi) \cap H^1_0(0, \pi)) \times R$ being the domain of A).

Let $B = B_1 \times B_2 \subset X^{\frac{1}{2}}$ be a bounded set and consider elements $(\phi, U), (\psi, \tilde{U}) \in B$. The nonlinear term

$$F(U, \phi) = \left(U\phi - \frac{1}{2}(\phi^2)_x, P - \int_0^\pi \phi^2(x)dx\right)$$

is globally Lipschitz in B, since we have

$$\begin{aligned}
\|F(U, \phi) - F(\tilde{U}, \psi)\|_H &\leq \|U\phi - \tilde{U}\psi\|_{L^2(0,\pi)} \\
&\quad + \left\|\frac{1}{2}(\phi^2)_x - \frac{1}{2}(\psi^2)_x\right\|_{L^2(0,\pi)} + \left|\int_0^\pi (\psi^2 - \phi^2)dx\right| \\
&\leq |U|\|\phi - \psi\|_{L^2(0,\pi)} + |U - \tilde{U}|\|\psi\|_{L^2(0,\pi)} \\
&\quad + \|\phi\|_{L^\infty(0,\pi)}\|\phi_x - \psi_x\|_{L^2(0,\pi)} + \|\phi - \psi\|_{L^\infty(0,\pi)}\|\psi_x\|_{L^2(0,\pi)} \\
&\leq c\left(\|\phi\|_{H^1(0,\pi)}, \|\psi\|_{H^1(0,\pi)}, |U|\right)\left(\|\phi - \psi\|_{H^1(0,\pi)} + |U - \tilde{U}|\right) \\
&\leq L_B\|(U, \phi) - (\tilde{U}, \psi)\|_{H^{\frac{1}{2}}},
\end{aligned}$$

where the embedding $H^1(0, \pi) \subset L^\infty(0, \pi)$ was used.

Therefore Proposition 7.1.1 with $\mathcal{W} = H^{\frac{1}{2}}$ is applicable, ensuring backward uniqueness of the $H^{\frac{1}{2}}$ solutions of the Burgers problem. For $p \geq 2$ and $\alpha \in [\frac{1}{2}, 1)$ global X^α solutions obtained in Chapter 6 are automatically $H^{\frac{1}{2}}$ solutions as considered above so that these considerations cover all such solutions as well.

In the above two examples nonlinearities allowed us to choose X as a Hilbert space, and verify the Lipschitz condition between $H^{\frac{1}{2}}$ and H, so that $H^{\frac{1}{2}}$ solutions could be considered. However, even for the simplest case of a scalar parabolic equation the choice $X := H$ and $\mathcal{W} := H^{\frac{1}{2}}$ may not be satisfactory. Such an approach may lead to certain growth restrictions for the right hand side term f or may limit the space dimension n, which is not reasonable from the point of view of the existence of local solutions.

Example 7.1.3. Consider a scalar parabolic equation

$$u_t = \Delta u + f(u), \; t > 0, \; x \in \Omega \subset R^n \tag{7.1.10}$$

(Ω a bounded domain), with either Neumann or Dirichlet boundary condition. Assume that the function $f : R \to R$ is locally Lipschitz continuous and let F be a substitution operator for f (or $f + I$ in the Neumann case). If $X = L^2(\Omega)$ (or, more generally, $X = H^k(\Omega)$) then, without additional assumptions on f, or restriction on n, we are rather unable to verify the Lipschitz condition required in Assumption 2.1.1 (see [CAR], [C-R]).

However, for $p > n$ and $X = L^p(\Omega)$, the existence of a local semigroup of X^α solutions, $\alpha \in [\frac{1}{2}, 1)$, is immediate (here $D(A) = W^{2,p}(\Omega) \cap W_0^{1,p}(\Omega)$ for Dirichlet and $D(A) = W^{2,p}_{\{\frac{\partial}{\partial N}\}}(\Omega)$ for Neumann data). Furthermore, (7.1.10) possesses the backward uniqueness property. The latter will be proved in Theorem 7.1.1 below for the general class of regular parabolic initial boundary value problems.

Backward uniqueness theorem. Consider a $2m$-th order semilinear parabolic problem

$$u_t = -Au + f(x, d^{m_0}u), \; t > 0, \; x \in \Omega \subset R^n, \tag{7.1.11}$$

$$B_j u = 0, \; t > 0, \; x \in \partial\Omega, \; j = 0, \dots, m-1, \; u(0, x) = u_0, \; x \in \Omega, \tag{7.1.12}$$

with $\partial\Omega \in C^{2m}$ and the notation of A and $d^{m_0}u$ as in (5.1.2) and (5.1.3) respectively.

Assume that

(*i*) $m \geq m_0, \; p > \max\{n, 2\}$,
(*ii*) A considered with the domain $D(A) = W^{2m,p}_{\{B_j\}}(\Omega)$ is a sectorial, positive operator in $L^p(\Omega)$,
(*iii*) $f : \overline{\Omega} \times R^{d_0} \to R$ is continuous and locally Lipschitz continuous with respect to each of its functional arguments uniformly for $x \in \Omega$,
(*iv*) A considered with the domain $H^{2m}_{2,\{B_j\}}(\Omega)$ is simultaneously a self-adjoint and positive definite operator in $L^2(\Omega)$.

The model equation we are going to discuss here was studied in Chapter 5 and corresponds to a regular elliptic boundary value problem $(A, \{B_j\}, \Omega)$. Note here the correspondence between condition (A-II) of Chapter 5 and the assumption (iv).

Below we shall prove that the Cauchy problem (2.1.1),

$$\dot{u} + Au = F(u),\ t > 0,\ u(0) = u_0,$$

generated by (7.1.11),(7.1.12) in a Banach space $X = L^p(\Omega)$ has the property of backward uniqueness.

Theorem 7.1.1. *Let the assumptions (i)-(iv) be satisfied and take $\alpha \in [\frac{1}{2}, 1)$ such that*

$$2m\alpha - \frac{n}{p} > m_0. \tag{7.1.13}$$

If $u(\cdot, u_0)$, $\tilde{u}(\cdot, \tilde{u}_0)$ are any local X^α solutions of (7.1.11),(7.1.12), defined on maximal intervals of existence $[0, \tau_{u_0})$ and $[0, \tau_{\tilde{u}_0})$ respectively, and there exists $\tau \in (0, \tau_{max})$ (where $\tau_{max} := \min\{\tau_{u_0}, \tau_{\tilde{u}_0}\}$) such that

$$u(\tau, u_0) = \tilde{u}(\tau, \tilde{u}_0), \tag{7.1.14}$$

then $\tau_{u_0} = \tau_{\tilde{u}_0} =: \tau_0$ and

$$u(t, u_0) = \tilde{u}(t, \tilde{u}_0) \ \ \textit{for all} \ \ t \in [0, \tau_0). \tag{7.1.15}$$

Proof. Fix any $t_0 \in (\tau, \tau_{max})$ and consider a function $w(t) := u(t, u_0) - \tilde{u}(t, \tilde{u}_0)$ on the interval $(0, t_0)$. Let

$$M := C\left(\sup_{t\in[0,t_0]} \|u(t, u_0)\|_{X^\alpha} + \sup_{t\in[0,t_0]} \|\tilde{u}(t, \tilde{u}_0)\|_{X^\alpha}\right),$$

where C denotes the embedding constant corresponding to the inclusion

$$X^\alpha \subset C^m(\overline{\Omega})$$

which follows from (7.1.13). Since $m_0 \le m$ and f restricted to the set $\Omega \times [-M, M]^{d_0}$ (d_0 being the length of the vector $d^{m_0}u$) is globally Lipschitz continuous with respect to functional variables uniformly for $x \in \Omega$, we obtain the estimate

$$\begin{aligned} &\int_\Omega |f(x, d^{m_0}u) - f(x, d^{m_0}\tilde{u})|^2 dx \\ &\le L_M \sum_{0\le|\alpha|\le m_0} \int_\Omega |D^\alpha(u - \tilde{u})|^2 dx. \end{aligned} \tag{7.1.16}$$

To take advantage of the Hilbert approach of Proposition 7.1.1 choose

$$\mathcal{W} := X^\alpha,\ H := L^2(\Omega)$$

and, based on the inclusion $W^{2m,p}_{\{B_j\}}(\Omega) \subset H^{2m}_{2,\{B_j\}}(\Omega)$, set

$$G := A,\ D(G) := H^{2m}_{2,\{B_j\}}(\Omega).$$

Then we have

$$H^{\frac{1}{2}} = [H, D(G)]_{\frac{1}{2}} \subset H^m(\Omega)$$
$$X \subset H, \ D(A) \subset D(G),$$

and hence (7.1.4) holds with $\tau = t_0$. Furthermore, w satisfies in H the equation

$$\dot{w} + Gw = f(\cdot, d^{m_0}u) - f(\cdot, d^{m_0}\tilde{u}) := h(t, w)$$

where, by (7.1.16), $\|h(t,w)\|_H \leq \sqrt{L_M}\|w\|_{H^{\frac{1}{2}}}$. Therefore Proposition 7.1.1 is applicable and

$$w \equiv 0 \ \text{ on } [0, \tau_0].$$

In particular $u_0 = \tilde{u}_0$, so that the condition (7.1.15) follows as a consequence of the uniqueness property for X^α solutions of (2.1.1). The proof is complete. □

Remark 7.1.4. The above result remains true also when f depends on t (under an appropriate Hölder continuity assumption). This, however, will not be pursued here since we are focused on the autonomous problems.

Remark 7.1.5. Theorem 7.1.1 extends further in a natural way to weakly coupled parabolic systems. A simple example is the system

$$(u_i)_t = \Delta u_i + f_i(x, u, \nabla u), \ i = 1, 2, \ t > 0, \ x \in \Omega, \tag{7.1.17}$$

with either Dirichlet or Neumann type boundary conditions. Here $u = (u_1, u_2)$, $\Omega \subset R^n$ is a bounded C^2 domain and functions f_i $(i = 1, 2)$ are locally Lipschitz continuous with respect to functional arguments uniformly for $x \in \Omega$.

Focusing on the Dirichlet problem we take $A = (-\Delta, -\Delta)$ in $[L^p(\Omega)]^2$, $p > n$, with the domain $D(A) := [W^{2,p}(\Omega) \cap W_0^{1,p}(\Omega)]^2$. For $\alpha \in (\frac{1}{2} + \frac{n}{2p}, 1)$ we have $X^\alpha \subset [C^1(\overline{\Omega})]^2$. Also, the substitution operator $F : X^\alpha \to X$ corresponding to (f_1, f_2) is Lipschitz continuous on bounded sets. The existence of local X^α solutions to (7.1.17) is thus straightforward.

For a pair u, $\tilde{u}$ of such solutions and each positive $\tau < \min\{\tau_{u_0}, \tau_{\tilde{u}_0}\}$ there is a bounded set $\mathcal{U}_\tau \subset X^\alpha$ such that

$$u(t), \tilde{u}(t) \in \mathcal{U}_\tau \ \text{ for } \ t \in [0, \tau].$$

Therefore

$$\forall_{0<\tau<\min\{\tau_{u_0}, \tau_{\tilde{u}_0}\}} \exists_{L_{\mathcal{U}_\tau}>0} \ \|(f_1, f_2)\|_{[L^2(\Omega)]^2} \leq L_{\mathcal{U}_\tau}\|u - \tilde{u}\|_{[H_0^1(\Omega)]^2}$$

and Proposition 7.1.1 may be used with $\mathcal{W} = X^\alpha$, $H = [L^2(\Omega)]^2$, $G = A$ on $D(G) = [H^2(\Omega) \cap H_0^1(\Omega)]^2$, $H^{\frac{1}{2}} = [H_0^1(\Omega)]^2$ similarly as in the proof of Theorem 7.1.1.

The remark above will be continued in the following considerations of the Navier-Stokes system (8.3.1).

Example 7.1.4. In this example we refer fully to considerations of Section 8.3 including basic facts concerning notation, abstract setting, and local solvability of the n-dimensional Navier-Stokes equation.

As shown in [FU-MO, Theorem 3], $A_2 = -P_2(\Delta, \dots, \Delta)_{n\times}$ with the domain $D(A_2) = \mathcal{W}^{2,2}(\Omega) \cap \mathcal{W}_0^{1,2}(\Omega) \cap X_2$ is a self-adjoint operator in $X_2 = cl_{\mathcal{L}^2(\Omega)}\{\phi \in [C_0^\infty(\Omega)]^n : div\,\phi = 0\}$. Furthermore, as a consequence of the Poincaré inequality and [FU-MO, Lemma 5] (in which $P_2 : \mathcal{L}^2(\Omega) \to X_2 \subset \mathcal{L}^2(\Omega)$ was proved to be self-adjoint) the operator A_2 is positive definite. Also, by (8.3.8),

$$X_2^{\frac{1}{2}} = \mathcal{W}_0^{1,2}(\Omega) \cap X_2$$

(see considerations in [TE 1, pp. 104-105]).

Fix $r > \max\{n, 2\}$ and choose $\alpha \in [0, 1)$ satisfying $2\alpha - \frac{n}{r} > 1$. Then for each $u_0 \in X_r^\alpha$ there exists an X_r^α solution $u(t, u_0)$ of the abstract equation (8.3.10) corresponding to (8.3.1)-(8.3.3). Moreover, recalling (8.3.9), for such α and r, we have

$$X_r^\alpha \subset C^1(\overline{\Omega}). \tag{7.1.18}$$

Consider a pair u, $\tilde{u}$ of such solutions, assuming that $u(t_0) = \tilde{u}(t_0)$ for a certain $t_0 \in [0, \tau_{max})$ ($\tau_{max} = \min\{\tau_{u_0}, \tau_{\tilde{u}_0}\}$) fix $\tau \in (t_0, \tau_{max})$ and define $w = u - \tilde{u}$ on $[0, \tau]$. Since w is a difference of two X_r^α solutions, we have

$$\dot{w} + P_r\Delta w = -(P_r(u, \nabla)u - P_r(\tilde{u}, \nabla)\tilde{u}), \; t \in (0, \tau), \tag{7.1.19}$$

where

$$w \in C([0, \tau], X_r^\alpha) \cap C^1((0, \tau), X_r) \cap C((0, \tau), D(A_r)). \tag{7.1.20}$$

In the presence of the embeddings

$$\begin{aligned}
&X_r^\alpha \subset \mathcal{W}^{2\alpha,r}(\Omega) \cap \mathcal{W}_0^{1,r}(\Omega) \cap X_r \subset \mathcal{W}_0^{1,2}(\Omega) \cap X_2 = X_2^{\frac{1}{2}},\\
&D(A_r) = \mathcal{W}^{2,r}(\Omega) \cap \mathcal{W}_0^{1,r}(\Omega) \cap X_r \subset \mathcal{W}^{2,2}(\Omega) \cap \mathcal{W}_0^{1,2}(\Omega) \cap X_2 = D(A_2),\\
&X_r \subset X_2,
\end{aligned}$$

the condition (7.1.20) ensures in particular that

$$w \in C([0, \tau], X_r^\alpha) \cap C^1((0, \tau), X_2) \cap C((0, \tau), D(A_2)). \tag{7.1.21}$$

Now, recalling the construction of $P_r : \mathcal{L}^r(\Omega) \to X_r \subset \mathcal{L}^r(\Omega)$ (see [FU-MO, §3]), it follows that

$$P_r v = P_2 v \;\text{ for }\; v \in \mathcal{L}^r(\Omega), \; r \geq 2.$$

Indeed, $P_r v := v - \nabla(\phi_1 + \phi_2)$, where

- ϕ_1 is a unique solution in $\mathcal{W}_0^{1,r}(\Omega)$ of

$$\begin{cases} \Delta\phi_1 = div\, v \;\text{ in }\; \Omega,\\ \phi_1 = 0 \;\text{ in }\; \partial\Omega, \end{cases}$$

- ϕ_2 is a unique solution of

$$\begin{cases} \Delta\phi_2 = 0 \ \text{ in } \ \Omega, \\ \frac{\partial \phi_2}{\partial N} = \frac{\partial v}{\partial N} - \frac{\partial \phi_1}{\partial N} \ \text{ in } \partial\Omega, \end{cases}$$

for which the estimate $\|\phi_2\|_{W^{1,r}(\Omega)} \leq \|\frac{\partial v}{\partial N} - \frac{\partial \phi_1}{\partial N}\|_{\mathcal{W}^{-1,r}(\Omega)}$ holds.

Let us note that $div\, v \in \mathcal{W}^{-1,r}(\Omega)$, whereas $v - \nabla\phi_1 \in \mathcal{L}^r(\Omega)$; i.e., from [FU-MO, Lemma 1], the normal component of $(u - \nabla\phi_1)$ has boundary value in $\mathcal{W}^{-\frac{1}{r},r}(\partial\Omega)$.

Note that for $r \geq 2$ and $v \in \mathcal{L}^r(\Omega)$, $P_r v = P_2 v$ by the above definition of $P_r v$. In consequence we obtain that w satisfying (7.1.19), (7.1.20) also fulfills (7.1.21) and

$$\dot{w} + P_2\Delta w = -(P_2(u,\nabla)u - P_2(\tilde{u},\nabla)\tilde{u}), \ t \in (0,\tau). \tag{7.1.22}$$

Next we shall use Proposition 7.1.1 with $H = X_2$, $G = A_2$, $D(G) = D(A_2)$, $H^{\frac{1}{2}} = X_2^{\frac{1}{2}}$ and $\mathcal{W} = X_r^{\alpha}$. For this we need yet to verify that $h(w) = -(P_2(u,\nabla)u - P_2(\tilde{u},\nabla)\tilde{u})$ satisfies the estimate required in (7.1.5). However, based on (7.1.18), we get

(7.1.23)

$$\begin{aligned}
&\|P_2(u,\nabla)u - P_2(\tilde{u},\nabla)\tilde{u}\|_{\mathcal{L}^2(\Omega)} \leq \|P_2\|_{\mathcal{L}(\mathcal{L}^2(\Omega),\mathcal{L}^2(\Omega))}\|(u,\nabla)u - (\tilde{u},\nabla)\tilde{u}\|_{\mathcal{L}^2(\Omega)} \\
&\leq \|P_2\|_{\mathcal{L}(\mathcal{L}^2(\Omega),\mathcal{L}^2(\Omega))} \sum_{k=1}^{n}\sum_{j=1}^{n} \|u_j\frac{\partial u_k}{\partial x_j} - \tilde{u}_j\frac{\partial \tilde{u}_k}{\partial x_j}\|_{L^2(\Omega)} \\
&\leq \|P_2\|_{\mathcal{L}(\mathcal{L}^2(\Omega),\mathcal{L}^2(\Omega))} \\
&\quad \times \sum_{k=1}^{n}\sum_{j=1}^{n}\left(\left\|(u_j - \tilde{u}_j)\frac{\partial u_k}{\partial x_j}\right\|_{L^2(\Omega)} + \left\|\tilde{u}_j\Big(\frac{\partial u_k}{\partial x_j} - \frac{\partial \tilde{u}}{\partial x_j}\Big)\right\|_{L^2(\Omega)}\right) \\
&\leq const._{r_0}\|u - \tilde{u}\|_{H^{\frac{1}{2}}},
\end{aligned}$$

where $r_0 = \sup_{t\in[0,\tau]}\|u(t)\|_{X_r^\alpha} + \sup_{t\in[0,\tau]}\|\tilde{u}(t)\|_{X_r^\alpha}$. Using Proposition 7.1.1 we thus obtain that $u \equiv \tilde{u}$ on $[0,\tau]$. The backward uniqueness property for the Navier-Stokes equations is thus justified.

7.2. $X^{s+\alpha}$ solutions; $s \geq 0$, $\alpha \in [0,1)$

Based on the results of Chapter 2 it is easy to obtain the existence of $X^{s+\alpha}$ solutions to (2.1.1) with $s \geq 0$, $\alpha \in [0,1)$ (see [HE 1, Exercises 1, 2, Section 3.5], [RB 1]). We shall consider here (2.1.1),

$$\begin{cases} \dot{u} + Au = F(u), \ t > 0, \\ u(0) = u_0, \end{cases}$$

under the following $(X^{s+\alpha}, X^s)$ *condition,* being a natural generalization of Assumption 2.1.1.

$(X^{s+\alpha}, X^s)$ condition. *Assume that A is a sectorial, positive operator in a Banach space X and, for some $\alpha \in [0,1)$, $s \in [0,+\infty)$, a function $F : X^{s+\alpha} \to X^s$ is Lipschitz continuous on bounded sets.*

Based on Definition 2.1.1 and Corollary 2.3.1 the notion of X^α solution of (2.1.1) will be extended as follows.

Definition 7.2.1. *Let u_0 be an element of $X^{s+\alpha}$ and $\tau > 0$. If a function $u : [0,\tau) \to X^{s+\alpha}$ satisfies*

$$
\begin{aligned}
&u(0) = u_0,\\
&u \in C([0,\tau), X^{s+\alpha}) \cap C^1((0,\tau), X^s) \cap C((0,\tau), X^{s+1}),\\
&\textit{the first equation in } (2.1.1) \textit{ holds in } X^s \textit{ for all } t \in (0,\tau),\\
&\dot{u} \in C((0,\tau), X^{s+\gamma}),\ \gamma \in [0,1),
\end{aligned}
$$

then u is called a ***local $X^{s+\alpha}$ solution*** *of* (2.1.1).

Remark 7.2.1. As before, $\dot{u}$ denotes here the strong time derivative of u in X (i.e. the limit of the difference quotient $\frac{1}{h}(u(t+h) - u(t))$ in the norm of X). However, since X^s is continuously embedded in X for $s \geq 0$ it may be seen that $\dot{u}$ coincides with the strong derivative in the X^s norm whenever the latter exists. In particular, $\dot{u}$ in Definition 7.2.1 above is thus the derivative of u in the X^s norm.

Existence of $X^{s+\alpha}$ solutions. Let us now show the following consequence of Theorem 2.1.1 (see the results of [RB 1] which are essentially exploited further under this heading).

Theorem 7.2.1. *Under the $(X^{s+\alpha}, X^s)$ condition there is a unique $X^{s+\alpha}$ solution of the problem* (2.1.1) *defined on its maximal interval of existence (see* (7.2.2)*).*

Proof. First note that, as a consequence of the definition of fractional power operators and property (1.3.45), A^s is an isomorphic map from X^s onto X and from $X^{s+\alpha}$ onto X^α (see [FR 1, p. 159], [TR, Section 1.15.2(e)]). Consider further the problem

$$
\begin{cases}
\dot{v} + Av = \tilde{F}(v), \ t > 0,\\
v(0) = v_0 =: A^s u_0,
\end{cases}
\tag{7.2.1}
$$

where $u_0 \in X^{s+\alpha}$ is fixed from now on and $\tilde{F} := A^s F A^{-s}$. By our assumptions $\tilde{F} : X^\alpha \to X$ is Lipschitz continuous on bounded sets. Indeed, if $w_1, w_2 \in B$, B a bounded subset of X^α, then $A^{-s}w_1, A^{-s}w_2 \in A^{-s}B$, where $A^{-s}B$ is bounded in $X^{s+\alpha}$, and we have

$$
\begin{aligned}
\|\tilde{F}(w_1) - \tilde{F}(w_2)\|_X &= \|A^s\big(FA^{-s}(w_1) - FA^{-s}(w_2)\big)\|_X\\
&= \|F(A^{-s}w_1) - F(A^{-s}w_2)\|_{X^s} \leq L_{A^{-s}B}\|A^{-s}w_1 - A^{-s}w_2\|_{X^{s+\alpha}}\\
&= L_{A^{-s}B}\|w_1 - w_2\|_{X^\alpha}.
\end{aligned}
$$

Therefore, from Theorem 2.1.1 and Corollary 2.3.1, (7.2.1) has a unique X^α solution $v \in C([0,\tau), X^\alpha)$ defined on a maximal interval of existence $[0,\tau)$ such that

$$v \in C([0,\tau), X^\alpha) \cap C^1((0,\tau), X) \cap C((0,\tau), X^1),$$
$$\dot{v} \in C((0,\tau), X^\gamma), \ \gamma \in [0,1).$$

Note that, since A^{-s} is a bounded linear operator in X and $\dot{v}$ exists, $A^{-s}v$ is differentiable and A^{-s} commutes with the differentiation in X, i.e.

$$\dot{\overline{A^{-s}v}} = A^{-s}\dot{v}.$$

From the above conditions, for $u := A^{-s}v$ we get

$$u \in C([0,\tau), X^{s+\alpha}) \cap C^1((0,\tau), X^s) \cap C((0,\tau), X^{s+1}),$$
$$\dot{u} \in C((0,\tau), X^{s+\gamma}), \ \gamma \in (0,1);$$

also, the function u satisfies (2.1.1). This justifies that u is a local $X^{s+\alpha}$ solution to (2.1.1). Moreover, recalling the equality $A^{s+\alpha}u = A^\alpha v$ and using Theorem 2.1.1 again we obtain that

$$\text{(7.2.2)} \qquad \text{either } \tau = +\infty \text{ or } \limsup_{t \to +\infty} \|u\|_{X^{s+\alpha}} = +\infty;$$

hence $[0,\tau)$ is a maximal interval of existence of u. The uniqueness of u is guaranteed by the uniqueness of v. Note here that $A^s u$ is an X^α solution to (7.2.1) whenever u is an $X^{s+\alpha}$ solution to (2.1.1); also $\dot{\overline{A^s u}} = A^s\dot{u}$ provided that $\dot{\overline{A^s u}}$ exists (see Remark 7.2.1). The proof is complete. □

Remark 7.2.2. Let r, s be two nonnegative numbers such that $r - s \in [0,1)$. Then $r = s + \alpha$ with $\alpha = r - s$ and hence the notion of (X^r, X^s) condition can be introduced as a simple analogy to the $(X^{s+\alpha}, X^s)$ condition previously defined. In this context, if an (X^r, X^s) condition is satisfied, Definition 7.2.1 also determines the notion of a local X^r solution to (2.1.1).

Space regularity. Based on [RB 1, Proposition 2.1] consider further two sets of numbers (finite or infinite) $\{\alpha_k\}$, $\{\beta_k\}$ such that, for all k,

$$\text{(7.2.3)} \qquad \begin{cases} 0 \leq \beta_k \leq \alpha_k < \alpha_{k+1} \leq \beta_k + 1, \\ \beta_k < \beta_{k+1} < \beta_k + 1. \end{cases}$$

We shall prove inductively for $m \in N$ the following implication.

Corollary 7.2.2. *If the $(X^{\alpha_k}, X^{\beta_k})$ conditions hold for $k = 0, \ldots, m$, then to any $u_0 \in X^{\alpha_0}$ corresponds on $[0, \tau_{u_0})$ a unique X^{α_0} solution of* (2.1.1) *such that*

$$\text{(7.2.4)} \qquad \begin{cases} u \in C([0,\tau_{u_0}), X^{\alpha_0}) \cap C^1((0,\tau_{u_0}), X^{\beta_m}) \cap C((0,\tau_{u_0}), X^{\beta_m+1}), \\ \dot{u} \in C((0,\tau_{u_0}), X^{\beta_m+\gamma}), \ \gamma \in (0,1). \end{cases}$$

Proof. For $m = 0$ the conclusion follows directly from Theorem 7.2.1 (with $s = \beta_0$ and $\alpha = \alpha_0 - \beta_0$). Next, whenever this result is true for a number m, we know from the inductive hypothesis (7.2.4) that

$$u(t + t_0, u_0) = u(t, u(t_0, u_0)) \in X^{\beta_m+1},\ 0 < t_0 \le t_0 + t < \tau_{u_0}. \tag{7.2.5}$$

Then, under the $(X^{\alpha_{m+1}}, X^{\beta_{m+1}})$ condition (note that, by (7.2.3), $\alpha_{m+1} \le \beta_m + 1$), there exists (see Theorem 7.2.1) a unique $X^{\alpha_{m+1}}$ solution $\tilde{u}$ of the problem

$$\begin{cases} \dot{u} + Au = F(u),\ t > 0, \\ u(0) = u(t_0, u_0). \end{cases} \tag{7.2.6}$$

Obviously $\tilde{u}$ is an X^{α_0} solution of (7.2.6) and, by uniqueness, the equality

$$\tilde{u}(t) = u(t + t_0, u_0)$$

holds at least in a certain right hand side neighborhood of zero. Since $t_0 \in (0, \tau_{u_0})$ was arbitrary, this shows that for each $t \in (0, \tau_{u_0})$ the function $u = u(\cdot, u_0)$ coincides in some neighborhood of t with the $X^{\alpha_{m+1}}$ solution of (7.2.6) and, in consequence, we have

$$\begin{cases} u \in C^1((0, \tau_{u_0}), X^{\beta_{m+1}}) \cap C((0, \tau_{u_0}), X^{\beta_{m+1}+1}), \\ \dot{u} \in C((0, \tau_{u_0}), X^{\beta_{m+1}+\gamma}),\ \gamma \in (0, 1). \end{cases} \tag{7.2.7}$$

The proof is complete. □

Remark 7.2.3. Let numbers α_i', β_i' $(i = 1, 2)$ satisfy (7.2.3), $u_0 \in X^{\alpha_2'}$ and, for $i = 1, 2$, the $(X^{\alpha_i'}, X^{\beta_i'})$ condition hold. For such a u_0 one may consider both $X^{\alpha_1'}$ and $X^{\alpha_2'}$ solutions of (2.1.1) on the maximal interval of existence $[0, \tau_1)$, $[0, \tau_2)$ respectively. Then, naturally, $\tau_1 = \tau_2$ and these solutions coincide. Indeed, the inequality $\tau_2 \le \tau_1$ follows immediately from the uniqueness property, whereas $\tau_1 \le \tau_2$ is a consequence of Corollary 7.2.2.

If, generally, $u_0 \in X^{\alpha_1'}$, then for arbitrarily small $t_1 > 0$ the corresponding $X^{\alpha_1'}$ solution enters $X^{\beta_2'+1} \subset X^{\alpha_2'}$ as follows from Corollary 7.2.2. The above reasoning shows that the $X^{\alpha_1'}$ and $X^{\alpha_2'}$ solutions originating at time t_1 in $u(t_1)$ have common interval of existence $[t_1, \tau_{u_0})$.

Time regularity. When Corollary 7.2.2 provides enough space regularity for the X^α solution it is easy to formulate a corresponding result concerning time regularity.

Denote further by $\frac{d^k}{dt^k}$ the Fréchet derivative operator of order $k \in N$. Obviously $\frac{du}{dt} = \dot{u}$ and, whenever Au can be differentiated with respect to t, we observe from (2.1.1) that $\frac{d^2u}{dt^2}$ exists if and only if there exists $\frac{d}{dt}F(u(t))$. Based on (2.1.1) we then obtain

$$\begin{aligned} \frac{d^2u}{dt^2} &= \frac{d}{dt}\big(-Au + F(u)\big) = -A\frac{du}{dt} + \frac{d}{dt}\big(F(u)\big) \\ &= (-A)^2u + (-A)F(u) + \frac{d}{dt}\big(F(u)\big), \end{aligned}$$

provided that u and F have values in X^2 and X^1 respectively. If we have still enough regularity to repeat the above calculations, we will obtain in the r-th step the formula

$$\frac{d^r u}{dt^r} = (-A)^r u + \sum_{j=0}^{r-1} \frac{d^j}{dt^j}\big((-A)^{r-j-1}F(u)\big). \tag{7.2.8}$$

Assume thus that the $(X^{\frac{2m-1+k}{2m}}, X^{\frac{k}{2m}})$ condition holds for each $k = 0, \ldots, 2mr$. Under this assumption Corollary 7.2.2 (used with $\beta_m = r$) ensures that $(-A)^r u$ is continuously differentiable so that, as seen from (7.2.8), $\frac{d^{r+1}u}{dt^{r+1}}$ exists and is continuous provided that there exists a continuous derivative with respect to t of the term $\sum_{j=0}^{r-1} \frac{d^j}{dt^j}\big((-A)^{r-j-1}F(u)\big)$.

The above considerations justify the following conclusion.

Corollary 7.2.3. *Let $m, r \in \{1, 2, \ldots\}$. If the $(X^{\frac{2m-1+k}{2m}}, X^{\frac{k}{2m}})$ condition holds for $k = 0, \ldots, 2mr$, then to each element $u_0 \in X^{\frac{2m-1}{2m}}$ corresponds the unique $X^{\frac{2m-1}{2m}}$ solution of* (2.1.1) *such that*

$$u \in C([0, \tau_{u_0}), X^{\frac{2m-1}{2m}}) \cap C^1((0, \tau_{u_0}), X^r) \cap C((0, \tau_{u_0}), X^{r+1}),$$
$$\dot{u} \in C((0, \tau_{u_0}), X^{r+\gamma}),\ \gamma \in (0,1).$$

If, in addition,

$$A^{r-j-1}F \in C^{j+1}(X^{r-j-1+\frac{2m-1}{2m}}, X),\ j = 0, \ldots, r-1, \tag{7.2.9}$$

then (7.2.8) *is satisfied and, consequently,*

$$u \in C^{r+1}((0, \tau_{u_0}), X). \tag{7.2.10}$$

Remarks on applications. Considering in Chapter 5 equation

$$u_t = -Au + f(x, d^{m_0}u),\ t > 0,\ x \in \Omega \subset R^n, \tag{7.2.11}$$

in $L^p(\Omega)$ spaces (see (5.1.2) and (5.1.3) for the notation of A and $d^{m_0}u$)) we have shown that, for $\varepsilon - \frac{n}{p} > 0$, the substitution operator F (which corresponds to f) maps $W^{m_0+\varepsilon,p}(\Omega)$ into $L^p(\Omega)$ and is Lipschitz continuous on bounded subsets of $W^{m_0+\varepsilon,p}(\Omega)$ (see (5.1.8)).

Let us now observe, by analogy, that if f has partial derivatives up to order r locally Lipschitz continuous with respect to functional arguments uniformly for $x \in \Omega$, then a similar result will be that

(7.2.12)

F takes $W^{m_0+r+\varepsilon,p}(\Omega)$ into $W^{r,p}(\Omega)$ and is Lipschitz continuous on bounded subsets of $W^{m_0+r+\varepsilon,p}(\Omega)$, where the set $\Omega \subset R^n$ may be unbounded under the additional assumption that $\frac{\partial^{|\sigma|} f}{\partial x_1^{\sigma_1} \dots x_n^{\sigma_n}}(\cdot, 0, \dots, 0) \in L^p(\Omega)$ for $0 \le |\sigma| \le r$. The condition (7.2.12) is a consequence of the Sobolev embedding,

$$W^{m_0+r+\varepsilon,p}(\Omega) \subset W^{m_0+r,p}(\Omega) \cap W^{m_0+r,\infty}(\Omega),\ \varepsilon - \frac{n}{p} > 0,$$

and Lemma 7.2.1 below.

Remark 7.2.4. The condition

$$\frac{\partial^{|\sigma|} f}{\partial x_1^{\sigma_1} \dots \partial x_n^{\sigma_n}}(\cdot, 0, \dots, 0) \in L^p(\Omega) \text{ for } 0 \le |\sigma| \le r,$$

assumed additionally in the case of unbounded Ω (e.g. for $\Omega = R^n$), is necessary and sufficient to guarantee that the values of the Lipschitz continuous substitution operator on the arguments taken from $W^{m_0+r+\varepsilon,p}(\Omega)$ be the elements of $W^{r,p}(\Omega)$. In particular, for $r = 0$, to conclude that F takes $W^{m_0+\varepsilon,p}(R^n)$ into $L^p(R^n)$ we estimate as follows:

$$\begin{aligned} \|F(\varphi)\|_{L^p(R^n)} &\le \|F(\varphi) - F(0)\|_{L^p(R^n)} + \|F(0)\|_{L^p(R^n)} \\ &\le \|f(\cdot, d^{m_0}\varphi) - f(\cdot, 0)\|_{L^p(R^n)} + \|f(\cdot, 0)\|_{L^p(R^n)} \\ &\le L_\varphi \|\varphi\|_{W^{m_0,p}(R^n)} + \|f(\cdot, 0)\|_{L^p(R^n)}, \end{aligned}$$

Note that when $r = 0$ and f does not depend of x, such a condition has the form $f(0, \dots, 0) = 0$ (see Example 6.7.1).

Lemma 7.2.1. *For each $r \in N$ and any choice of*

- *numbers $k_0,\ k \in N$,*
- *a function $g : R^{n+k_0} \to R$ having partial derivatives up to order r locally Lipschitz continuous with respect to functional arguments uniformly for $x \in \Omega$,*
- *multi-indices $\sigma^{(1)}, \dots, \sigma^{(k_0)} \in N^n$ with $\max_{i=1,\dots,k_0}\{|\sigma^{(i)}|\} = k$,*

there is a locally bounded function $L : [0, +\infty) \to [0, +\infty)$ such that the estimate

(7.2.13)

$$\begin{aligned} &\|g(x, D^{\sigma^{(1)}}\phi, \dots, D^{\sigma^{(k_0)}}\phi) - g(x, D^{\sigma^{(1)}}\psi, \dots, D^{\sigma^{(k_0)}}\psi)\|_{W^{r,p}(\Omega)} \\ &\le L\Big(\max\{\|\phi\|_{W^{k+r,\infty}(\Omega)}, \|\psi\|_{W^{k+r,\infty}(\Omega)}\}\Big) \|\phi - \psi\|_{W^{k+r,p}(\Omega)} \end{aligned}$$

holds for all $\phi, \psi \in W^{k+r,p}(\Omega) \cap W^{k+r,\infty}(\Omega)$.

Proof. Induction with respect to $r \in N$ will be used. The case $r = 0$ follows after calculations similar to (5.1.8). Assuming further that the lemma holds for $r = s$ we are going to show that it is true for $r = s + 1$. Let us take $k, k_0 \in N$, a function $g : R^{n+k_0} \to R$ having Lipschitz derivatives up to order $s + 1$ and finite sequence of multi-indices $\sigma^{(1)},\dots,\sigma^{(k_0)} \in N^n$ with $\max_{i=1,\dots,k_0}\{|\sigma^{(i)}|\} = k$. For $\phi,\psi \in W^{k+s+1,p}(\Omega) \cap W^{k+s+1,\infty}(\Omega)$ we have

(7.2.14)

$$\begin{aligned}
&\|g(x, D^{\sigma^{(1)}}\phi,\dots,D^{\sigma^{(k_0)}}\phi) - g(x, D^{\sigma^{(1)}}\psi,\dots,D^{\sigma^{(k_0)}}\psi)\|_{W^{s+1,p}(\Omega)} \\
&= \|g(x, D^{\sigma^{(1)}}\phi,\dots,D^{\sigma^{(k_0)}}\phi) - g(x, D^{\sigma^{(1)}}\psi,\dots,D^{\sigma^{(k_0)}}\psi)\|_{L^p(\Omega)} \\
&\quad + \sum_{0\leq|\sigma|\leq s}\sum_{i=1}^{n}\Bigg\| D^\sigma\left[\frac{\partial}{\partial x_i}\left(g(x, D^{\sigma^{(1)}}\phi,\dots,D^{\sigma^{(k_0)}}\phi)\right)\right] \\
&\quad - D^\sigma\left[\frac{\partial}{\partial x_i}\left(g(x, D^{\sigma^{(1)}}\psi,\dots,D^{\sigma^{(k_0)}}\psi)\right)\right]\Bigg\|_{L^p(\Omega)} \\
&\leq \|g(x, D^{\sigma^{(1)}}\phi,\dots,D^{\sigma^{(k_0)}}\phi) - g(x, D^{\sigma^{(1)}}\psi,\dots,D^{\sigma^{(k_0)}}\psi)\|_{L^p(\Omega)} \\
&\quad + const. \sum_{i=1}^{n}\left\|\frac{\partial}{\partial x_i}\left(g(x, D^{\sigma^{(1)}}\phi,\dots,D^{\sigma^{(k_0)}}\phi) - g(x, D^{\sigma^{(1)}}\psi,\dots,D^{\sigma^{(k_0)}}\psi)\right)\right\|_{W^{s,p}(\Omega)}.
\end{aligned}$$

Note further that

(7.2.15)

$$\begin{aligned}
&\frac{\partial}{\partial x_i}\left(g(x, D^{\sigma^{(1)}}\phi,\dots,D^{\sigma^{(k_0)}}\phi)\right) = \frac{\partial g}{\partial x_i}(x, D^{\sigma^{(1)}}\phi,\dots,D^{\sigma^{(k_0)}}\phi) \\
&+ \frac{\partial g}{\partial(D^{\sigma^{(1)}}\phi)}(x, D^{\sigma^{(1)}}\phi,\dots,D^{\sigma^{(k_0)}}\phi)\frac{\partial(D^{\sigma^{(1)}}\phi)}{\partial x_i} \\
&+ \dots + \frac{\partial g}{\partial(D^{\sigma^{(k_0)}}\phi)}(x, D^{\sigma^{(1)}}\phi,\dots,D^{\sigma^{(k_0)}}\phi)\frac{\partial(D^{\sigma^{(k_0)}}\phi)}{\partial x_i} \\
&=: \tilde{g}_i(x, D^{\sigma^{(1)}}\phi,\dots,D^{\sigma^{(k_0)}}\phi, D^{\sigma^{(k_0+1)}}\phi,\dots,D^{\sigma^{(\tilde{k}_0)}}\phi),\ i = 1,\dots,n,
\end{aligned}$$

where

$$\tilde{k}_0 := 2k_0 \ \text{ and } \ \tilde{k} := \max\{|\sigma^{(j)}| : j = 1,\dots,\tilde{k}_0\} = k+1.$$

Also, by our assumptions on g,ϕ,ψ, composite functions $\tilde{g}_i : R^{n+\tilde{k}_0} \to R$ ($i = 1,\dots,n$) have Lipschitz continuous partial derivatives up to the order s. Therefore we may apply the inductive hypothesis to functions $\tilde{g}_i$ and parameters $\tilde{k}_0, \tilde{k}$.

With the use of this hypothesis, for $\phi,\psi \in W^{\tilde{k}+s,p}(\Omega) \cap W^{\tilde{k}+s,\infty}(\Omega)$ and $i = 1,\dots,n$, we obtain

(7.2.16)

$$\begin{aligned}
&\|\tilde{g}_i(x, D^{\sigma^{(1)}}\phi,\dots,D^{\sigma^{(k_0)}}\phi, D^{\sigma^{(k_0+1)}}\phi,\dots,D^{\sigma^{(\tilde{k}_0)}}\phi) \\
&- \tilde{g}_i(x, D^{\sigma^{(1)}}\psi,\dots,D^{\sigma^{(k_0)}}\psi, D^{\sigma^{(k_0+1)}}\psi,\dots,D^{\sigma^{(\tilde{k}_0)}}\psi)\|_{W^{s,p}(\Omega)} \\
&\leq L_i\left(\max\left\{\|\phi\|_{W^{\tilde{k}+s,\infty}(\Omega)}, \|\psi\|_{W^{\tilde{k}+s,\infty}(\Omega)}\right\}\right)\|\phi-\psi\|_{W^{\tilde{k}+s,p}(\Omega)}.
\end{aligned}$$

Combining finally the inequalities (7.2.14)-(7.2.16), we get

(7.2.17)
$$
\begin{aligned}
&\|g(x, D^{\sigma(1)}\phi, \dots, D^{\sigma(k_0)}\phi) - g(x, D^{\sigma(1)}\psi, \dots, D^{\sigma(k_0)}\psi)\|_{W^{s+1,p}(\Omega)} \\
&\le \|g(x, D^{\sigma(1)}\phi, \dots, D^{\sigma(k_0)}\phi) - g(x, D^{\sigma(1)}\psi, \dots, D^{\sigma(k_0)}\psi)\|_{L^p(\Omega)} \\
&\quad + const. \sum_{i=1}^{n} \|\tilde{g}_i(x, D^{\sigma(1)}\phi, \dots, D^{\sigma(k_0)}\phi, D^{\sigma(k_0+1)}\phi, \dots, D^{\sigma(\tilde{k}_0)}\phi) \\
&\qquad - \tilde{g}_i(x, D^{\sigma(1)}\psi, \dots, D^{\sigma(k_0)}\psi, D^{\sigma(k_0+1)}\psi, \dots, D^{\sigma(\tilde{k}_0)}\psi)\|_{W^{s,p}(\Omega)} \\
&\le L\big(\max\{\|\phi\|_{W^{\tilde{k}+s,\infty}(\Omega)}, \|\psi\|_{W^{\tilde{k}+s,\infty}(\Omega)}\}\big)\|\phi - \psi\|_{W^{\tilde{k}+s,p}(\Omega)}.
\end{aligned}
$$

Since $\tilde{k} = k + 1$, the proof is complete. □

Remark 7.2.5. Let the regular elliptic boundary value problem $(A, \{B_j\}, \Omega)$ be such that A is a positive, sectorial operator on $L^p(\Omega)$ with $D(A) = W^{2m,p}_{\{B_j\}}(\Omega)$ $(p > 1, \Omega \subset R^n)$. Suppose further that $p > n$, $\varepsilon > \frac{n}{p}$ and, in addition,

$$
(7.2.18) \qquad F \text{ takes } X^{\frac{m_0+r}{2m}+\frac{\varepsilon}{2m}} \text{ into } X^{\frac{r}{2m}},
$$

where $r \in N$ and F, as before, is an abstract counterpart of the right hand side term f in (7.2.11). The condition (7.2.12) will ensure that $F : X^{\frac{m_0+r}{2m}+\frac{\varepsilon}{2m}} \to X^{\frac{r}{2m}}$ is Lipschitz continuous on bounded subsets of $X^{\frac{m_0+r}{2m}+\frac{\varepsilon}{2m}}$ $(r \in N)$ since we know that

(i) $X^{\frac{r}{2m}} \subset W^{r,p}(\Omega)$,
(ii) the $W^{r,p}(\Omega)$ norm is an equivalent norm in $X^{\frac{r}{2m}}$,
(iii) $X^{\frac{m_0+r}{2m}+\frac{\varepsilon}{2m}} \subset W^{m_0+r,p}(\Omega) \cap W^{m_0+r,\infty}(\Omega)$.

Indeed, for the fractional power spaces connected with regular elliptic boundary value problems having C^∞ smooth data, conditions (i)-(iii) are known to hold (see interpolation arguments of Section 1.3.3). Nevertheless, (7.2.18) may provide certain limitations for applications. Roughly speaking, in (7.2.18) we require of the composite function $f(x, d^{m_0}u)$ to fulfill suitable boundary conditions, which is similar to the requirement of fulfillment of appropriate compatibility conditions known from the classical approach (see Remark 8.4.2). We point out this problem when considering the simple scalar parabolic equation in Example 7.2.2 below. However, sometimes (7.2.18) may not be restrictive at all, as e.g. for $\Omega = R^n$ and $A = (I - \Delta)$ acting in $L^p(\Omega)$, which will be exploited in Example 7.2.1.

Remark 7.2.6. Note also that if we assume additionally in Remark 7.2.5 that $1 \le m_0 + \varepsilon \le 2m - 1$, then the $(X^{\frac{m_0+r}{2m}+\frac{\varepsilon}{2m}}, X^{\frac{r}{2m}})$ condition holds for $r \in N$ and, moreover, sequences $\{\alpha_r\}_{r\in N} = \{\frac{m_0+r}{2m} + \frac{\varepsilon}{2m}\}_{r\in N}$ and $\{\beta_r\}_{r\in N} = \{\frac{r}{2m}\}_{r\in N}$ have the properties (7.2.3) so that Corollary 7.2.2 is applicable (see also Corollary 7.2.3).

Example 7.2.1. Consider the Cauchy problem

$$\text{(7.2.19)} \quad \begin{cases} u_t = \Delta u + f(x,u), \ t \geq 0, \ x \in R^n, \\ u(0,x) = u_0(x), \ x \in R^n, \end{cases}$$

where $f \in C^j(R^{n+1})$ has j-th order partial derivative with respect to u locally Lipschitz continuous uniformly for $x \in R^n$. Assume that $p > n$, $\frac{\partial^{|\sigma|} f}{\partial x_1^{\sigma_1} \ldots x_n^{\sigma_n}}(\cdot, 0) \in L^p(R^n)$ for $0 \leq |\sigma| \leq j$ and consider $A = (I - \Delta)$ in $L^p(R^n)$ on the domain $D(A) = W^{2,p}(R^n) = W_0^{2,p}(R^n)$. Such an A is sectorial and positive on $L^p(R^n)$. Furthermore, (7.2.12) is satisfied as a consequence of Lemma 7.2.1. From the results of Chapter 1, we have

$$\text{(7.2.20)} \quad \begin{cases} X^{\frac{1+k}{2}} \subset W^{1+k,p}(R^n) \cap C^{k+\nu}(R^n), & k = 0, \ldots, j, \ 0 < \nu \leq 1 - \frac{n}{p}, \\ X^{\frac{k}{2}} = H_p^k(R^n) = W^{k,p}(R^n), & k = 0, \ldots, j. \end{cases}$$

Therefore, the function $F(u)$ (corresponding to $(f(x,u) + u)$) takes $X^{\frac{1+k}{2}}$ into $X^{\frac{k}{2}}$ and is Lipschitz continuous on bounded subsets of $X^{\frac{1+k}{2}}$ for each $k = 0, \ldots, j$. Corollary 7.2.2 guarantees that to each $u_0 \in X^{\frac{1}{2}} = W^{1,p}(R^n)$ (see (1.3.62)) corresponds a unique $X^{\frac{1}{2}}$ solution u of (7.2.19) satisfying (the conditions (7.2.4) and (7.2.20) are combined here)

$$u \in C([0, \tau_{u_0}), W^{1,p}(R^n)) \cap C^1((0, \tau_{u_0}), W^{j,p}(R^n)) \cap C((0, \tau_{u_0}), W^{j+2,p}(R^n)),$$
$$\dot{u} \in C((0, \tau_{u_0}), H_p^{j+2\gamma}(R^n)), \ \gamma \in (0,1),$$

so that, in particular, by (1.2.36),

$$u(t, \cdot) \in C^{j+1+\nu}(R^n), \ t \in (0, \tau_{u_0}), \ \nu \in \left(0, 1 - \frac{n}{p}\right],$$

$$\dot{u}(t, \cdot) \in C^{j+1+\nu}(R^n), \ t \in (0, \tau_{u_0}), \ \nu \in \left(0, 1 - \frac{n}{p}\right).$$

If $j = 2$, we obtain from Corollary 7.2.3 with $r = 1$ that

$$\text{(7.2.21)} \quad u \in C^2((0, \tau_{u_0}), L^p(R^n)).$$

For the justification of the property (7.2.9), required for the validity of (7.2.21) note that by the assumption on f, there is a continuous Fréchet derivative F' of the map

$$W^{1,p}(R^n) \ni u \xrightarrow{F} F(u) = f(\cdot, u) + u \in L^p(R^n)$$

given by the formula

$$F'(u)v = \left(\frac{\partial f}{\partial u}(\cdot, u) + 1\right) v, \ u, v \in X^{\frac{1}{2}}.$$

The procedure described above can then be continued as long as sufficient regularity of the right hand side term f is given.

Example 7.2.2. As another simple application of the above regularity theory consider the second order equation

$$u_t = \Delta u + f(u),\ t > 0,\ x \in \Omega \subset R^n, \tag{7.2.22}$$

where Ω is a bounded C^∞ domain and u satisfies the homogeneous Dirichlet condition. Assume that $j \geq 2$ is even and $f \in C^j(\overline{\Omega})$ has locally Lipschitz j-th order derivative. Moreover, let

$$f(0) = f''(0) = f'''(0) = \ldots = f^j(0) = 0; \tag{7.2.23}$$

in particular $f(u)$ will be a polynomial of the form

$$a_1 u + \text{ terms of order higher than } j.$$

Consider the above problem in $X = L^p(\Omega)$ with $p > n$ and $D(-\Delta) = W^{2,p}(\Omega) \cap W_0^{1,p}(\Omega)$. We choose next the sequences $\{\alpha_k\}, \{\beta_k\}, k \leq j$, in (7.2.3), setting

$$\alpha_0 = \frac{1}{2}, \beta_0 = 0, \ldots, \alpha_j = \frac{j+1}{2}, \beta_j = \frac{j}{2}. \tag{7.2.24}$$

Evidently the inequalities required in the condition (7.2.3) are satisfied. It now follows from (7.2.12) with $m_0 = 0, \varepsilon = 1$ that the substitution operator F corresponding to f is Lipschitz on bounded subsets as a map between the spaces

$$W^{1,p}(\Omega) \to X, \ldots, W^{j+1,p}(\Omega) \to W^{j,p}(\Omega). \tag{7.2.25}$$

Moreover, as a consequence of (7.2.23), one can check that for every φ satisfying (in the sense of traces) the conditions

$$\varphi = \Delta\varphi = \ldots = \Delta^i \varphi = 0 \ \text{ on } \ \partial\Omega, \tag{7.2.26}$$

where $i \in \{0, 1, \ldots, \frac{j}{2}\}$, one has

$$f(\varphi) = \Delta(f(\varphi)) = \ldots = \Delta^i(f(\varphi)) = 0 \text{ on } \ \partial\Omega. \tag{7.2.27}$$

The observations above justify that $(X^{\alpha_0}, X^{\beta_0}), \ldots, (X^{\alpha_j}, X^{\beta_j})$ conditions are satisfied (see Remark 1.3.7). It thus follows from Corollary 7.2.2, that the solution u of (7.2.22) fulfills the condition (7.2.4); i.e.

$$u \in C([0, \tau_{u_0}), X^{\frac{1}{2}}) \cap C^1((0, \tau_{u_0}), X^{\frac{j}{2}}) \cap C((0, \tau_{u_0}), X^{\frac{j}{2}+1}),$$
$$\dot{u} \in C((0, \tau_{u_0}), X^{\frac{j}{2}+\gamma}),\ \gamma \in (0, 1).$$

Classical solutions. Most of the basic properties of solutions to parabolic equations, like the maximum principle, have been originally formulated for *classical solutions* of such problems. Let Ω be a bounded domain in R^n with C^∞ boundary and (A, B_j, Ω) be a regular elliptic boundary value problem with C^∞ data. Let the highest order of the operators B_j be equal to m_1, the order of A equal to $2m$ and $0 \leq m_1 \leq 2m - 1$.

Definition 7.2.2. *A continuous function $u : [0,T] \times \overline{\Omega} \to R$ will be called the* ***classical solution*** *of an initial-boundary value problem,*

$$\begin{cases} \dot{u} + Au = f(x, d^{m_0}u) & in \ (0,T] \times \Omega, \\ B_j u = 0 & in \ (0,T] \times \partial\Omega, \\ u(0,x) = u_0(x) & in \ \Omega \end{cases} \tag{7.2.28}$$

$(0 \leq m_0 \leq 2m-1)$, provided that $\dot{u} = \frac{\partial u}{\partial t}$, $\frac{\partial^{|\alpha|}u}{\partial x_1^{\alpha_1}\ldots\partial x_n^{\alpha_n}}$, $0 \leq |\alpha| \leq 2m$, exist and are continuous for $(t,x) \in (0,T] \times \Omega$. Moreover, the spatial derivatives of u of order less than or equal to m_1 are continuous in $(0,T] \times \overline{\Omega}$ and the equation and boundary conditions are fulfilled on prescribed sets as equalities of continuous functions.

The above definition of the classical solution is well suited for the purposes of the (classical) setting of the problem; however, it is inconvenient in further studies. In particular, the known techniques of proving the existence of solutions to (7.2.28) do not operate inside the class of classical solutions. Typically we need to construct Hölder continuous solutions (see [L-S-U], [FR 2]) which are, in particular, classical solutions. Another access is through sufficiently regular solutions to (7.2.28) in Sobolev spaces. The last approach will be exploited below.

If we fix

$$p > n, \ \varepsilon \in \left(\frac{n}{p}, 1\right), \tag{7.2.29}$$

then, as a consequence of Lemma 7.2.1 considered with $r = 1$ and $k = m_0$ (see also Remark 7.2.5), we see that the nonlinear term F corresponding to f in (7.2.28) is Lipschitz continuous on bounded sets as a map between $X^{\frac{1+m_0+\varepsilon}{2m}}$ and $X^{\frac{1}{2m}}$. Of course, our implicit assumption here is that

(7.2.30)

for any $\phi \in X^{\frac{1+m_0+\varepsilon}{2m}}$ the composite function $f(x, d^{m_0}\phi)$ satisfies the boundary conditions required for an element of $X^{\frac{1}{2m}}$.

It is next a consequence of the embedding (1.3.73) that

$$X^{1+\frac{1}{2m}} \subset C^{2m+\mu}(\overline{\Omega}), \ 0 \leq \mu \leq 1 - \frac{n}{p}. \tag{7.2.31}$$

For $u_0 \in X^{\frac{1+m_0+\varepsilon}{2m}}$, the Lipschitz continuity of F between $X^{\frac{1+m_0+\varepsilon}{2m}}$ and $X^{\frac{1}{2m}}$ together with Theorem 7.2.1 allows us to consider the $X^{\frac{1+m_0+\varepsilon}{2m}}$ solution $u = u(\cdot, u_0)$ to (7.2.28) having the following properties:

$$\begin{cases} u \in C([0,\tau), X^{\frac{1+m_0+\varepsilon}{2m}}) \cap C^1((0,\tau), X^{\frac{1}{2m}}) \cap C((0,\tau), X^{1+\frac{1}{2m}}), \\ \dot{u} \in C((0,\tau), X^{\frac{1}{2m}+\gamma}), \ \gamma \in [0,1) \end{cases} \tag{7.2.32}$$

(here τ may depend on u_0). As a result of the embedding (7.2.31) and of (7.2.32), we find that

$$u \in C((0,\tau), C^{2m+\mu}(\overline{\Omega})) \cap C([0,\tau), C^{m_0}(\overline{\Omega})), \tag{7.2.33}$$

which shows, in particular, that u is continuous on $[0,\tau) \times \overline{\Omega}$, the spatial partial derivatives of u up to order $2m$ are continuous in $(0,\tau)\times\overline{\Omega}$ and the initial-boundary conditions in (7.2.28) are satisfied in the classical sense. Also, $\dot{u}$ is a derivative in the classical sense and, moreover,

$$\dot{u} \in C((0,\tau), C^{2m+\nu}(\overline{\Omega})), \ \nu \in \left[0, 1 - \frac{n}{p}\right). \tag{7.2.34}$$

By (7.2.33) and (7.2.34), u possesses all the properties of a classical solution to (7.2.28) on any set $(0,T]\times\overline{\Omega}$ with $T \in (0,\tau)$. We have thus justified the following.

Observation 7.2.1. *Let* (7.2.29) *hold and* $u_0 \in X^{\frac{1+m_0+\varepsilon}{2m}}$. *Whenever the non-linear term* $f : \overline{\Omega} \times R^{d_0} \to R$ *in* (7.2.28) *is continuous and has first order partial derivatives with respect to functional arguments locally Lipschitz continuous uniformly for* $x \in \Omega$ *and* (7.2.30) *is satisfied, the* $X^{\frac{1+m_0+\varepsilon}{2m}}$ *solution to* (7.2.28) *constructed in Theorem* 7.2.1 *is a classical solution of this problem.*

Remark 7.2.7. When $m_0 = 2m-1$ we usually need to take the initial data from the space $X^{\frac{1+m_0+\varepsilon}{2m}} = X^{1+\frac{\varepsilon}{2m}}$ ($\varepsilon \in (\frac{n}{p}, 1)$). However, in the case when $m_0 \leq 2m-2$, Observation 7.2.1 holds also for the initial data $u_0 \in X^{\frac{m_0+\varepsilon}{2m}}$.

Indeed, under both $(X^{\frac{1+m_0+\varepsilon}{2m}}, X^{\frac{1}{2m}})$ and $(X^{\frac{m_0+\varepsilon}{2m}}, X)$ conditions, there exists an $X^{\frac{m_0+\varepsilon}{2m}}$ solution $u = u(\cdot, u_0)$ to (7.2.28) with the properties

$$u \in C([0,\tau), X^{\frac{m_0+\varepsilon}{2m}}) \cap C((0,\tau), X^{1+\frac{1}{2m}}) \cap C^1((0,\tau), X^{\frac{1}{2m}}),$$
$$\dot{u} \in C((0,\tau), X^{\frac{1}{2m}+\gamma}), \ \gamma \in [0,1)$$

(see Corollary 7.2.2). Given the Sobolev embeddings such a u is thus the classical solution to (7.2.28).

Example 7.2.3. To emphasize the requirements of the above method consider a simple Dirichlet problem on a bounded domain $\Omega \subset R^n$:

$$\begin{cases} u_t = \Delta u + f(u), \ t > 0, \ x \in \Omega, \\ u = 0 \ \text{ on } \ \partial\Omega, \ u(0,x) = u_0(x) \ \text{ in } \ \Omega. \end{cases} \tag{7.2.35}$$

Here $\partial\Omega \in C^\infty$, f is assumed to have locally Lipschitz continuous first order derivative and to satisfy the condition $f(0) = 0$. If $u_0 \in X^{\frac{\varepsilon}{2}} = \{\phi \in H^\varepsilon_p(\Omega) : \phi_{|\partial\Omega} = 0\}$, $\varepsilon \in (\frac{n}{p}, 1)$ (see (1.3.65)), then the corresponding local $X^{\frac{\varepsilon}{2}}$ solution to (7.2.35) is, in particular, the classical solution of that problem. Note that both $(X^{\frac{1+\varepsilon}{2}}, X^{\frac{1}{2}})$ and $(X^{\frac{\varepsilon}{2}}, X)$ conditions are satisfied in this example.

Another glance at classical solutions. It is often possible to obtain classical solutions of various special parabolic problems under weaker regularity assumptions on $\partial\Omega$ and the coefficients of the elliptic operator using the following simple procedure. Consider the initial-boundary value problem (7.2.28) and assume that:

(i) The corresponding triple $(A, \{B_j\}, \Omega)$ forms a regular elliptic boundary value problem such that the triple $(A^2, \{B_j, A \circ B_j\}, \Omega)$ also forms a regular elliptic boundary value problem (see Definition 1.2.1).

(ii) The nonlinear term $F : X^{\frac{1}{2m}+\gamma} \to X^{\frac{1}{2m}}$ is Lipschitz continuous on bounded sets for some $\gamma \in (0,1)$.

It is clear that assumption (i) is satisfied when $A = -\Delta$ with homogeneous Dirichlet boundary condition, $\Omega \subset R^n$ a bounded domain with $\partial\Omega \in C^4$. Assumption (i) justifies also Henry's characterization (see Remark 1.3.8) of the fractional order spaces corresponding to the square operator A^2 considered on $L^p(\Omega)$, $p > n$. We denote by $X^{\alpha^{(2)}}$ the α-th order fractional power space corresponding to A^2. Then

$$X^{\alpha^{(2)}} \subset C^k(\overline{\Omega}) \text{ if } 4m\alpha - \frac{n}{p} > k, \ k \in N. \tag{7.2.36}$$

Remark 7.2.8. We have the equality

$$X^{\alpha^{(2)}} = X^{2\alpha}, \ \alpha \geq 0, \tag{7.2.37}$$

where $X^{2\alpha}$ are the spaces corresponding to the operator A on $L^p(\Omega)$, $p > n$. Indeed, formula (1.3.49) (see Proposition 1.3.7 with $\beta = 2$) yields that $(A^2)^\alpha = A^{2\alpha}$ in this case. Therefore, (7.2.37) holds.

Consider now the problem (7.2.28) in the abstract formulation

$$\dot{u} + Au = F(u), \ t > 0, \ u(0) = u_0, \tag{7.2.38}$$

on the phase space $X^{\frac{1}{2m}} = X^{\frac{1}{4m}^{(2)}}$. Based on Proposition 1.3.8 the problem (7.2.38) will be considered as an equation in $X^{\frac{1}{2m}}$ with sectorial operator $A = A_{|_{X^{\frac{1}{2m}}}}$.

As shown in [AM 5, p. 260],

$$D\left(\left(A_{|_{X^{\frac{1}{2m}}}}\right)^\gamma\right) = X^{\frac{1}{2m}+\gamma}.$$

Consequently, by Theorem 2.1.1 (see also Corollary 2.3.1), there exists a unique $X^{\frac{1}{2m}+\gamma}$ solution to that problem such that

$$\begin{cases} u \in C([0,\tau), X^{\frac{1}{2m}+\gamma}) \cap C^1((0,\tau), X^{\frac{1}{2m}}) \cap C((0,\tau), X^{\frac{1}{2m}+1}), \\ \dot{u} \in C((0,\tau), X^{\frac{1}{2m}+\beta}), \ \beta \in [0,1). \end{cases} \tag{7.2.39}$$

If $p > n$ and assumption (ii) holds, it is then a consequence of the inclusion (7.2.36) and Remark 7.2.8 that

(7.2.40)
$$\begin{cases} u \in C([0,\tau), C^{2m\gamma+\mu}(\overline{\Omega})) \cap C((0,\tau), C^{2m+\mu}(\overline{\Omega})), \ \mu \in \left[0, 1 - \frac{n}{p}\right), \\ \dot{u} \in C((0,\tau), C^0(\overline{\Omega})). \end{cases}$$

Therefore, both the space derivatives appearing in the nonlinear term of (7.2.38) and the boundary operators can be understood in the classical sense. Also, the time derivative $\dot{u}$ (strong derivative in $X^{\frac{1}{2m}}$) coincides, because of the inclusion $X^{\frac{1}{2m}} \subset C^0(\overline{\Omega})$, with a pointwise classical time derivative of $u(t,x)$. Thus we deal with the *classical solution* of (7.2.28).

Remark 7.2.9. The necessity of introducing the square operator A^2 is forced by the fact that Henry's estimates (see Remark 1.3.8) are formulated only for fractional power spaces intermediate bewteen X and the domain of the operator considered. It seems however, that the assumption (i) is quite commonly satisfied by elliptic operators.

Remark 7.2.10. In general we remark that, by [KO 1, Theorem 10.5], if A is sectorial and positive then A^α also possesses these properties provided that $0 < \alpha < \frac{\pi}{2\phi}$. As before, ϕ denotes here half of the angle opening of the sector $\mathcal{S}_{0,\phi}$ corresponding to A (see Remark 1.3.3). Further, for $0 < \alpha < \frac{\pi}{\phi}$, the resolvent set $\rho(A^\alpha)$ of the power operator A^α always contains the sector $\mathcal{S}_{0,\alpha\phi}$ and, in particular,

$$\|\lambda(\lambda I + A^\alpha)^{-1}\|_{\mathcal{L}(X,X)} \leq M, \ \lambda > 0$$

(see [KO 1, Proposition 10.2 and Theorem 10.3]). In this latter case the powers of A^α may be defined as in [KO 1, p. 299]. Note also that half of the angle opening $\alpha\phi$ of the sector corresponding to A^α increases with respect to $\alpha \in (0, \frac{\pi}{\phi})$.

Bibliographical notes.

- **Section 7.1.** The proof of Proposition 7.1.1, crucial for the considerations of Section 7.1, comes from [TE 1] and is based on the log-convexity method of [A-N].
- **Section 7.2.** The section originates from [HE 1] and uses the ideas developed further in [RB 1]. Also, mention should be made of the original paper [HA-SC] where smoothness of the solutions on the attractors was studied.
- **Classical solutions.** These were considered in Section 7.2 within the technique of X^α spaces. We thus do not use Calderón-Zygmund type estimates as in [HE 1, Example 3.6, p. 75]. The classical approach to higher regularity of the solutions to parabolic equations in *Hölder spaces* has been presented in [L-S-U]. With the use of the semigroup approach, analyticity of solutions of the linear $2m$-th order parabolic equations has been studied in [FR 1, p. 212]. We refer also to the interesting recent paper [L-S-W] devoted to $2m$-th order linear parabolic problems with time dependent coefficients, and to [WA 5] in which nonlinear problems have been considered.

Iteration formulas for calculating higher order derivatives of the composite functions were strongly exploited in [KL]. If one wants to study within the theory of the present monograph very smooth (or analytic) solutions to semilinear parabolic problems then, as seen in Example 7.2.3, a number of troublesome conditions, like (7.2.23), have to be imposed on the nonlinear term f.

• **Acknowledgement.** We are grateful to Doctor Anibal Rodriguez-Bernal for the discussion concerning $X^{s+\alpha}$ solutions reported in this chapter.

CHAPTER 8

Extensions

8.1. Non-Lipschitz nonlinearities

Some problems originating in applied sciences admit the formulation in a sectorial form (2.1.1) with $F : X^\alpha \to X$ ($\alpha \in [0,1)$) continuous, but nevertheless, F not satisfying the Lipschitz condition. The diffusion equation with strong absorption mentioned in Remark 7.1.1 may serve here as an example. It is thus reasonable to extend the previous considerations to problems of the latter type. We shall do that based on the results of [L-M] concerning *mild solutions* to (2.1.1). For the purpose of brevity we here stay away from the generality of [L-M], [MAT], focusing on some consequences of their approach. The results mentioned here may be generalized in several directions.

Existence result. In the following, consider the problem (2.1.1),

$$\begin{cases} \dot{u} + Au = F(u), \ t > 0, \\ u(0) = u_0, \end{cases}$$

under the usual assumptions that A is a sectorial, positive operator in a Banach space X and the resolvent of A is compact.

Definition 8.1.1. *Let $\alpha \in (0,1)$ and $\tau > 0$. A function $u \in C([0,\tau), X) \cap C((0,\tau), X^\alpha)$ is called a* **local mild X^α solution of** *(2.1.1)* **through** *$u_0 \in X$ if u fulfills in X the Cauchy integral formula*

$$u(t) = e^{-At}u_0 + \int_0^t e^{-A(t-s)}F(u(s))ds, \quad for \ \ t \in [0,\tau).$$

As a consequence of the general theorem of [L-M] the following result holds.

Proposition 8.1.1. *Let $\alpha \in (0,1)$. If $F : X^\alpha \to X$ is a continuous map taking bounded subsets of X^α into bounded subsets of X, then for each $u_0 \in X^\alpha$ there exists at least one local mild X^α solution u to (2.1.1) and $u(t,u_0) \to u_0$ in X^α when $t \to 0^+$.*

Proof. Recall that, since the resolvent of A is compact, $e^{-At} : X \to X$ is a compact map for each $t > 0$ (see Remark 3.3.1). Therefore, the assumptions $(C1)$-

($C5$) of [L-M, Theorem 1] hold where, in the notation of [L-M], one should specify $D = X$, $L = A$, $B(t, v) \equiv F(v)$, and $T(t) \equiv e^{-At}$. This completes the proof. □

If u fulfill the requirements of the above definition with $\tau = +\infty$, then u is called a *global mild X^α solution* to (2.1.1). Of course, global solutions will exist if (3.0.1) holds:

$$\|F(v)\|_X \leq const.(1 + \|v\|_{X^\alpha}), \quad \text{for} \quad v \in X^\alpha.$$

More precisely:

Proposition 8.1.2. *Let $\alpha \in (0,1)$ and $F : X^\alpha \to X$ be a continuous function satisfying the restriction* (3.0.1). *Then to each $u_0 \in X$ corresponds at least one global mild X^α solution to* (2.1.1). *Furthermore, the set $\{u(t, u_0) : u_0 \in B\}$ is precompact in X for each bounded set $B \subset X$ and each $t > 0$.*

Proof. The assertion of the proposition follows from [L-M, Theorem 2] (see also [L-M, Remark 2]). □

A glance at the stability. Let us strengthen (3.0.1) to the condition

$$\|F(v)\|_X \leq const.(1 + \|v\|^\theta_{X^\alpha}), \quad \text{for} \quad v \in X^\alpha, \tag{8.1.1}$$

where $\theta \in [0, 1)$ is fixed from now on and *const.* is independent of v. We then obtain the following conclusion.

Corollary 8.1.1. *Let $\alpha \in (0,1)$ and $F : X^\alpha \to X$ be a continuous function satisfying* (8.1.1). *Let, in addition, each mild solution to* (2.1.1) *resulting from Proposition* 8.1.2 *be also unique for $u_0 \in V$, where $V \subset X$ is closed and positively invariant. Then the problem* (2.1.1) *generates a C^0 semigroup $T(t) : V \to V$, $t \geq 0$, of global mild X^α solutions which has a global attractor in V.*

Proof. By our assumptions the existence of a compact C^0 semigroup $\{T(t)\}$ on V (in the sense of Definition 1.1.1) is straightforward (see [L-M, Remark 2]). Continuity of this semigroup with respect to u_0 (as mentioned in [L-M, Remark 2]) is a consequence of the uniqueness of solutions. Thus, to finish the proof, it suffices to justify that $\{T(t)\}$ is point dissipative. For $u_0 \in V$, Cauchy's formula reads

$$T(t)T(1)u_0 = e^{-At}T(1)u_0 + \int_0^t e^{-A(t-s)}F(T(s)T(1)u_0)ds, \quad \text{for} \quad t > 0.$$

Similar calculations as in Theorems 3.1.1 and 4.1.1 show that each orbit enters and remains inside a certain fixed ball in X^α. In particular $\{T(t)\}$ is point dissipative in V, which completes the proof. □

Remarks on applications. We first describe a large class of problems for which Proposition 8.1.1 is applicable. There will be the initial boundary value problem (5.1.1),(5.1.4) of Chapter 5 with the function $f : \overline{\Omega} \times R^{d_0} \to R$ being merely continuous. As shown in Chapter 5 equation (5.1.1) admits the abstract formulation (2.1.1) with A sectorial, positive in $X = L^p(\Omega)$ ($p \in (1, +\infty)$) and such that the resolvent of A is compact.

Let $\alpha \in [0,1)$ and $2m\alpha - \frac{n}{p} > m_0$. For such values of parameters $X^\alpha \subset C^{m_0}(\overline{\Omega})$, so that the assumptions of Proposition 8.1.1 hold, leading to the existence of local mild X^α solutions to (5.1.1), (5.1.4).

Example 8.1.1. Let us now come back to the equation of type (7.1.2) with strong absorption term. Consider the generalized problem

$$
\begin{cases} u_t = \Delta u - \lambda |u|^\theta, \ \lambda > 0, \ \theta \in (0,1), \\ u = 0 \ \text{ on } \ \partial\Omega, \\ u(0,x) = u_0(x) \ \text{ for } \ x \in \Omega. \end{cases} \tag{8.1.2}
$$

Choosing $\alpha \in [0,1)$, $p \in [2,+\infty)$ such that $2\alpha > \frac{n}{p}$ we may apply Proposition 8.1.2 to obtain the existence of the global mild X^α solution $u(\cdot,u_0)$ for each $u_0 \in L^p(\Omega)$. As follows from the proof of [L-M, Theorem 2], such a solution is constructed as a limit of a sequence $u_{z_n} = u(\cdot,z_n)$ of the global X^α solutions, when $\|z_n - u_0\|_{L^p(\Omega)} \to 0$ and $\{z_n\} \subset X^\alpha$.

Define $D^+ = \{\phi \in C^{2+\epsilon}(\overline{\Omega}) : \phi_{|\partial\Omega} = 0, \Delta\phi_{|\partial\Omega} = 0, \phi \geq 0\}$ and assume that $\partial\Omega$ is of class $C^{2+\epsilon}$. It was shown in [DL 6, Theorem 1] that, for $z \in D^+$, there exists a unique positive Hölder solution $v_z = v_z(t,x)$ to (8.1.2); i.e. a unique $v_z \geq 0$ satisfying (8.1.2) in a classical sense and such that $v_z \in C^{1+\frac{\epsilon}{2},2+\epsilon}([0,\tau]\times\overline{\Omega})$ for each $\tau > 0$. The latter implies that v_z may be considered as an element of $C^1(0,+\infty),X) \cap C([0,+\infty),X^\alpha)$. In particular, v_z is a global mild X^α solution to (8.1.2).

Consider $V = cl_{L^p(\Omega)}D^+$ and choose $u_0 \in V$. Based on the proof of [L-M, Theorem 2] a global mild X^α solution through u_0 may be thus obtained as the limit of the sequence $\{v_{z_n}(t,\cdot)\}$ of Hölder solutions corresponding to $z_n \in D^+$ when $z_n \to u_0$ in $L^p(\Omega)$. It follows from (8.1.2) that for any $z_n, w_n \in D^+$ and the corresponding nonnegative Hölder solutions v_{z_n}, v_{w_n}

$$
(v_{z_n} - v_{w_n})_t = \Delta(v_{z_n} - v_{w_n}) - \lambda(v_{z_n}^\theta - v_{w_n}^\theta).
$$

Multiplying the above equation in $L^2(\Omega)$ by $(v_{z_n} - v_{w_n})$, integrating by parts and using the equality

$$
sgn(v_{z_n}^\theta - v_{w_n}^\theta) = sgn(v_{z_n} - v_{w_n}),
$$

we obtain an estimate

$$
\|v_{z_n}(t,\cdot) - v_{w_n}(t,\cdot)\|_{L^2(\Omega)} \leq \|z_n - w_n\|_{L^2(\Omega)}, \ z_n, w_n \in D^+, \ t \geq 0.
$$

The last estimate guarantees uniqueness of the limit solution $v(\cdot,u_0)$ corresponding to $u_0 \in V$. We have thus justified that (8.1.2) generates on V a C^0 semigroup of global mild X^α solutions. From Corollary 8.1.1, this semigroup possesses a global attractor $\mathcal{A}$ in V. Here we have

$$
\|F(v)\|_{L^p(\Omega)} = \||v|^\theta\|_{L^p(\Omega)} = \|v\|^\theta_{L^{p\theta}(\Omega)}, \ v \in L^p(\Omega),
$$

so that (8.1.1) holds as a result of the Sobolev inclusions $X^\alpha \subset L^p(\Omega) \subset L^{p\theta}(\Omega)$, $\theta \in (0,1)$, $\alpha \in [0,1)$.

It is worth noting that $V = cl_{L^p(\Omega)}D^+$ is the cone of nonnegative elements of $L^p(\Omega)$. We also remark that the attractor $\mathcal{A}$ is trivial and the *extinction phenomenon* for (8.1.2) concerns orbits of bounded sets, that is, bounded subsets of V are absorbed by $\{0\}$ in a finite time (see [DL 6, Theorem 2] for details).

Remark 8.1.1. Example 8.1.1 can be treated within the theory presented in [PAZ 1]. In particular, [PAZ 1, Theorem 5.2] guarantees that the mild solutions constructed above are actually solutions of the abstract sectorial differential equation corresponding to (8.1.2).

8.2. Application of the principle of linearized stability

When the nonlinear term in the equation grows too fast there is no hope, in general, that the global solutions exist and hence one can hardly expect to construct a global attractor. However, just for fast growing nonlinearities it may happen that there will be a subset of the phase space producing solutions global in time which, moreover, will approach asymptotically an equilibrium point or, more generally, a local attractor.

Following the presentation of [HE 1, Chapter 5] we shall formulate a theorem concerning *local stability of equilibrium points.* The example for this result will be the celebrated n-dimensional Navier-Stokes equation.

We start with some classical definitions from the *Lyapunov stability theory.*

Let A be a sectorial operator in a Banach space X and $\mathcal{U}$ be an open subset of X^α for some $\alpha \in [0,1)$. For $f : \mathcal{U} \to X$ consider the equation

$$\dot{u} + Au = f(u), \; t > 0. \tag{8.2.1}$$

Definition 8.2.1. *An element $u_0 \in \mathcal{U} \cap D(A)$ will be called an* **equilibrium point** *of equation* (8.2.1) *if*

$$Au_0 = f(u_0). \tag{8.2.2}$$

Remembering the general notions of stability of invariant sets introduced in Definition 1.1.2, we shall formulate explicitly the definition of *stability of an equilibrium point.* We shall refer partially to [HE 1, Chapter 5], although the considerations below are simpler since equation (8.2.1) is autonomous (that is, f is independent of t).

Definition 8.2.2. *Assume that there exists an open neighborhood $\mathcal{U}$ of the equilibrium point u_0 in X^α such that the solution of* (8.2.1) *originating in $\mathcal{U}$ exists globally for $t \geq 0$. We say that an equilibrium point u_0 is* **stable** *in X^α, if*

$$\forall_{\varepsilon>0}\, \exists_{\delta>0}\, \|u(0) - u_0\|_{X^\alpha} < \delta \Longrightarrow \forall_{t\geq 0}\, \|u(t, u(0)) - u_0\|_{X^\alpha} < \varepsilon, \tag{8.2.3}$$

where $u(t, u(0))$ denotes a global X^α solution of (8.2.1) *with the initial data $u(0)$ (i.e. when the one point set $\{u_0\}$ is stable in the sense of Definition* 1.1.2*).*

Definition 8.2.3. *An equilibrium point u_0 is* ***uniformly asymptotically stable*** *if it is stable and*

$$\lim_{t\to+\infty} \|u(t,u(0)) - u_0\|_{X^\alpha} \to 0 \quad as \quad t \to +\infty, \tag{8.2.4}$$

uniformly for any $u(0)$ with $\|u(0)-u_0\|_{X^\alpha} \le \delta_0$ for some fixed $\delta_0 > 0$ (that is when u_0 attracts some open neighborhood of itself; see Definition 1.1.2).

The proposition below concerning stability of equilibrium point of equation (8.2.1) is a version of [HE 1, Theorem 5.1.1].

Proposition 8.2.1. *Let A be a sectorial operator in a Banach space X and let, for fixed $\alpha \in [0,1)$, the function $f : \mathcal{U} \to X$ be Lipschitz continuous on bounded subsets of an open set $\mathcal{U} \subset X^\alpha$. Suppose further that $u_0 \in \mathcal{U}$ is an equilibrium point of (8.2.1) and*

$$f(u_0 + v) = f(u_0) + Bv + g(v), \tag{8.2.5}$$

where B is a bounded linear map from X^α to X and g fulfills the estimate

$$\|g(v)\|_X \le C\|v\|_{X^\alpha}^{1+\delta}, \tag{8.2.6}$$

with some positive constants C and δ.

Under the above assumptions $A - B$ is sectorial in X and, if $Re\,\sigma(A-B) > a$ for some $a > 0$, then u_0 is uniformly asymptotically stable in X^α. More precisely, there exists constant ρ, such that any solution $u(t,u(0))$ of (8.2.1) corresponding to initial data $u(0) \in B_{X^\alpha}(u_0,\rho)$ exists globally in time and, fulfills the estimate

$$\|u(t,u(0)) - u_0\|_{X^\alpha} \le 2M\rho e^{-a't} \tag{8.2.7}$$

with any $a' \in (0,a)$ and M as in condition (8.2.8).

Remark 8.2.1. As will be seen from the proof of Proposition 8.2.1 the assumption (8.2.6) may be weakened to the form

$$\frac{\|g(v)\|_X}{\|v\|_{X^\alpha}} \to 0 \quad \text{as} \quad \|v\|_{X^\alpha} \to 0.$$

In that case, however, the estimates obtained throughout the proof will not have so explicit a form.

Proof. To show that the operator $\mathcal{L} := A - B$ is sectorial we shall use Proposition 1.3.2. By our assumptions

$$\|Bv\|_X \le const.\|v\|_{X^\alpha}, \; v \in X^\alpha.$$

Using the *moments inequality* (1.3.60) and the Young inequality, for any $\varepsilon > 0$, we get further

$$\begin{aligned}\|Bv\|_X &\le const.'\|Av\|_X^\alpha \|v\|_X^{1-\alpha} \\ &\le \varepsilon\|Av\|_X + C_\varepsilon\|v\|_X, \; v \in D(A).\end{aligned}$$

In consequence, all the assumptions of Proposition 1.3.2 are satisfied, guaranteeing sectoriality of $\mathcal{L}$ in X.

Next, since the spectrum of $\mathcal{L}$ is contained in a half plane $\{Re\,\lambda > a\}$, for a certain constant $M \geq 1$, we have the estimates (see [HE 1, p. 99])

$$\begin{cases} \|e^{-\mathcal{L}t}v\|_{X^\alpha} \leq Me^{-at}\|v\|_{X^\alpha}, & t \geq 0,\ v \in X^\alpha, \\ \|e^{-\mathcal{L}t}v\|_{X^\alpha} \leq M\frac{e^{-at}}{t^\alpha}\|v\|_X, & t > 0,\ v \in X. \end{cases} \tag{8.2.8}$$

We should also note here the equality (see [HE 1, p. 29])

$$D(\mathcal{L}^\alpha) = D\big((A-B)^\alpha\big) = D(A^\alpha) = X^\alpha,\ \alpha \in (0,1).$$

Assume that $B_{X^\alpha}(u_0, \rho_0) \subset \mathcal{U}$ and write

$$\Gamma := \frac{\Gamma(1-\alpha)}{a^{1-\alpha}} = \int_0^\infty \frac{e^{-ar}}{r^\alpha}dr,\ K_0 := \min\{\rho_0, (CM\Gamma)^{-\frac{1}{\delta}}\}.$$

We shall first show that the equilibrium u_0 is *stable*, that is, for any $K \in (0, K_0)$ there is $\rho(K) > 0$ such that any solution $u(t, u(0))$ of (8.2.1) with

$$u(0) \in B_{X^\alpha}(u_0, \rho(K)) \tag{8.2.9}$$

exists globally for $t \geq 0$ and satisfies the estimate

$$\|u(t, u(0)) - u_0\|_{X^\alpha} < K.$$

Fix $K < K_0$ and set

$$\rho = \rho(K) < \frac{K}{M}(1 - CM\Gamma K^\delta). \tag{8.2.10}$$

Since u_0 is an equilibrium, we have

$$Au_0 = f(u_0) \text{ and } u_0 \in D(A).$$

Let $u(t, u(0))$ be a local solution to (8.2.1) corresponding to $u(0) \in B_{X^\alpha}(u_0, \rho)$. The difference $v(t) = u(t, u(0)) - u_0$ is thus a local X^α solution of the problem

$$\begin{cases} \dot{v} + \mathcal{L}v = g(v), \\ v(0) = u(0) - u_0. \end{cases} \tag{8.2.11}$$

We need to show that $\|v(t)\|_{X^\alpha} < K$ for $t \geq 0$.

Since $M \geq 1$, we have $\rho < K$ by (8.2.10). Therefore, by continuity, the local solution v originating in $B_{X^\alpha}(u_0, \rho)$ fulfills

$$\|v(t)\|_{X^\alpha} < K,\ t \in [0, \tau],$$

for some $\tau > 0$ (see the proof of Proposition 2.3.1). Using (8.2.6), (8.2.8), (8.2.11) we get

(8.2.12)

$$\begin{aligned}\|v(t)\|_{X^\alpha} &= \|e^{-\mathcal{L}t}(u(0)-u_0) + \int_0^t e^{-\mathcal{L}(t-s)}g(v(s))ds\|_{X^\alpha} \\ &\leq Me^{-at}\|u(0)-u_0\|_{X^\alpha} + \int_0^t M\frac{e^{-a(t-s)}}{(t-s)^\alpha}\|g(v(s))\|_X ds \\ &\leq M\|u(0)-u_0\|_{X^\alpha} + MC\int_0^t \frac{e^{-a(t-s)}}{(t-s)^\alpha}\|v(s)\|_{X^\alpha}^{1+\delta}ds.\end{aligned}$$

Now, as long as $\|v(s)\|_{X^\alpha} \leq K$, we have

$$\|v(t)\|_{X^\alpha} \leq M\rho + MCK^\delta\Gamma \sup_{s\in[0,t]}\|v(s)\|_{X^\alpha}.$$

Hence, writing $y(t) := \sup_{s\in[0,t]}\|v(s)\|_{X^\alpha}$ and using the definition of K_0 and (8.2.10), we find that

$$y(t)(1-MCK^\delta\Gamma) \leq M\rho \tag{8.2.13}$$

or

$$y(t) \leq \frac{M\rho}{1-MC\Gamma K^\delta} < \frac{K(1-MC\Gamma K^\delta)}{1-MC\Gamma K^\delta} = K.$$

The estimate (8.2.13) shows that any local solution $v(t)$ satisfying (8.2.9) varies *inside* the ball $B_{X^\alpha}(u_0, K)$, never reaching its boundary. Hence, according to Theorem 2.1.1, such a solution exists globally in time and u_0 is a stable equilibrium.

We shall prove further the *uniform asymptotic stability* of the equilibrium u_0.

Fix any number $a' \in (0, a)$ and consider a constant $K \in (0, K_0)$ satisfying an additional restriction:

$$K^\delta < \frac{1}{2}\left(CM\int_0^\infty \frac{e^{-(a-a')r}}{r^\alpha}dr\right)^{-1} = \frac{1}{2}\left(CM\frac{\Gamma(1-\alpha)}{(a-a')^{1-\alpha}}\right)^{-1}. \tag{8.2.14}$$

Let $\rho = \rho(K)$ be given by (8.2.10) and take initial data $v(0)$ fulfilling

$$\|v(0)\|_{X^\alpha} = \|u(0)-u_0\|_{X^\alpha} \leq \rho(K). \tag{8.2.15}$$

As follows from the previous considerations

$$\|v(t)\|_{X^\alpha} < K \quad \text{for} \quad t \geq 0,$$

so that extending (8.2.12) we find

(8.2.16)

$$\begin{aligned}\|v(t)\|_{X^\alpha} &\leq M\rho e^{-at} + MC\int_0^t \frac{e^{-a(t-s)}}{(t-s)^\alpha}\|v(s)\|_{X^\alpha}^{1+\delta}ds \\ &\leq M\rho e^{-at} + MCK^\delta\int_0^t \frac{e^{-a(t-s)}}{(t-s)^\alpha}\|v(s)\|_{X^\alpha}ds.\end{aligned}$$

Multiplying (8.2.16) by $e^{a't}$ and substituting $z(t) := \|v(t)\|_{X^\alpha} e^{a't}$, we obtain

$$
\begin{aligned}
(8.2.17) \qquad z(t) &\le M\rho e^{-(a-a')t} + MCK^\delta \int_0^t \frac{e^{-(a-a')(t-s)}}{(t-s)^\alpha} z(s)ds \\
&\le M\rho + MCK^\delta \sup_{s\in[0,t]} z(s) \int_0^\infty \frac{e^{-(a-a')r}}{r^\alpha} dr.
\end{aligned}
$$

Since the right hand side of (8.2.17) dominates $z(s)$ for any $s \in [0,t]$, with the use of (8.2.14) we get

$$
(8.2.18) \qquad \sup_{s\in[0,t]} z(s) \le M\rho + \frac{1}{2} \sup_{s\in[0,t]} z(s).
$$

The last estimate implies next that

$$
\sup_{s\in[0,t]} \left(\|v(s)\|_{X^\alpha} e^{a's} \right) \le 2M\rho
$$

and, consequently,

$$
(8.2.19) \qquad \|u(t,u(0)) - u_0\|_{X^\alpha} = \|v(t)\|_{X^\alpha} \le 2M\rho e^{-a't}, \; t \ge 0.
$$

The proof is complete. □

8.3. The n-dimensional Navier-Stokes system

We shall aplly the preceding stability result to examine the stationary solutions of the Navier-Stokes system in dimension $n \ge 3$.

Formulation of the problem. Notation. We deal with the n-dimensional Navier-Stokes equation

$$
(8.3.1) \qquad u_t = \nu\Delta u - \nabla p - (u,\nabla)u + h, \quad div\, u = 0, \;\text{ for }\; t > 0,\; x \in \Omega,
$$

where $n \ge 3$, $\nu > 0$ is a constant viscosity, $u = \big(u_1(t,x), \dots, u_n(t,x)\big)$ denotes velocity, $p = p(t,x)$ pressure and $h = \big(h_1(x), \dots, h_n(x)\big)$ external force. Here Ω is a bounded subdomain of R^n with boundary $\partial\Omega$ of class $C^{2+\rho}$, $\rho \in (0,1)$, whereas $(\cdot,\cdot)$ stands for the standard inner product in R^n.

Equation (8.3.1) is studied with a boundary condition of Dirichlet type

$$
(8.3.2) \qquad u = 0, \; t > 0, \; x \in \partial\Omega,
$$

and subject to an initial condition

$$
(8.3.3) \qquad u(0,x) = u_0(x), \;\text{ for }\; x \in \Omega.
$$

For simplicity of the notation let us introduce the following list of vector function spaces:

$$\mathcal{L}^r(\Omega) := [L^r(\Omega)]^n,$$
$$\mathcal{W}^{2,r}(\Omega) := [W^{2,r}(\Omega)]^n,$$
$$X_r := cl_{\mathcal{L}^r(\Omega)}\{\phi \in [C_0^\infty(\Omega)]^n : div\,\phi = 0\},$$
$$\mathcal{C}^\mu(\overline{\Omega}) := [C^\mu(\overline{\Omega})]^n,$$

and define in X_r an unbounded product operator A_r by the formula

$$A_r = -\nu P_r(\Delta, \dots, \Delta)_{n\times}. \tag{8.3.4}$$

Here P_r is the linear, continuous projection from $\mathcal{L}^r(\Omega)$ to X_r which is given by the decomposition of $\mathcal{L}^r(\Omega)$ onto the spaces of divergence free vector fields and scalar function gradients (see [FU-MO, Theorem 2]) so that

$$X_r = P_r\mathcal{L}^r(\Omega),\ \mathcal{L}^r(\Omega) = X_r \oplus \{\nabla\phi : \phi \in \mathcal{W}^{1,r}(\Omega)\}\ \text{(direct sum)}$$

and (see [FU-MO, p. 694])

$$\begin{cases} P_r v = v \ \text{ for } v \in X_r, \\ \|P_r v\|_{\mathcal{L}^r(\Omega)} \le const. \|v\|_{\mathcal{L}^r(\Omega)}, \quad v \in \mathcal{L}^r(\Omega). \end{cases} \tag{8.3.5}$$

Generically, P_r is thus a counterpart of the orthogonal projection in $\mathcal{L}^2(\Omega)$ (see [KA]).

From [GI-MI, Lemma 1.1] we know that:

Proposition 8.3.1. *The operator $-A_r$ considered on the domain*

$$D(A_r) := X_r \cap \{\phi \in \mathcal{W}^{2,r}(\Omega) : \phi|_{\partial\Omega} = 0\}, \tag{8.3.6}$$

generates an analytic semigroup $\{e^{-tA_r}\}$ in X_r $(1 < r < \infty)$.

It is well known (see [S-W, p. 31]) that $Re\,\sigma(A_r) > a > 0$. Therefore, fractional powers A_r^α ($\alpha \in [0,1]$) of A_r may be defined on the domains $X_r^\alpha := D(A_r^\alpha)$ and for each $r \in (1,\infty)$, $\alpha \in [0,1]$ (where we write $A^0 := I$),

$$\|A_r^\alpha e^{-tA_r}\|_{\mathcal{L}(X_r,X_r)} \le C_{\alpha,r} t^{-\alpha} e^{-at},\ t > 0. \tag{8.3.7}$$

Since, from [S-W, p. 29] (see also [GI-MI, Lemma 3.1]),

$$\|u\|_{\mathcal{W}^{2,r}(\Omega)} \le c(r)\|A_r u\|_{\mathcal{L}^r(\Omega)},\ r > 1, u \in D(A_r),$$

the operator $(\lambda I - A_r)^{-1}$ is compact for $\lambda = 0$ as a result of the Rellich-Kondrachov theorem (see the conditions below formula (1.2.36)). Therefore the resolvent of A_r is compact (see (1.2.1)) and, consequently, also compact are the embeddings

$$X_r^\beta \subset X_r^\alpha,\ 0 \le \alpha < \beta,\ 1 < r < \infty$$

(see Proposition 1.3.5). As follows further from [GI 2, Theorem 1], imaginary powers A_r^{it} are bounded in the $\mathcal{L}(X_r, X_r)$ norm uniformly for $t \in [-\varepsilon, \varepsilon]$ so that the interpolation formula of Corollary 1.3.4 reads (see [GI 2, Theorem 2])

$$X_r^\alpha = [X_r, D(A_r)]_\alpha, \ \alpha \in (0,1).$$

The above interpolation result allows us further to characterize X_r^α ($\alpha \in [0,1]$) spaces by the equality (see [GI 2, Theorem 3])

$$X_r^\alpha = D(B_r^\alpha) \cap X_r \tag{8.3.8}$$

where $B_r = -\Delta$ with $D(B_r) := \{\phi \in \mathcal{W}^{2,r}(\Omega) : \phi|_{\partial\Omega} = 0\}$ from which, using Proposition 1.3.10, we get the embeddings

(8.3.9)
$$X_r^\alpha \subset \begin{cases} \mathcal{W}^{t,r}(\Omega) \cap X_r, & 2\alpha \geq t, \ r \in (2, +\infty), \\ \mathcal{C}^\mu(\overline{\Omega}) \cap X_r, & 2\alpha - \frac{n}{r} \geq k + \mu, \ k \in N, \ \mu \in (0,1), \ r \in (2,+\infty). \end{cases}$$

Here the spaces $\mathcal{W}^{t,r}(\Omega) \cap X_r$ and $\mathcal{C}^\mu(\overline{\Omega}) \cap X_r$ are considered with the $\mathcal{W}^{t,r}(\Omega)$ and $\mathcal{C}^\mu(\overline{\Omega})$ topologies, respectively. Note that the second inclusion holds also for $\mu = 0$, provided that strict inequality $2\alpha - \frac{n}{r} > k$ is satisfied.

Local solvability of (8.3.1)-(8.3.3). For any $h \in \mathcal{L}^r(\Omega)$ the system (8.3.1)-(8.3.3) may thus be studied as an abstract Cauchy problem in X_r:

$$u_t + A_r u = F_r u + P_r h, \ t > 0, \ u_{|t=0} = u(0), \tag{8.3.10}$$

where A_r considered with the domain (8.3.6) is sectorial in X_r and $F_r u = -P_r(u, \nabla)u$. Moreover, for $\alpha \in [\frac{1}{2}, 1)$ and $r > n$ the nonlinear term F_r, acting from X_r^α into X_r, is Lipschitz continuous on bounded sets. Indeed, the estimate [GI-MI, Lemma 3.3, (iii)] reads

$$\begin{aligned} &\|P_r(w, \nabla)v\|_{\mathcal{L}^r(\Omega)} \\ &\leq c_r \|w\|_{\mathcal{W}^{1,r}(\Omega)} \|v\|_{\mathcal{W}^{1,r}(\Omega)}, \ w, v \in \mathcal{W}^{1,r}(\Omega), \ r > n. \end{aligned} \tag{8.3.11}$$

Therefore, when $\phi, \psi \in \mathcal{U}$ and $\mathcal{U}$ is bounded in X_r^α, we have

(8.3.12)
$$\begin{aligned} \|F_r\phi - F_r\psi\|_{X_r} &\leq c_r \|\phi\|_{\mathcal{W}^{1,r}(\Omega)} \|\phi - \psi\|_{\mathcal{W}^{1,r}(\Omega)} + c_r \|\phi - \psi\|_{\mathcal{W}^{1,r}(\Omega)} \|\psi\|_{\mathcal{W}^{1,r}(\Omega)} \\ &\leq c_r \max\{\|\phi\|_{\mathcal{W}^{1,r}(\Omega)}, \|\psi\|_{\mathcal{W}^{1,r}(\Omega)}\} \ \|\phi - \psi\|_{\mathcal{W}^{1,r}(\Omega)} \\ &\leq c_{r,\mathcal{U}} \|\phi - \psi\|_{\mathcal{W}^{1,r}(\Omega)}. \end{aligned}$$

Note that $X_r^\alpha \subset X_r^{\frac{1}{2}}$ for $\alpha \in [\frac{1}{2}, 1)$, whereas $X_r^{\frac{1}{2}}$ is continuously embedded in $X_r \cap \mathcal{W}^{1,r}(\Omega)$ (see (8.3.9)).

Stability of equlibrium. We are now ready to apply Proposition 8.2.1 to the Navier-Stokes system. We have

Lemma 8.3.1. *For any $r > n$ and $\alpha \in [\frac{1}{2}, 1)$ the equilibrium solution u_0 of* (8.3.1)-(8.3.3) *is asymptotically stable in X_r^α provided that $\|u_0\|_{X_r^\alpha}$ is restricted by* (8.3.18) *below.*

Proof. We need to justify applicability of Proposition 8.2.1. Fix r and α as prescribed. Defining

$$f(v) := F_r(v) + P_r h,\ v \in X_r^\alpha,$$

we then obtain easily that

$$f(u_0 + v) - f(u_0) = B_r v + g(v),$$

where

(8.3.13)
$$\begin{cases} g(v) = F_r(v), \\ B_r v = -P_r\left(\sum_{i=1}^n \left(\frac{\partial u_0^1}{\partial x_i} v^i + \frac{\partial v^1}{\partial x_i} u_0^i\right), \dots, \sum_{i=1}^n \left(\frac{\partial u_0^n}{\partial x_i} v^i + \frac{\partial v^n}{\partial x_i} u_0^i\right)\right). \end{cases}$$

Therefore the decomposition (8.2.5) is justified. Next, as a consequence of (8.3.11), we have

$$\|g(v)\|_{X_r} = \|F_r(v)\|_{X_r} \le C_r \|v\|^2_{X_r^{\frac{1}{2}}} \le C_r' \|v\|^2_{X_r^\alpha},\ v \in X_r^\alpha, \tag{8.3.14}$$

with $C_r' = C_r'(\alpha)$, so that the condition (8.2.6) is satisfied with $\delta = 1$. The Lipschitz continuity of the nonlinear term f on bounded subsets of X_r^α has been already discussed in (8.3.11)-(8.3.12). Hence, we need only make precise conditions for the equilibrium u_0 ensuring that $Re\big(\sigma(A_r - B_r)\big) > a$ for some $a > 0$.

Let us first observe that

$$\|B_r v\|_{X_r} \le const. \|u_0\|_{X_r^\alpha} \|A_r v\|_{X_r},\ v \in D(A_r). \tag{8.3.15}$$

Indeed, from (8.3.13), (8.3.5) and the embeddings $X_r^\alpha \subset C^\mu(\overline{\Omega})$ ($r > n$, $\alpha \in [\frac{1}{2}, 1)$), $X_r^\alpha \subset W^{1,r}(\Omega)$ (see (8.3.9)) we have the estimates (see [GI-MI, p. 277])

$$\begin{aligned} \|B_r v\|_{X_r} &= \left\| -P_r\left(\sum_{i=1}^n \left(\frac{\partial u_0^1}{\partial x_i} v^i + \frac{\partial v^1}{\partial x_i} u_0^i\right), \dots, \sum_{i=1}^n \left(\frac{\partial u_0^n}{\partial x_i} v^i + \frac{\partial v^n}{\partial x_i} u_0^i\right)\right)\right\|_{L^r(\Omega)} \\ &\le const. \left\| \left(\sum_{i=1}^n \left(\frac{\partial u_0^1}{\partial x_i} v^i + \frac{\partial v^1}{\partial x_i} u_0^i\right), \dots, \sum_{i=1}^n \left(\frac{\partial u_0^n}{\partial x_i} v^i + \frac{\partial v^n}{\partial x_i} u_0^i\right)\right)\right\|_{L^r(\Omega)} \\ &\le const.' \sum_{i=1}^n \left(\|v\|_{X_r^\alpha} \|\nabla u_0^i\|_{L^r(\Omega)} + \|u_0\|_{X_r^\alpha} \|\nabla v^i\|_{L^r(\Omega)} \right) \\ &\le const.'' \|u_0\|_{X_r^\alpha} \|v\|_{X_r^\alpha}. \end{aligned}$$

Let us recall now the following result concerning the resolvent of A_r (see [S-W, p. 31]):

$$\begin{cases} \{Re\,\lambda \le 0\} \subset \rho(A_r), \\ \|\lambda(\lambda I - A_r)^{-1}\|_{\mathcal{L}(X_r, X_r)} \le M_r, \quad \text{for each } Re\,\lambda \le 0. \end{cases} \tag{8.3.16}$$

Using (8.3.15), (8.3.16) we obtain further

(8.3.17)
$$\begin{aligned}
&\|B_r(\lambda I - A_r)^{-1}v\|_{X_r} \leq const.\|u_0\|_{X_r^\alpha}\|A_r(\lambda I - A_r)^{-1}v\|_{X_r} \\
&\leq const.\|u_0\|_{X_r^\alpha}\|(\lambda I - A_r)(\lambda I - A_r)^{-1}v\|_{X_r} + \|\lambda(\lambda I - A_r)^{-1}v\|_{X_r} \\
&\leq const.\|u_0\|_{X_r^\alpha}(1 + M_r)\|v\|_{X_r}, \ v \in X_r, \ Re\,\lambda \leq 0.
\end{aligned}$$

Condition (8.3.17) ensures that, for each $Re\,\lambda \leq 0$, $1 \in \rho\Big(B_r(\lambda I - A_r)^{-1}\Big)$ provided that

$$const.(1 + M_r)\|u_0\|_{X_r^\alpha} < 1. \tag{8.3.18}$$

The equality

$$[\lambda I - (A_r + B_r)]^{-1} = (\lambda I - A_r)^{-1}\left(I - B_r(\lambda I - A_r)^{-1}\right)^{-1}, \ Re\,\lambda \leq 0,$$

shows finally (see (1.3.28), (1.3.29)) that $A_r - B_r$ is sectorial in X_r and $\{Re\,\lambda \leq 0\} \subset \rho(A_r - B_r)$, which completes the proof of Lemma 8.3.1. □

Stationary solutions of the Navier-Stokes equation. Let Ω be a bounded domain in R^n, $n \geq 3$, with $\partial\Omega \in C^{2+\rho}$. We are looking for a vector function $v : \Omega \to R^n$, $v = (v_1, \dots, v_n)$, and a scalar function $p : \Omega \to R$ satisfying the problem

$$\begin{cases} -\nu\Delta v_k + \sum_{i=1}^n v_i \frac{\partial v_k}{\partial x_i} + \frac{\partial p}{\partial x_k} = h_k, \ k = 1, \dots, n, \ x \in \Omega, \\ div\, v = 0 \ \text{ in } \ \Omega, \\ v = 0 \ \text{ on } \ \partial\Omega. \end{cases} \tag{8.3.19}$$

For $n \in N$ existence of a weak solution to (8.3.19) and a priori estimates are given in [TE 2]. Set

$$\begin{cases} \mathcal{H}_0^1(\Omega) = [H_0^1(\Omega)]^n, \\ \mathcal{V} = \{\phi \in [C_0^\infty(\Omega)]^n : div\,\phi = 0\}, \\ V = cl_{\mathcal{H}_0^1(\Omega)}\mathcal{V}, \\ \tilde{V} = cl_{\mathcal{H}_0^1(\Omega)\cap\mathcal{L}^n(\Omega)}\mathcal{V}. \end{cases} \tag{8.3.20}$$

Note that $\tilde{V} \subset V$ but $\tilde{V} = V$ only if $n \leq 4$. In [TE 2, Chapter II, §1] existence of a solution (v, p) to (8.3.19) is shown in the following weak sense.

Definition 8.3.1. *For arbitrarily given $h \in \mathcal{H}^{-1}(\Omega)$ a pair $(v, p) \in V \times \mathcal{L}^1_{loc}(\Omega)$ is a **weak solution** of (8.3.19) if v solves the variational problem*

(8.3.21)
$$\nu \sum_{i,k=1}^n \int_\Omega \frac{\partial v_k}{\partial x_i}\frac{\partial \phi_k}{\partial x_i}dx + \sum_{i,k=1}^n \int_\Omega v_i \frac{\partial v_k}{\partial x_i}\phi_k dx = \sum_{k=1}^n \int_\Omega h_k \phi_k dx \ \ for\ all\ \phi \in \tilde{V},$$

and p is an element of $\mathcal{L}^1_{loc}(\Omega)$ such that the first equation in (8.3.19) holds in the distributional sense on Ω (such a p, determined up to an additive constant, *may always be chosen for v if* (8.3.21) *is satisfied, as shown in* [TE 2, Chapter I, Section 1.4]*).*

In this book we are mostly interested in smooth solutions of the problems considered. Therefore we decided to give a proof fully independent of Temam's results of the existence of X^α stationary solutions to the Navier-Stokes system. We require that the right side $P_r h$ is sufficiently small in the $\mathcal{L}^r(\Omega)$ norm. Our main tool will be the following variant of the Schauder fixed point theorem, which can be found in [GI-TR, Lemma 11.7].

Proposition 8.3.2. *Let $B = B_{\mathcal{B}}(0,\rho)$ be an open ball centered at zero with radius ρ in a Banach space $\mathcal{B}$ and let T be a continuous mapping of $cl_{\mathcal{B}}B$ into $\mathcal{B}$ such that the image $T(cl_{\mathcal{B}}B)$ is precompact and $T(\partial B) \subset B$. Then T has a fixed point in the interior of B.*

Proof. The proof is given in [GI-TR, p. 286] except that we need slightly to modify the definition of the improved mapping there, setting

$$T^*z = \begin{cases} Tz & \text{if } \|Tz\|_{\mathcal{B}} \le \rho, \\ \rho\frac{Tz}{\|Tz\|_{\mathcal{B}}} & \text{if } \|Tz\|_{\mathcal{B}} \ge \rho. \end{cases}$$

□

Using the projection P_r we shall reformulate the problem (8.3.19) into the following abstract setting.

Abstract setting of (8.3.19). *For $r \ge 2$ and given $h \in \mathcal{L}^r(\Omega)$ find a function $u_0 \in D(A_r) = \{\phi \in \mathcal{W}^{2,r}(\Omega) : \phi_{|\partial\Omega} = 0\} \cap X_r$ satisfying the equation*

$$A_r u_0 = P_r[h - (u_0, \nabla)u_0]. \tag{8.3.22}$$

We are now able to formulate

Theorem 8.3.1. *When $r > n$ and $\|P_r h\|_{\mathcal{L}^r(\Omega)}$ satisfies the smallness restriction* (8.3.29) *below, there exists a stationary solution $u_0 \in D(A_r)$ of the problem* (8.3.22)*. Moreover $\|u_0\|_{W^{2,r}(\Omega)}$ tends to zero when $\|P_r h\|_{\mathcal{L}^r(\Omega)}$ tends to zero. The solution u_0 is unique provided that* (8.3.33) *below is satisfied.*

Proof. Proposition 8.3.2 will be used in the proof. For a given $h \in \mathcal{L}^r(\Omega)$, $r > n$, from (8.3.8) $X_r^{\frac{1}{2}} = \mathcal{W}_0^{1,r}(\Omega) \cap X_r$. Consider $X_r^{\frac{1}{2}}$ with the $\mathcal{W}^{1,r}(\Omega)$ norm. We define a mapping $T : X_r^{\frac{1}{2}} \to X_r^{\frac{1}{2}}$ as follows:

for given $w \in X_r^{\frac{1}{2}}$ we set $v = T(w)$ as the unique $X_r^{\frac{1}{2}}$ solution of the linear problem

$$A_r v = P_r[h - (w, \nabla)w], \tag{8.3.23}$$

$$v_{|\partial\Omega} = 0, \ div\, v = 0 \tag{8.3.24}$$

(condition (8.3.24) is implicitly included in the definition of the space $X_r^{\frac{1}{2}}$ and we added it only for the clarity). To justify correctness of the definition of T note that, according to (8.3.16), $0 \in \rho(A_r)$ and for $h \in \mathcal{L}^r(\Omega)$ and $w \in X_r^{\frac{1}{2}}$ the right hand side of (8.3.23) belongs to X_r,

$$\|P_r(w,\nabla)w\|_{\mathcal{L}^r(\Omega)} \leq C\|w\|_{\mathcal{L}^\infty(\Omega)}\|w\|_{\mathcal{W}^{1,r}(\Omega)} < +\infty, \tag{8.3.25}$$

since $\mathcal{W}^{1,r}(\Omega) \subset \mathcal{L}^\infty(\Omega)$ for $r > n$. Therefore

$$v = A_r^{-1}\Big(P_r[h-(w,\nabla)w]\Big)$$

is a uniquely determined element of $D(A_r)$ and the operator T is well defined. Taking the $\mathcal{L}^r(\Omega)$ norm in (8.3.23) it follows from (8.3.25) and Proposition 1.3.5 that

(8.3.26)
$$\begin{aligned}\|Tw\|_{D(A_r)} = \|v\|_{D(A_r)} &= \|A_r v\|_{\mathcal{L}^r(\Omega)} = \|P_r[h-(w,\nabla)w]\|_{\mathcal{L}^r(\Omega)}\\ &\leq C'\Big(\|P_r h\|_{\mathcal{L}^r(\Omega)} + \|(w,\nabla)w\|_{\mathcal{L}^r(\Omega)}\Big)\\ &\leq C'_{2,r}\Big(\|P_r h\|_{\mathcal{L}^r(\Omega)} + \|w\|^2_{\mathcal{W}^{1,r}(\Omega)}\Big),\end{aligned}$$

where the estimate (8.3.11) has been used. Compactness of the inclusion $D(A_r) \subset \mathcal{W}^{1,r}(\Omega)$ and (8.3.26) show that T is *compact*.

Continuity of T is a consequence of the following simple estimate: for $v_1 = Tw_1$ and $v_2 = Tw_2$,

$$\begin{cases} A_r(v_1-v_2) = -P_r[(w_1,\nabla)w_1 - (w_2,\nabla)w_2],\\ (v_1-v_2)_{|\partial\Omega} = 0, \ div(v_1-v_2) = 0.\end{cases}$$

Hence estimating as in (8.3.25), (8.3.26) we find that

(8.3.27)
$$\begin{aligned}\|Tw_1 - Tw_2\|_{D(A_r)} &= \|A_r(v_1-v_2)\|_{\mathcal{L}^r(\Omega)} = \|P_r[(w_1,\nabla)w_1 - (w_2,\nabla)w_2]\|_{\mathcal{L}^r(\Omega)}\\ &\leq C'_{2,r}\Big(\|w_1-w_2\|_{\mathcal{W}^{1,r}(\Omega)}\|w_1\|_{\mathcal{W}^{1,r}(\Omega)} + \|w_1-w_2\|_{\mathcal{W}^{1,r}(\Omega)}\|w_2\|_{\mathcal{W}^{1,r}(\Omega)}\Big).\end{aligned}$$

Finally we will set admissible values of the radii ρ to ensure that

$$T\Big(\partial B_{X_r^{\frac{1}{2}}}(0,\rho)\Big) \subset B_{X_r^{\frac{1}{2}}}(0,\rho).$$

As a consequence of (8.3.26) and the inclusion (8.3.9)

$$\begin{aligned}\|v\|_{X_r^{\frac{1}{2}}} = \|v\|_{\mathcal{W}^{1,r}(\Omega)} &\leq c_1\|v\|_{D(A_r)}\\ &\leq C''_{2,r}\Big(\|P_r h\|_{\mathcal{L}^r(\Omega)} + \|w\|^2_{X_r^{\frac{1}{2}}}\Big),\end{aligned} \tag{8.3.28}$$

where $C''_{2,r} := c_1 C'_{2,r}$ is a constant independent of h and w. Assume from now on that $P_r h$ satisfies the following *smallness restriction*:

$$(C''_{2,r})^{-2} - 4\|P_r h\|_{\mathcal{L}^r(\Omega)} =: \delta > 0, \tag{8.3.29}$$

which guarantees that the quadratic algebraic equation

$$C''_{2,r}(\|P_r h\|_{\mathcal{L}^r(\Omega)} + s^2) = s$$

has two different positive roots

$$s_{1,2} = \frac{1}{2}(C''_{2,r})^{-1} \pm \sqrt{\frac{1}{4}(C''_{2,r})^{-2} - \|P_r h\|_{\mathcal{L}^r(\Omega)}}.$$

Then, for any number $\rho \in (s_1, s_2)$, taking $\|w\|_{X_r^{\frac{1}{2}}} = \rho$ we obtain from (8.3.28) that

$$\|Tw\|_{X_r^{\frac{1}{2}}} = \|v\|_{X_r^{\frac{1}{2}}} \leq C''_{2,r}(\|P_r h\|_{\mathcal{L}^r(\Omega)} + \rho^2) < \rho.$$

Therefore T transforms the boundary of $B_{X_r^{\frac{1}{2}}}(0, \rho)$ into interior of this ball.

All the assumptions of Proposition 8.3.2 are thus fulfilled and hence T has *at least* one fixed point $u_0 \in B_{X_r^{\frac{1}{2}}}(0, \rho)$, where $\rho \in (s_1, s_2)$ is arbitrary. It is thus clear that $u_0 \in \bigcap_{\rho \in (s_1, s_2)} B_{X_r^{\frac{1}{2}}}(0, \rho) = cl_{X_r^{\frac{1}{2}}} B_{X_r^{\frac{1}{2}}}(0, s_1)$, which provides an estimate:

$$\|u_0\|_{X^{\frac{1}{2}}} \leq s_1 = \frac{1}{2}(C''_{2,r})^{-1} - \sqrt{\frac{1}{4}(C''_{2,r})^{-2} - \|P_r h\|_{\mathcal{L}^r(\Omega)}}. \tag{8.3.30}$$

It is evident from (8.3.30) and the elementary equality

$$\sqrt{c} - \sqrt{c - \Delta c} = \frac{1}{2\sqrt{c - \theta \Delta c}} \Delta c, \ \theta = \theta(c) \in (0, 1),$$

that

$$\|u_0\|_{X^{\frac{1}{2}}} \leq \frac{1}{\sqrt{\delta}} \|P_r h\|_{\mathcal{L}^r(\Omega)}, \tag{8.3.31}$$

with δ introduced in (8.3.29). Hence convergence to zero of $\|P_r h\|_{\mathcal{L}^r(\Omega)}$ forces $\|u_0\|_{X_r^{\frac{1}{2}}}$ to decay to zero. Moreover, as a result of (8.3.26) with $w = v = u_0$ we find that $u_0 \in D(A_r)$ and also obtain the estimate

$$\|u_0\|_{D(A_r)} \leq C'_{2,r} \left(1 + \frac{\|P_r h\|_{\mathcal{L}^r(\Omega)}}{\delta}\right) \|P_r h\|_{\mathcal{L}^r(\Omega)}. \tag{8.3.32}$$

Finally, *uniqueness* of the equilibrium u_0 for small $\|P_r h\|_{\mathcal{L}^r(\Omega)}$ follows from (8.3.27) (with $v_1 = u_0$, $v_2 = \overline{u}_0$, $Tu_0 = u_0$, $T\overline{u}_0 = \overline{u}_0$) and (8.3.31):

$$\|u_0 - \overline{u}_0\|_{D(A_r)} \leq C'_{2,r} \frac{2}{\sqrt{\delta}} \|P_r h\|_{\mathcal{L}^r(\Omega)} c_1 \|u_0 - \overline{u}_0\|_{D(A_r)}$$

provided that

$$C'_{2,r} \frac{2}{\sqrt{\delta}} \|P_r h\|_{\mathcal{L}^r(\Omega)} c_1 < 1. \tag{8.3.33}$$

The proof of Theorem 8.3.1 is complete. □

Theorem 8.3.1 provides all the information necessary to justify by Lemma 8.3.1 asymptotic stability of u_0 in each of the spaces X_r^α, $r > n$, $\alpha \in [\frac{1}{2}, 1)$. The restriction (8.3.18) of Lemma 8.3.1 follows from (8.3.32) provided $\|P_r h\|_{\mathcal{L}^r(\Omega)}$ is sufficiently small.

This justifies the following conclusion.

Corollary 8.3.2. *For any $r > n$, $\alpha \in [\frac{1}{2}, 1)$, if only $h \in \mathcal{L}^r(\Omega)$ is restricted by*

$$\|P_r h\|_{\mathcal{L}^r(\Omega)} \leq \min\left\{\frac{\sqrt{\delta}}{2C'_{2,r}}, \frac{1}{(2C''_{2,r})^2}, \frac{\sqrt{\delta}}{const.(1+M_r)}\right\},$$

there is a unique, uniformly asymptotically stable equilibrium point of the Navier-Stokes system (8.3.10) *considered in X_r^α.*

8.4. Parabolic problems in Hölder spaces

Considering semigroups generated by p.d.e.'s there are a number of possible choices of phase spaces and consequently properties of the attractor may often be displayed in the context of various topologies. It might sometimes be important to get an attractor on a possibly 'small' space (i.e. in a strong topology) or, when it is known in a 'larger' space, where its existence is usually easier to prove, to show that it is the same set in a 'smaller' one. For partial differential equations, this imposes the need to study semigroups of very smooth solutions, which usually leads to additional technical difficulties. Nevertheless, more regularity of the semigroup still seems to be worth considering, since it provides better control over the asymptotic behavior of solutions.

In this section our aim is to show the existence of a global attractor for the semigroup generated on Hölder type spaces (i.e. in a possibly strong topology) by the initial boundary value problem for an autonomous semilinear $2m$-th order parabolic equation introduced in Chapter 5:

(8.4.1)

$$\begin{cases} u_t = -\sum_{|\alpha|\leq 2m} a_\alpha(x) D^\alpha u + f(x, d^{m_0}u), \ t > 0, \ x \in \Omega \subset R^n, \\ \sum_{|\beta|\leq m_j} b_{j\beta}(x) D^\beta u = 0, \ m_j \leq 2m-1, \ j = 1, \dots, m, \ t > 0, \ x \in \partial\Omega, \\ u(0,x) = u_0(x), \ x \in \Omega. \end{cases}$$

Here Ω is a bounded domain in R^n, $n \geq 2$, with $C^{2m+\mu}$ boundary $\partial\Omega$, $m_0 \in \{0, \dots, 2m-1\}$ and $d^{m_0}u$ is a vector as in (5.1.3) of partial derivatives with respect to x of order not exceeding m_0.

As follows from Corollary 4.2.2 a necessary and sufficient condition for existence of a global attractor for the semigroup generated by (8.4.1) on a fractional power space, under some growth restrictions for a nonlinear term f, is the asymptotic independence of any 'weak' a priori estimate for solutions of (8.4.1) of the initial data. Using our abstract approach and basing on the linear theory of parabolic equations (see [L-S-U], [L-S-W]) dissipativeness of semigroups generated by (8.4.1) on smoother phase spaces will be shown under the same assumptions as in Chapter 5 except that we naturally need the data of (8.4.1) to possesses Hölder type regularity and satisfy compatibility conditions necessary for the existence of solutions in the classes considered further of smooth Hölder functions.

Technicalities. We start by listing the assumptions on the data of (8.4.1) and basic tools from the theory of linear parabolic equations.

Assumption 8.4.1. *Let f be locally Lipschitz continuous with respect to all its arguments separately, $A = \sum_{|\alpha|\le 2m} a_\alpha(x)D^\alpha$ be a uniformly strongly elliptic operator, and additionally let both A and the boundary operators $B_j = \sum_{|\beta|\le m_j} b_{j\beta}(x)D^\beta$ ($m_i \ne m_j$ for $i \ne j$, $m_j \in \{0,\dots,2m-1\}$) fulfill the following regularity assumptions (see* [L-S-W, p. 1287]*); for*

(8.4.2)

$$\mu_0 = \frac{\nu_0^{m_0+1}}{(1+\nu_0)^{m_0}}, \text{ where } \nu_0 = 1 - \frac{n+1}{p_0}, \; p_0 > n+1 \text{ and } n \ge 2 \text{ are fixed,}$$

$$(8.4.3) \quad \begin{cases} (i) & \Omega \subset R^n \text{ is a bounded domain with } C^{2m+\mu_0} \text{ boundary } \partial\Omega, \\ (ii) & \text{each } a_\alpha \; (|\alpha| \le 2m) \text{ is an element of } C^{\mu_0}(\overline{\Omega}), \\ (iii) & b_{j\beta} \; (|\beta| \le m_j,\ j = 1,\dots,m) \text{ belongs to } C^{2m-m_j+\mu_0}(\partial G), \\ (iv) & \text{the complementing condition of [L-S-W, p. 1287] holds.} \end{cases}$$

Assumption (iv) is slightly stronger than the strong complementary condition of Subsection 1.2.4. Note also, that Assumption 8.4.1 ensures validity of the requirements of [L-S-W, conditions (2), (3), p. 1287].

Observation 8.4.1. *As a consequence of the requirements introduced above, the triple $(A, \{B_j\}, \Omega)$ forms a regular elliptic boundary value problem (see Example 1.3.8), so that (8.4.1) falls into a class of abstract evolutionary equations:*

$$(8.4.4) \qquad \begin{cases} \dot u + A_p u = F_p(u), \; t > 0, \\ u(0) = u_0, \end{cases}$$

where for each $p \in (1, +\infty)$ an operator

$$(8.4.5) \qquad A_p = A + k_0 I$$

(here $k_0 > 0$ is sufficiently large and independent of p) considered on the domain

$$D(A_p) = W^{2m,p}_{\{B_j\}}(\Omega) = \{\phi \in W^{2m,p}(\Omega) : \; \forall_{j=1,\dots,m-1} \; B_j\phi = 0 \text{ on } \partial\Omega\}$$

is sectorial in $X_p = L^p(\Omega)$. Also

(8.4.6) *A_p is positive, so that $Re(\sigma(A_p)) > a$ for some $a > 0$,*

and, by Proposition 1.2.3,

(8.4.7) *A_p has compact resolvent.*

In addition, (1.2.65) and (1.2.66) imply that

(8.4.8) *A_p is an isomorphism from $D(A_p)$ onto $L^p(\Omega)$.*

Assumption 8.4.2. *Let $p_0 \in (n+1,\infty)$ and $\alpha_0 \in (0,1)$ satisfy the inequalities*

$$m_0 < 2m\alpha_0 - \frac{n}{p_0}, \ \alpha_0 > 1 - \frac{1}{p_0}. \tag{8.4.9}$$

Observation 8.4.2. *The above assumption ensures that the space $X^{\alpha_0}_{p_0}$ is embedded in $W^{k,\infty}(\Omega)$, $X^{\alpha_0}_{p_0} \subset B^{2m-\frac{2m}{p_0}}_{p_0}(\Omega)$, and the nonlinear term $F_{p_0} : X^{\alpha_0}_{p_0} \longrightarrow L^{p_0}(\Omega)$ (which is an abstract counterpart of a locally Lipschitz continuous function $f + k_0 I$ in (8.4.1)) is Lipschitz continuous on bounded subsets of $X^{\alpha}_{p_0}$ for each $\alpha \in [\alpha_0, 1)$. Therefore (8.4.4) is known to have a unique $X^{\alpha}_{p_0}$ solution on a maximal interval of existence $[0, \tau_{u_0})$ (see Section 5.1).*

To justify global existence of the $X^{\alpha}_{p_0}$ solutions of (8.4.4) we take further

Assumption 8.4.3. *Assume that for some $l \geq 0$, $r \geq 1$,*

$$D(A_{p_0}) \subset W^{l,r}(\Omega), \tag{8.4.10}$$

and let the following a priori $W^{l,r}(\Omega)$ estimate of $X^{\alpha_0}_{p_0}$ solutions be known:

$$\|u(t)\|_{W^{l,r}(\Omega)} \leq c_1(\|u_0\|_{X^{\alpha_0}_{p_0}}), \ t \in (0, \tau_{u_0}), \tag{8.4.11}$$

where $c_1 : [0,+\infty) \to [0,+\infty)$ is locally bounded. Let additionally the abstract counterpart F_{p_0} of the function $f = f(x, d^{m_0}u)$ satisfy the subordination condition

$$\begin{aligned} &\|F(u(t,u_0))\|_{L^{p_0}(\Omega)} \\ &\leq g\left(\|u(t,u_0)\|_{W^{l,r}(\Omega)}\right)\left(1 + \|A^{\alpha_0}u(t,u_0)\|^{\eta}_{L^{p_0}(\Omega)}\right), \ t \in (0,\tau_{u_0}), \end{aligned} \tag{8.4.12}$$

with a certain constant $\eta \in [0,1)$ and some nondecreasing function $g : [0,+\infty) \to [0,+\infty)$.

Remark 8.4.1. Recall that an estimate of type (8.4.12) holds whenever f has a growth limited by Restriction 1 (see Lemma 5.2.1, Remark 5.2.2).

Observation 8.4.3. *Under Assumptions* 8.4.1-8.4.3 *we obtain from Theorems* 3.1.1, 3.3.1 *that, for each $\alpha \in [\alpha_0, 1)$, the solution u of the problem* (8.4.4) *corresponding to $u_0 \in X^{\alpha}_{p_0}$ exists globally for $t \geq 0$ and* (8.4.1) *generates a compact C^0 semigroup $T(t) : X^{\alpha}_{p_0} \longrightarrow X^{\alpha}_{p_0}$, $t \geq 0$, which has bounded orbits of bounded sets.*

We provide below two illustrative problems fulfilling Assumption 8.4.3.

Example 8.4.1. The first example is the Dirichlet problem for the *single second order dissipative equation* (6.2.1) considered with the *dissipativeness condition* (6.2.5) for the nonlinear term. In this case (8.4.11) has the form

$$\|u(t,u_0)\|_{L^\infty(\Omega)} \leq const. \max\left\{1, \max\left\{\|u_0\|_{L^2(\Omega)}, \frac{2m_{\varepsilon_0}|\Omega|}{c_\Omega C}\right\}\right\}, \ t > 0,$$

following from the estimates (6.2.8) and (6.2.11). Consequently, the counterpart of (8.4.12) is given in the formula (6.2.12).

Example 8.4.2. As another example will serve the second order equation with nonlinearity dependent on the gradient considered earlier in (5.3.1) under homogeneous Dirichlet data. Assuming that $f(x,u,0)=0$, as a consequence of the maximum principle, we have

$$\|u(t,u_0)\|_{L^\infty(\Omega)} \leq \|u_0\|_{L^\infty(\Omega)},$$

which is a version of (8.4.11) for that problem. Assuming next the *subquadratic growth restriction* on f as in (5.3.7) we obtain, through Lemma 5.2.1 and Remark 5.2.2, the estimate (5.2.15) corresponding to (8.4.12).

It is possible to give examples of higher order parabolic problems satisfying Assumption 8.4.3. Such is, for instance, the Kuramoto-Velarde equation considered in [RB 3] (under special conditions on parameters as in [RB 3, Proposition 8]). Another fourth order problem is the Cahn-Hilliard equation that will be studied in detail in Example 8.5.1 at the end of Section 8.5.

We take finally

Assumption 8.4.4. *There is a closed subset V of $X^{\alpha_0}_{p_0}$ such that $T(t)V \subset V$ for $t \geq 0$ and the a priori estimate* (8.4.11) *is asymptotically independent of $u_0 \in V$.*

Observation 8.4.4. *The last condition, when added to Assumptions* 8.4.1-8.4.3, *allows us to fulfill the requirements of Theorem* 4.2.1 *(see also Theorem* 5.3.1*) and hence, for each $\alpha \in [\alpha_0, 1)$, the semigroup $T(t)$ generated by* (8.4.1) *on a complete metric subspace $V \cap X^\alpha_{p_0}$ of $X^\alpha_{p_0}$ has a global attractor $\mathcal{A}_\alpha \subset V \cap X^\alpha_{p_0}$.*

Let us introduce the space of Hölder functions $C^{2m+\mu}_{\{B_j\}}(\overline{\Omega})$,

$$C^{2m+\mu}_{\{B_j\}}(\overline{\Omega}) = \{\phi \in C^{2m+\mu}(\overline{\Omega}) : \ \forall_{j=1,\dots,m} \ B_j\phi = 0 \ \text{ on } \ \partial\Omega\}, \tag{8.4.13}$$

and define for $\mu \in (0,\mu_0)$ a Banach space

$$\mathcal{X}^\mu = cl_{C^{2m+\mu}(\overline{\Omega})}\left(\bigcup_{\mu<\nu} C^{2m+\nu}_{\{B_j\}}(\overline{\Omega})\right), \tag{8.4.14}$$

where a norm in $\mathcal{X}^\mu$ is the usual $C^{2m+\mu}(\overline{\Omega})$ norm.

It may be helpful to give the characterization of elements of $\mathcal{X}^\mu$ in the language of *little Hölder continuous functions.* For $\theta \in (0,1)$ and $k \in N$ denote by $h^{k+\theta}(\overline{\Omega})$ the set of $(k+\theta)$*-little Hölder continuous functions:*

$$h^{k+\theta}(\overline{\Omega}) := \left\{\phi \in C^{k+\theta}(\overline{\Omega}) : \forall_{|\sigma|=k} \lim_{r\to 0^+} \sup_{\substack{x,\overline{x}\in\Omega \\ 0<|x-\overline{x}|<r}} \frac{|D^\sigma\phi(x) - D^\sigma\phi(\overline{x})|}{|x-\overline{x}|^\theta} = 0\right\}$$

(see [LU 1, Section 0.2], [A-T, §1]). As is known, (see [LU 1, p. 8], [LU 2, p. 305]), $h^{k+\theta}(\overline{\Omega})$ equipped with the $C^{k+\theta}(\overline{\Omega})$ norm is a Banach space, a subspace of $C^{k+\theta}(\overline{\Omega})$. Moreover, $C^{l+\nu}(\overline{\Omega})$ is dense in $h^{k+\theta}(\overline{\Omega})$ for any $l=k,k+1,\dots$ and any $\nu \in [0,1)$ such that $l+\nu > k+\theta$ (Ω is of class C^{l+1}, or C^l if $\nu=0$).

In Observation 8.5.1 it will be proved that

(8.4.15)
$$\mathcal{X}^\mu = h^{2m+\mu}_{\{B_j\}}(\overline{\Omega}) := \{\phi \in h^{2m+\mu}(\overline{\Omega}) : \forall_{j=1,\dots,m}\ B_j\phi = 0 \text{ on } \partial\Omega\},\ \mu \in (0,\mu_0).$$

Thus, $\mathcal{X}^\mu$ consists of those elements $\phi \in C^{2m+\mu}_{\{B_j\}}(\overline{\Omega})$ whose $2m$-th order partial derivatives are μ-little Hölder continuous functions.

Overview. We shall show in Theorems 8.4.1, 8.4.2 that, under Assumptions 8.4.1-8.4.3, the problem (8.4.1) generates a C^0 semigroup of global solutions both on $X^\alpha_{p_0}$ ($\alpha_0 \leq \alpha < 1$) and on $\mathcal{X}^\mu$ ($0 < \mu < \mu_0$) phase spaces. If, in addition, Assumption 8.4.4 is satisfied, then there exist global attractors $\mathcal{A}_\alpha$, $\mathcal{A}_\mu$ corresponding to the semigroup considered on $V \cap X^\alpha_{p_0}$ or $V \cap \mathcal{X}^\mu$ respectively. Furthermore, these attractors coincide, that is

$$\mathcal{A}_\alpha = \mathcal{A}_\mu = \mathcal{A} \text{ for } \alpha \in [\alpha_0, 1),\ \mu \in (0,\mu_0).$$

Remark 8.4.2. The space $C^{2m+\mu}_{\{B_j\}}(\overline{\Omega})$ consists of $C^{2m+\mu}(\overline{\Omega})$ functions satisfying the boundary conditions stated in (8.4.1). Hence, if only the set of boundary conditions $\{B_j\}$ does not contain a zero order operator then for initial functions from $C^{2m+\mu}_{\{B_j\}}(\overline{\Omega})$ the compatibility conditions (see [L-S-W, p. 1288]), necessary for existence of Hölder $C^{1+\frac{\mu}{2m},2m+\mu}([0,\tau]\times\overline{\Omega})$ solutions of (8.4.1), are satisfied. In the case when B_1 has order $m_1 = 0$, additionally to the usual compatibility conditions any initial function u_0 should also satisfy

$$B_1[-Au_0 + f(x, d^{m_0}u_0)] = 0 \text{ on } \partial\Omega. \tag{8.4.16}$$

Therefore, in the case discussed when B_1 has order $m_1 = 0$, we shall also require

$$B_1[A\phi] = 0 \text{ on } \partial\Omega \text{ for } \phi \in C^{2m+\mu}_{\{B_j\}}(\overline{\Omega}) \tag{8.4.17}$$

and that the function f satisfies the following implication:

$$B_1[f(x, d^{m_0}\phi)] = 0 \text{ on } \partial\Omega \text{ whenever } B_1\phi = 0 \text{ on } \partial\Omega. \tag{8.4.18}$$

Remark 8.4.3. Observations 8.4.1-8.4.4 show that to get the existence of the global attractor in a strong $C^{2m+\mu}(\overline{\Omega})$ topology it suffices to have, apart the assumptions that are needed in the case of fractional power spaces, merely the extra μ_0 Hölder regularity of the data expressed in conditions (i)-(iii) of Assumption 8.4.1.

Regular parabolic initial boundary value problems. Problems of the type (8.4.1) for which Assumption 8.4.1 holds are called *regular parabolic initial boundary value problems.* Whenever u_0 and a composite function f in (8.4.1) are shown to have appropriate regularity and satisfy the required compatibility conditions, the linear theory of [L-S-U, Chapter VII, §10, Th.10.4] and [L-S-W, Theorem 4.1] can be applied to (8.4.1) and the following estimates become valid:

$$\|u\|_{C^{1+\frac{\mu}{2m},2m+\mu}([0,\tau]\times\overline{\Omega})} \leq c(\tau)\left(\|f\|_{C^{\frac{\mu}{2m},\mu}([0,\tau]\times\overline{\Omega})} + \|u_0\|_{C^{2m+\mu}(\overline{\Omega})}\right), \tag{8.4.19}$$

$$\|u\|_{W_q^{1,2m}((0,\tau)\times\Omega)} \leq const. \left(\|f\|_{L^q((0,\tau)\times\Omega)} + \|u_0\|_{B_q^{2m-\frac{2m}{q}}(\Omega)} \right), \tag{8.4.20}$$

where c in (8.4.19) is a continuous and increasing function of τ. Moreover, according to [L-S-W, Th. 5.2], for each $\mu \in (0, \mu_0)$ the set

$$\mathcal{D} = \left\{ \phi \in \bigcap_{p>1} W^{2m,p}_{\{B_j\}}(\Omega) : A\phi \in C^\mu(\overline{\Omega}) \right\} \tag{8.4.21}$$

is contained in $C^{2m+\mu}(\overline{\Omega})$ and

$$\|\phi\|_{C^{2m+\mu}(\overline{\Omega})} \leq const.'(\|A\phi\|_{C^\mu(\overline{\Omega})} + \|\phi\|_{C(\overline{\Omega})}), \phi \in \mathcal{D}. \tag{8.4.22}$$

These basic tools will be very useful for further study of the regular solutions of the problem (8.4.1).

Regularity result. To simplify further notation take it that $\mathcal{Y}$ denotes a space (together with the topology) selected from the family of Banach spaces

$$\mathcal{F} = \{\mathcal{X}^\mu : \mu \in (0, \mu_0)\} \cup \{X^\alpha_{p_0} : \alpha \in [\alpha_0, 1)\}, \tag{8.4.23}$$

where $\mathcal{X}^\mu$ was introduced in (8.4.14). Additionally, when $V \cap \mathcal{Y}$ appears in considerations below as a metric space, the topology in $V \cap \mathcal{Y}$ will always be induced from $\mathcal{Y}$:

Let Assumptions 8.4.1-8.4.3 be satisfied from now on. Then, for each $\mathcal{Y} \in \mathcal{F}$:

Theorem 8.4.1. *A C^0 semigroup $\{T(t)\}$ of solutions of the problem (8.4.1) is generated on $\mathcal{Y}$. Moreover, $T(t) : \mathcal{Y} \longrightarrow \mathcal{Y}$ is completely continuous for $t > 0$ and if, in addition, for some $V \subset \mathcal{Y}$ the a priori estimate (8.4.11) is asymptotically independent of $u_0 \in V$, i.e. the condition*

$$\exists_{const.>0} \forall_{u_0 \in V} \limsup_{t\to+\infty} \|u(t, u_0)\|_{W^{l,r}(\Omega)} \leq const., \tag{8.4.24}$$

holds, then there is a bounded subset $\mathcal{P}_1$ of $\mathcal{Y}$ attracting each point of V with respect to $T(t) : \mathcal{Y} \longrightarrow \mathcal{Y}$, i.e.

$$dist_{\mathcal{Y}}(T(t)u_0, \mathcal{P}_1) \overset{t\to+\infty}{\longrightarrow} 0, \ u_0 \in V.$$

As a consequence of the above theorem we have immediately

Theorem 8.4.2. *(Smoothing property) If for some $\mathcal{Y}' \in \mathcal{F}$ a subset $V \subset \mathcal{Y}'$ exists such that*

(8.4.25)

(a) the a priori estimate (8.4.11) is asymptotically independent of $u_0 \in V$,

(b) V is closed in $\mathcal{Y}'$,

(c) $T(t)V \subset V$, $t \geq 0$,

then the semigroup $\{T(t)\}$ restricted to the complete metric space $V \cap \mathcal{Y}'$ has a global attractor $\mathcal{A} \subset V \cap \mathcal{Y}'$ (in the topology of $\mathcal{Y}'$). Furthermore, for each $\mathcal{Y} \in \mathcal{F}$ such that $\mathcal{Y} \subset \mathcal{Y}'$, $\mathcal{A}$ is also the global attractor for $\{T(t)\}$ restricted to a complete metric space $V \cap \mathcal{Y}$ (in the topology of $\mathcal{Y}$).

8.5. Dissipativeness in Hölder spaces

This section is fully devoted to the proof of Theorems 8.4.1, 8.4.2 which will be given in a sequence of lemmas.

By our assumptions the problem (8.4.1) can be studied in the abstract form (8.4.4) with parameters $p_0 > n+1$ and $\alpha \in [\alpha_0, 1)$ satisfying (8.4.9). Also, to each $u_0 \in X^\alpha_{p_0}$ corresponds an $X^\alpha_{p_0}$ solution global in time u of (8.4.4) (see the considerations of Chapter 5) given by a variation of constants formula

$$u(t,u_0) = e^{-A_{p_0}t}u_0 + \int_0^t e^{-A_{p_0}(t-s)} F_{p_0}\big(u(s,u_0)\big)\,ds. \tag{8.5.1}$$

We shall consider further three families of mappings $T(t)$, $S(t)$, $U(t)$, where

$$T(t) = S(t) + U(t),\ t \geq 0, \tag{8.5.2}$$

$$T(t)u_0 = e^{-A_{p_0}t}u_0 + \int_0^t e^{-A_{p_0}(t-s)} F_{p_0}\big(u(s,u_0)\big)\,ds, \tag{8.5.3}$$

$$S(t)u_0 = e^{-A_{p_0}t}u_0, \tag{8.5.4}$$

$$U(t)u_0 = \int_0^t e^{-A_{p_0}(t-s)} F_{p_0}\big(u(s,u_0)\big)\,ds. \tag{8.5.5}$$

According to Observation 8.4.3, for each $\alpha \in [\alpha_0, 1)$, $T(t) : X^\alpha_{p_0} \longrightarrow X^\alpha_{p_0}$, $t \geq 0$, is a C^0 semigroup of $X^\alpha_{p_0}$ solutions of (8.4.4). However, since $S(t)$ is an analytic semigroup on $L^{p_0}(\Omega)$, $S(t) : X^\alpha_{p_0} \longrightarrow X^\alpha_{p_0}$, $t \geq 0$, is also a C^0 semigroup of $X^\alpha_{p_0}$ solutions to

$$\begin{cases} \dot v + A_{p_0} v = 0,\ t > 0, \\ v(0) = u_0, \end{cases} \tag{8.5.6}$$

and as a consequence, for each $u_0 \in X^\alpha_{p_0}$, $U(\cdot)u_0$ as a difference $T(\cdot)u_0 - S(\cdot)u_0$ must be an $X^\alpha_{p_0}$ solution of

$$\begin{cases} \dot w + A_{p_0} w = F_{p_0}(T(t)u_0),\ t > 0, \\ w(0) = 0. \end{cases} \tag{8.5.7}$$

Furthermore, the maps T, S, U introduced above can be considered on the product $[0, +\infty) \times X^\alpha_{p_0}$ and

$$S, T, U : [0, +\infty) \times X^\alpha_{p_0} \longrightarrow X^\alpha_{p_0} \text{ are continuous.} \tag{8.5.8}$$

It is also worth noting that (8.4.4), (8.5.6), (8.5.7) are abstract formulations of regular parabolic initial boundary value problems resulting from (8.4.1) and hence, to obtain a priori bounds, the linear theory of parabolic problems (see formulas (8.4.19)-(8.4.22)) is applicable whenever the necessary compatibility conditions are satisfied and suitable boundedness of right hand sides and initial functions is given.

Lemma 8.5.1. *$\mathcal{X}^\mu$ is compactly embedded and dense in $X^\alpha_{p_0}$ for each $\mu \in (0, \mu_0)$, $\alpha \in [\alpha_0, 1)$.*

Proof. Compactness of the embedding $\mathcal{X}^\mu \subset X^\alpha_{p_0}$ follows immediately from the inclusions

$$\mathcal{X}^\mu \underset{\text{cont.}}{\subset} C^{2m+\mu}_{\{B_j\}}(\overline{\Omega}) \underset{\text{cont.}}{\subset} D(A_{p_0}) \underset{\text{comp.}}{\subset} X^\alpha_{p_0}.$$

For the proof of denseness we recall that $D(A_p) = W^{2m,p}_{\{B_j\}}(\Omega)$ and $A_p = A + k_0 I$ where $k_0 > 0$ has been chosen independently of $p \in (0, +\infty)$ in such a way that $A + k_0 I$ is an isomorphism from $D(A_p)$ onto $L^p(\Omega)$ (Observation 8.4.1).

Let us fix an arbitrary $\psi_0 \in W^{2m,p_0}_{\{B_j\}}(\Omega)$, take $\phi_0 \in L^{p_0}(\Omega)$ such that $\psi_0 := (A+k_0I)^{-1}\phi_0$ and consider any sequence $\{\phi_n\} \subset C_0^\infty(\Omega)$ convergent to ϕ_0 in $L^{p_0}(\Omega)$. Since $\{\phi_n\}_{n=1}^\infty \subset \bigcap_{p>1} L^p(\Omega)$, $\psi_n := (A + k_0 I)^{-1}\phi_n$ also belongs to $\bigcap_{p>1} W^{2m,p}_{\{B_j\}}(\Omega)$ for each $n = 1, 2, \dots$, and

$$\psi_n \overset{D(A_{p_0})}{\longrightarrow} \psi_0. \tag{8.5.9}$$

Furthermore, since $(A + k_0 I)\psi_n = \phi_n$ is an element of $C_0^\infty(\Omega) \subset C^{\mu_0}(\overline{\Omega})$ for $n = 1, 2, \dots$, basing on the property (8.4.21) we obtain that

$$\psi_n \in C^{2m+\mu_0}_{\{B_j\}}(\overline{\Omega}), \ n = 1, 2, \dots,$$

which proves denseness of $\mathcal{X}^\mu$ in $D(A_{p_0})$ (note also that $(A + k_0 I)\psi_n = 0$ on $\partial\Omega$, which is important in the case when B_1 has order $m_1 = 0$; see Remark 8.4.2). Since it is known that $D(A_{p_0})$ is dense in $X^\alpha_{p_0}$ (see Proposition 1.3.5) the lemma is proved. □

Arguments similar to the proof of Lemma 8.5.1 allow us to justify

Observation 8.5.1. *For $\mu \in (0, \mu_0)$ the following characterization holds:*

$$\mathcal{X}^\mu = h^{2m+\mu}_{\{B_j\}}(\overline{\Omega}) := \{\phi \in h^{2m+\mu}(\overline{\Omega}) : \forall_{j=1,\dots,m}\ B_j\phi = 0 \text{ on } \partial\Omega\}. \tag{8.5.10}$$

Proof. To show (8.5.10) we proceed as in the proof of Lemma 8.5.1 but now we take $\psi \in h^{2m+\mu}_{\{B_j\}}(\overline{\Omega})$. Clearly $(A + k_0 I)\psi_0 \in h^\mu(\overline{\Omega})$. Therefore, since $C^1(\overline{\Omega})$ is densely embedded in $h^\mu(\overline{\Omega})$ (see [LU 2, p. 305]), there is $\{\phi_n\} \subset C^1(\overline{\Omega})$ such that

$$\phi_n \to \phi_0 \ \text{ in } \ h^\mu(\overline{\Omega}). \tag{8.5.11}$$

We then obtain that, for each $\nu \in (0,1)$, $\psi_n = (A + k_0 I)^{-1}\phi_n \in \{\phi \in \bigcap_{p>1} W^{2m,p}_{\{B_j\}}(\Omega) : (A + k_0 I)\phi \in C^\nu(\overline{\Omega})\}$, $n \in N$ (see (8.4.21)), and, consequently,

$$\psi_n \in C^{2m+\nu}(\overline{\Omega}),\ \nu \in (0,1),\ n \in N,$$

$$\begin{aligned}\|\psi_n - \psi_0\|_{C^{2m+\mu}(\overline{\Omega})} &\le const.'(\|\psi_n - \psi_0\|_{C^\mu(\overline{\Omega})} + \|\psi_n - \psi_0\|_{C^0(\overline{\Omega})} \\ &\le const.'(\|\psi_n - \psi_0\|_{C^\mu(\overline{\Omega})} + \|\psi_n - \psi_0\|_{D(A_{p_0})}.\end{aligned}$$

By (8.5.9) and (8.5.11) the right hand side of the above inequality tends to zero as $n \to +\infty$. It is thus seen that $\{\psi_n\} \subset \bigcup_{\mu<\nu} C^{2m+\nu}_{\{B_j\}}(\overline{\Omega})$ and $\psi_n \to \psi_0$ in the $C^{2m+\mu}(\overline{\Omega})$ norm. The proof is complete. □

Remark 8.5.1. Obvious modifications of the proof allow us to cover the case when B_1 has the order $m_1 = 0$, i.e. when (8.4.17) is required. Indeed, in this case [LU 2, Proposition 2.6] allows us to take $\{\phi_n\} \subset C_0^1(\overline{\Omega})$, $\phi_n \to \phi_0$ in $h^\mu(\overline{\Omega})$ so that $(A + k_0 I)\psi_n$ will be zero on $\partial\Omega$. It is also a particular consequence of [LU 2, Remark 2.8] that under our assumptions on Ω, $C^{k+1}(\overline{\Omega})$ is contained in $C^{k+\nu}(\overline{\Omega})$, $k \in N$, $\nu \in (0,1)$. The simple example of [GI-TR, Section 4.1] reminds us that this may not be true in general.

Lemma 8.5.2. *For each $\mathcal{Y} \in \mathcal{F}$, $S(t): \mathcal{Y} \longrightarrow \mathcal{Y}$ is a linear C^0 semigroup and*

$$\|S(t)\|_{\mathcal{L}(\mathcal{Y},\mathcal{Y})} \le K(t), \tag{8.5.12}$$

with $K(t) \longrightarrow 0$ for $t \longrightarrow +\infty$. Furthermore, if $\alpha \in [\alpha_0, 1)$ and B is bounded in $X^\alpha_{p_0}$, then

$$S(\tau)B \text{ is bounded in } C^{2m+\mu}(\overline{\Omega}) \text{ for each } \tau > 0,\ \mu \in (0,\mu_0). \tag{8.5.13}$$

Proof. If $\mathcal{Y}$ is a fractional power space the assertion follows from analyticity of $S(t)$ and the condition (1.3.47). Considering Hölder's $\mathcal{Y}$ and applying to (8.5.6) the linear theory (8.4.19) we observe that

$$S(t)\left(\bigcup_{\mu<\nu} C^{2m+\nu}_{\{B_j\}}(\overline{\Omega})\right) \subset \bigcup_{\mu<\nu} C^{2m+\nu}_{\{B_j\}}(\overline{\Omega}),\ \mu \in (0,\mu_0). \tag{8.5.14}$$

Also, for any $u_0 \in \mathcal{X}^\mu$ and a sequence $\{u_n\} \subset \bigcup_{\mu<\nu} C^{2m+\nu}_{\{B_j\}}(\overline{\Omega})$,

$$\|S(\cdot)u_n - S(\cdot)u_0\|_{C^{1+\frac{\mu}{2m},2m+\mu}([0,T]\times\overline{\Omega})} \le c(T)\|u_n - u_0\|_{C^{2m+\mu}(\overline{\Omega})},\ T > 0, \tag{8.5.15}$$

so that if $u_n \to u_0$ in $C^{2m+\mu}(\overline{\Omega})$ then according to (8.5.15)

$$S(t)u_n \overset{C^{2m+\mu}(\overline{\Omega})}{\longrightarrow} S(t)u_0,\ t \ge 0. \tag{8.5.16}$$

The conditions (8.5.16), (8.5.14) thus imply that, for each $\mu \in (0,\mu_0)$,

$$S(t)(\mathcal{X}^\mu) \subset \mathcal{X}^\mu,\ t \ge 0.$$

Hence, in order to verify that $S(t)$ is a linear C^0 semigroup when $\mathcal{Y}$ is a Hölder space, only continuity of $S(t)$ at zero needs to be shown.

Let us take any sequence $\{h_n\}_{n\in N} \subset (0,\tau]$ convergent to 0. Using on the problem (8.5.6) the linear estimate (8.4.19), for $u_0 \in \mathcal{X}^\mu$ and $v \in \bigcup_{\mu<\nu} C^{2m+\nu}_{\{B_j\}}(\overline{\Omega})$ we have the inequality

$$\|S(h_n)u_0 - S(h_n)v\|_{C^{2m+\mu}(\overline{\Omega})} \le c(\tau)\|u_0 - v\|_{C^{2m+\mu}(\overline{\Omega})}. \tag{8.5.17}$$

Therefore, it suffices to prove continuity of $S(t)$ on the dense subset of $\mathcal{X}^\mu$.

When $v \in \bigcup_{\mu<\nu} C^{2m+\nu}_{\{B_j\}}(\overline{\Omega})$ it is clear that $v \in C^{2m+\nu}_{\{B_j\}}(\overline{\Omega})$ for some $\nu \in (\mu,\mu_0)$. Applying (8.4.19) again, we get the estimate uniform with respect to $n \in N$,

$$\|S(h_n)v\|_{C^{2m+\nu}(\overline{\Omega})} \le c(\tau)\|v\|_{C^{2m+\nu}(\overline{\Omega})}. \tag{8.5.18}$$

In the presence of a compact embedding $C^{2m+\nu}(\overline{\Omega}) \subset C^{2m+\mu}(\overline{\Omega})$, $\nu > \mu$, (8.5.18) guarantees that a sequence $\{S(h_n)v\}_{n\in N}$ is precompact in $\mathcal{X}^\mu$. Since $S(t)$ is continuous on $X^\alpha_{p_0}$, we have

$$S(h_n)v \xrightarrow{X^\alpha_{p_0}} v, \quad \text{when} \quad n \to \infty, \tag{8.5.19}$$

so that, recalling (8.5.18), the sequence $\{S(h_n)v\}_{n\in N}$ (as precompact in $\mathcal{X}^\mu$ and convergent to v in $X^\alpha_{p_0}$) must converge to v in $\mathcal{X}^\mu$, i.e. continuity of $S(t)$ at zero is justified.

To prove (8.5.12) for Hölder's $\mathcal{Y}$ we fix any $\mu \in (0, \mu_0)$ and start from the Schauder type estimate (8.4.22) used on $S(t)u_0$ with fixed $t > 0$:

$$\|S(t)u_0\|_{C^{2m+\mu}(\overline{\Omega})} \le c_2(\|A_{p_0}S(t)u_0\|_{C^\mu(\overline{\Omega})} + \|S(t)u_0\|_{C(\overline{\Omega})}). \tag{8.5.20}$$

In consequence of an embedding $X^{\frac{1}{2m}}_{p_0} \subset C^\mu(\overline{\Omega})$ ($\mu \in (0, \mu_0)$) we have further

(8.5.21)
$$\|A_{p_0}S(t)u_0\|_{C^\mu(\overline{\Omega})} \le C\|A_{p_0}S(t)u_0\|_{X^{\frac{1}{2m}}_{p_0}}$$
$$= \|A^{\frac{1}{2m}}_{p_0}A_{p_0}S(t)u_0\|_{L^{p_0}(\Omega)} \le C_{1+\frac{1}{2m}}\frac{e^{-at}}{t^{1+\frac{1}{2m}}}\|u_0\|_{L^{p_0}(\Omega)},$$

and analogously

$$\|S(t)u_0\|_{C^0(\overline{\Omega})} \le C\|A^{\frac{1}{2m}}_{p_0}S(t)u_0\|_{L^{p_0}(\Omega)} \le C_{\frac{1}{2m}}\frac{e^{-at}}{t^{\frac{1}{2m}}}\|u_0\|_{L^{p_0}(\Omega)}. \tag{8.5.22}$$

Inserting (8.5.21), (8.5.22) into the right side of (8.5.20) we verify immediately that $\|S(t)\|_{\mathcal{L}(\mathcal{X}^\mu,\mathcal{X}^\mu)} \longrightarrow 0$ when $t \longrightarrow +\infty$.

For the proof of (8.5.13) it suffices to note that since $S(t)$ is analytic on $L^{p_0}(\Omega)$ we have $S(t)u_0 \in D(A^2_{p_0}) \subset D\left(A^{1+\frac{1}{2m}}_{p_0}\right)$ for each $u_0 \in X^\alpha_{p_0}$ (see Propositions 1.3.6, 1.3.5). By the embedding of Proposition 1.3.10 we have

$$A_{p_0}S(t)u_0 \in X^{\frac{1}{2m}}_{p_0} \subset C^\mu(\overline{\Omega}),\ t > 0,\ \mu \in (0, \mu_0),$$

and therefore $A_{p_0}S(t)u_0 \in \bigcap_{p>1} L^p(\Omega)$ for each $t > 0$. However, the latter implies the condition (note (8.4.8) and the choice of $k_0 > 0$ in (8.4.5))

$$S(t)u_0 = (A + k_0I)^{-1}(A_{p_0}S(t)u_0) \in \bigcap_{p>1} W^{2m,p}_{\{B_j\}}(\Omega), \tag{8.5.23}$$

so that, when $t > 0$ and $\mu \in (0, \mu_0)$ are fixed, the Schauder type estimate (8.4.22) can be used on $S(t)u_0$ with arbitrary $u_0 \in X^\alpha_{p_0}$ and inequality (8.5.20) follows. The condition (8.5.13) is then a consequence of the same estimates as in formulas (8.5.20)-(8.5.22). The proof is complete. □

Lemma 8.5.3. *For each $\mu \in (0, \mu_0)$ there are $\alpha \in [\alpha_0, 1)$ and $\varepsilon > 0$ such that when $u_0 \in X^\alpha_{p_0}$, we have*

$$F_{p_0}(T(\cdot)u_0) \in C^{\mu+\varepsilon,\mu+\varepsilon}([0, \tau] \times \overline{\Omega}),\ \tau > 0.$$

Moreover, for each $\tau > 0$, any bounded subset $\mathcal{B}$ of $X^{\alpha}_{p_0}$ and all $u_0 \in \mathcal{B}$,

$$\|F_{p_0}(T(\cdot)u_0)\|_{C^{\mu+\epsilon,\mu+\epsilon}([0,\tau]\times\overline{\Omega})} \leq const.(\tau,\mu,\varepsilon,\alpha,\mathcal{B}).$$

Proof. For arbitrary $u_0 \in X^{\alpha}_{p_0}$ there exists an $X^{\alpha}_{p_0}$ solution $u(t) = T(t)u_0$ of (8.4.4), with bounded trajectory, and condition (3.1.12) (see Theorem 3.1.1) reads

$$\sup_{t\in[0,\infty)} \|u(t)\|_{X^{\alpha}_{p_0}} \leq z_1(u_0) \tag{8.5.24}$$

(note that z_1 in (8.5.24) is locally bounded on $X^{\alpha}_{p_0}$). From (8.5.24) and an embedding $X^{\alpha}_{p_0} \subset C^{m_0+\nu}(\overline{\Omega})$, $\nu \in (0, 2m\alpha - \frac{n}{p_0} - m_0)$ (see Assumption 8.4.2 and note that $\alpha \geq \alpha_0$) we obtain

(8.5.25)

$$\sup_{t\in[0,\infty)} \|u(t)\|_{C^{m_0+\nu_\alpha}(\overline{\Omega})} \leq z_1(u_0),$$

$$\text{for } \nu_\alpha \in \left(0, 2m\alpha - \frac{n}{p_0} - m_0 - \left[2m\alpha - \frac{n}{p_0} - m_0\right]\right).$$

Thanks to (8.5.25) the nonlinear term F_{p_0} (being the counterpart of the sum $f + k_0 I$) becomes bounded on $(0,+\infty)\times\Omega$, so that when α is close to 1 the estimate (8.4.20) can be used on the $X^{\alpha}_{p_0}$ solution $u(t) = T(t)u_0$ of (8.4.4) with $q = p_0$ on each cylinder $(0,\tau)\times\Omega$ and we get

$$\|u\|_{W^{1,2m}_{p_0}((0,\tau)\times\Omega)} \leq C(\tau, u_0, p_0). \tag{8.5.26}$$

The equality above ensures that

$$\|u\|_{C^{\nu_0,\nu_0}([0,\tau]\times\overline{\Omega})} \leq C(\tau, u_0, p_0),\ \tau > 0, \tag{8.5.27}$$

where ν_0 was defined in (8.4.2).

It follows next from properties of Hölder functions (see [L-S-U, Chapt.II, §3, Lemma 3.1]) that if for any $h : [0,\tau]\times\overline{\Omega} \longrightarrow R$

$$\sup_{x\in\overline{\Omega}} \|h(\cdot,x)\|_{C^{\delta_1}([0,\tau])} \leq M_1 \tag{8.5.28}$$

and

$$\sup_{t\in[0,\tau]} \left\|\frac{\partial h}{\partial x_i}(t,\cdot)\right\|_{C^{\delta_2}(cl\Omega)} \leq M_2,\ i = 1,\dots,n, \tag{8.5.29}$$

then also

$$\left\|\frac{\partial h}{\partial x_i}\right\|_{C^{\frac{\delta_1\delta_2}{1+\delta_2},\frac{\delta_1\delta_2}{1+\delta_2}}([0,\tau]\times\overline{\Omega})} \leq const.(M_1, M_2, \delta_1, \delta_2),\ i = 1,\dots,n. \tag{8.5.30}$$

If $m_0 > 0$, then basing on (8.5.25), (8.5.27) and applying the above property m_0 times for $l = 1,\dots,m_0$, step by step we verify that for each $\tau > 0$

$$D^l u \in C^{\nu_{\nu_0\alpha},\nu_{\nu_0\alpha}}([0,\tau]\times\overline{\Omega}),\ l = 0,\dots,m_0,\ \nu_{\nu_0\alpha} = \frac{\nu_0\nu_\alpha^{m_0}}{(1+\nu_\alpha)^{m_0}}. \tag{8.5.31}$$

Since $\alpha \in [\alpha_0, 1)$ may be arbitrarily close to 1, increasing ν_α to ν_0 in (8.5.31) we observe further that the Hölder exponents $\nu_{\nu_0 \alpha}$ remain inside $(0, \mu_0)$ and become arbitrarily close to μ_0 defined in (8.4.2). Hence, when $\mu \in (0, \mu_0)$ is fixed it is possible to find $\alpha(\mu) \in [\alpha_0, 1)$ for which (8.5.31) holds with $\nu_{\nu_0 \alpha(\mu)} > \mu$ so that, writing $\varepsilon = \nu_{\nu_0 \alpha(\mu)} - \mu$, formula (8.5.31) may be rewritten in the form

$$D^l u \in C^{\mu+\varepsilon, \mu+\varepsilon}([0,\tau] \times \overline{\Omega}), \ l = 0, \ldots, k, \ \tau > 0. \tag{8.5.32}$$

Next, (8.5.32) and Lipschitz continuity of f lead to the condition

$$f(\cdot, d^{m_0} u) \in C^{\mu+\varepsilon, \mu+\varepsilon}([0,\tau] \times \overline{\Omega}), \ \tau > 0, \tag{8.5.33}$$

which gives immediately

$$F_{p_0}(T(\cdot)u_0) \in C^{\mu+\varepsilon, \mu+\varepsilon}([0,\tau] \times \overline{\Omega}), \ \tau > 0 \tag{8.5.34}$$

(recall that F_{p_0} is a substitution operator defined by the sum $f + k_0 I$ whereas $u(t)$ was used to denote $T(t)u_0$). Moreover, more precise calculations based on the Hölder property described in conditions (8.5.28)-(8.5.30) show that for each $\tau > 0$ the estimate of a composite function $f(\cdot, d^{m_0} u)$ (and hence also of $F_{p_0}(T(\cdot)u_0)$) in the $C^{\mu+\varepsilon,\mu+\varepsilon}([0,\tau] \times \overline{\Omega})$ norm is uniform with respect to u_0 varying in bounded subsets of $X^{\alpha(\mu)}_{p_0}$, i.e. for each $\tau > 0$ and arbitrary bounded subset $\mathcal{B} \subset X^{\alpha(\mu)}_{p_0}$

$$\|F_{p_0}(T(\cdot)u_0)\|_{C^{\mu+\varepsilon,\mu+\varepsilon}([0,\tau]\times\overline{\Omega})} \leq const.(\tau, \mu, \varepsilon, \alpha, \mathcal{B}), \ u_0 \in \mathcal{B}. \tag{8.5.35}$$

(This is in general a consequence of Assumption 8.4.3 which implies that z_1 appearing in the condition (8.5.24) is locally bounded on $X^\alpha_{p_0}$ for $\alpha \in [\alpha_0, 1)$; see Theorem 3.1.1). Lemma 8.5.3 is thus proved. □

Lemma 8.5.4. *For each $\mu \in (0, \mu_0)$ there is an $\alpha \in [\alpha_0, 1)$ such that $U : [0, +\infty) \times X^\alpha_{p_0} \longrightarrow C^{2m+\mu}_{\{B_j\}}(\overline{\Omega})$ is continuous. Moreover, for any $\tau > 0$ and any bounded subset B of $X^\alpha_{p_0}$ the set $\bigcup_{0 \leq t \leq \tau}\{U(t)B\}$ is precompact in $C^{2m+\mu}_{\{B_j\}}(\overline{\Omega})$.*

Proof. For each $u_0 \in X^\alpha_{p_0}$ we have an $X^\alpha_{p_0}$ solution $U(t)u_0 = T(t)u_0 - S(t)u_0$ of the problem

$$\begin{cases} \dot{w} + A_{p_0} w = F_{p_0}(T(t)u_0), \ t > 0, \\ w(0) = 0. \end{cases} \tag{8.5.36}$$

Making use of Lemma 8.5.3, for fixed $\mu \in (0, \mu_0)$ we can choose $\alpha(\mu) \in [\alpha_0, 1)$ sufficiently close to 1 to ensure that for some $\varepsilon > 0$,

$$F_{p_0}(T(\cdot)u_0) \in C^{\mu+\varepsilon,\mu+\varepsilon}([0,\tau] \times \overline{\Omega}), \ \tau > 0, \ u_0 \in X^{\alpha(\mu)}_{p_0}. \tag{8.5.37}$$

Since the problem (8.5.36) starts from zero, the linear estimate (8.4.19) applied to (8.5.36) guarantees, given (8.5.37), that for each $\tau > 0$ and $u_0 \in X^{\alpha(\mu)}_{p_0}$,

(8.5.38) $\quad U(\cdot)u_0$ is a $C^{1+\frac{\mu+\varepsilon}{2m}, 2m+\mu+\varepsilon}([0,\tau] \times \overline{\Omega})$ solution of (8.5.36)

(note the restrictions (8.4.17), (8.4.18) imposed on A and f in Remark 8.4.2 in the case when B_0 has order $m_0 = 0$). Furthermore, according to Lemma 8.5.3, for each $\tau > 0$ and any bounded subset $\mathcal{B}$ of $X^{\alpha(\mu)}_{p_0}$,

$$(8.5.39) \qquad \|U(\cdot)u_0\|_{C^{1+\frac{\mu+\varepsilon}{2m}, 2m+\mu+\varepsilon}([0,\tau]\times\overline{\Omega})} \leq const._1(\tau,\mu,\varepsilon,\alpha,\mathcal{B}), \; u_0 \in \mathcal{B}.$$

From (8.5.39),

$$(8.5.40) \qquad \bigcup_{0\leq t\leq \tau} \{U(t)\mathcal{B}\} \text{ is bounded in } C^{2m+\mu+\varepsilon}(\overline{\Omega}),$$

and hence, compactness of U follows immediately from compactness of the embedding $C^{2m+\mu+\varepsilon}(\overline{\Omega}) \subset C^{2m+\mu}(\overline{\Omega})$.

In order to prove continuity of $U : [0,+\infty) \times X^{\alpha}_{p_0} \longrightarrow C^{2m+\mu}_{\{B_j\}}(\overline{\Omega})$ it suffices to note that because of (8.5.8)

$$(8.5.41) \qquad U : [0,+\infty) \times X^{\alpha(\mu)}_{p_0} \longrightarrow X^{\alpha(\mu)}_{p_0} \text{ is continuous,}$$

and, as a consequence of the first part of this proof, for each $\tau > 0$ and any bounded subset $\mathcal{B}$ of $X^{\alpha(\mu)}_{p_0}$

$$(8.5.42) \qquad \bigcup_{0\leq t\leq \tau} \{U(t)\mathcal{B}\} \text{ is precompact in } C^{2m+\mu}(\overline{\Omega}).$$

Since the conditions (8.5.41), (8.5.42) ensure that $U(t_n,u_n) \to U(t_0,u_0)$ in $C^{2m+\mu}(\overline{\Omega})$ whenever $(t_n,u_n) \to (t_0,u_0)$ in $[0,+\infty)\times X^{\alpha(\mu)}_{p_0}$, the proof of Lemma 8.5.4 is complete. □

Because $\mathcal{X}^{\mu} \subset \bigcap_{\alpha\in(0,1)} X^{\alpha}_{p_0}$, from the condition (8.5.38) of Lemma 8.5.4 it follows further that for each $\mu \in (0,\mu_0)$ there is $\alpha(\mu) \in [\alpha_1, 1)$ such that

$$U(t)(\mathcal{X}^{\mu}) \subset U(t)(X^{\alpha(\mu)}_{p_0}) \subset C^{2m+\mu+\varepsilon}_{\{B_j\}}(\overline{\Omega})$$
$$\subset cl_{C^{2m+\mu}(\overline{\Omega})}\left(\bigcup_{\mu<\nu} C^{2m+\nu}_{\{B_j\}}(\overline{\Omega})\right) = \mathcal{X}^{\mu}, \; t \geq 0.$$

Since also for all $\mu \in (0,\mu_0)$, $\alpha \in [\alpha_1, 1)$ an embedding $\mathcal{X}^{\mu} \subset X^{\alpha}_{p_0}$ is compact and continuous, Lemma 8.5.4 implies immediately

Lemma 8.5.5. *For each $\mathcal{Y} \in \mathcal{F}$, $U(t) : \mathcal{Y} \longrightarrow \mathcal{Y}$ is completely continuous for $t \geq 0$.*

Recalling next that $T(t) = S(t) + U(t)$, $t \geq 0$, we obtain from Lemmas 8.5.2, 8.5.4 the required inclusions

$$T(t)(\mathcal{X}^{\mu}) \subset \mathcal{X}^{\mu}, \; t \geq 0, \; \mu \in (0,\mu_0).$$

Hence, since the families $S(t) : \mathcal{X}^{\mu} \longrightarrow \mathcal{X}^{\mu}$ and $U(t) : \mathcal{X}^{\mu} \longrightarrow \mathcal{X}^{\mu}$, $t \geq 0$, have already been proved to be continuous with respect to a pair of arguments (see Lemmas 8.5.2, 8.5.4), we obtain finally C^0 continuity of $T(t) : \mathcal{X}^{\mu} \longrightarrow \mathcal{X}^{\mu}$, $t \geq 0$. From this and Observation 8.4.3 we thus conclude:

Lemma 8.5.6. *For each $\mathcal{Y} \in \mathcal{F}$, $T(t) : \mathcal{Y} \longrightarrow \mathcal{Y}$, $t \geq 0$, is a C^0 semigroup of solutions of the problem* (8.4.1).

Lemma 8.5.7. *For each $\mathcal{Y} \in \mathcal{F}$, $T(t) : \mathcal{Y} \longrightarrow \mathcal{Y}$, $t \geq 0$, is completely continuous. Moreover, for each $\mu \in (0, \mu_0)$ there is $\alpha(\mu) \in [\alpha_0, 1)$ such that whenever $\alpha \in [\alpha(\mu), 1)$ and B is a bounded subset of $X^\alpha_{p_0}$, we have*

$$T(t)B \text{ is precompact in } \mathcal{X}^\mu \text{ for } t > 0. \tag{8.5.43}$$

Proof. Given Observation 8.4.3 only Hölder's $\mathcal{Y}$ needs to be considered. The linear estimate (8.4.19) and Lemma 8.5.3 ensure that for each $\mu \in (0, \mu_0)$, $\tau > 0$ and any bounded subset B of $\mathcal{X}^\mu$, $\bigcup_{0 \leq t \leq \tau} T(t)B$ is bounded in $\mathcal{X}^\mu$. Next, from the condition (8.5.40) and Lemma 8.5.2, for each $\mu \in (0, \mu_0)$ there are $\alpha(\mu) \in [\alpha_1, 1)$ and $\varepsilon > 0$ such that both $U(\tau)B$ and $S(\tau)B$, $\tau > 0$, are bounded in $C^{2m+\mu+\varepsilon}(\overline{\Omega})$ whenever B is bounded in $X^{\alpha(\mu)}_{p_0}$. Since $T(t)$ is the sum $S(t) + U(t)$ and the embedding $C^{2m+\mu+\varepsilon}(\overline{\Omega}) \subset C^{2m+\mu}(\overline{\Omega})$ is compact, the proof of Lemma 8.5.7 is complete. □

Lemma 8.5.8. *For each $\mu \in (0, \mu_0)$ and all $\alpha \in [\alpha(\mu), 1)$*

$$\omega_{X^\alpha_{p_0}}(u_0) = \omega_{\mathcal{X}^\mu}(u_0), \ u_0 \in X^\alpha_{p_0},$$

where $\omega_{\mathcal{Y}}(u_0)$ denotes an ω-limit set of u_0 with respect to the topology of $\mathcal{Y}$.

Proof. Since the $\mathcal{X}^\mu$ topology is stronger than that of $X^\alpha_{p_0}$ (see Lemma 8.5.1), we have $\omega_{\mathcal{X}^\mu}(u_0) \subset \omega_{X^\alpha_{p_0}}(u_0)$. Let us fix $\mu \in (0, \mu_0)$ and for $\alpha(\mu)$ as in Lemma 8.5.7 choose $\alpha \in (\alpha(\mu), 1)$ and take any $v_0 \in \omega_{X^\alpha_{p_0}}$. The positive orbit $\gamma^+(u_0)$ is bounded in $X^\alpha_{p_0}$ (see Theorem 3.1.1), so that $T(1)\gamma^+(u_0)$ is precompact in $\mathcal{X}^\mu$ as a result of Lemma 8.5.7. From the characterization of *ω-limit sets*, $T(t_j)u_0 \overset{X^\alpha_{p_0}}{\longrightarrow} v_0$ for some sequence $\{t_j\}_{j \in N}$, $t_j \geq 1$, $t_j \to +\infty$. Since $\{T(t_j)u_0\}_{j \in N} \subset T(1)\gamma^+(u_0)$, there is a subsequence $\{T(t_{j'})u_0\}$ convergent in $\mathcal{X}^\mu$. The only possible limit of this is v_0, therefore $v_0 \in \omega_{\mathcal{X}^\mu}(u_0)$. Lemma 8.5.8 is proved. □

Lemma 8.5.9. *For each $\mathcal{Y} \in \mathcal{F}$, if for some $V \subset \mathcal{Y}$ an a priori estimate* (8.4.11) *is asymptotically independent of $u_0 \in V$, then there is a bounded subset $\mathcal{P}_1$ of $\mathcal{Y}$ attracting (with respect to $T(t) : \mathcal{Y} \longrightarrow \mathcal{Y}$) each point of V, i.e.*

$$dist_{\mathcal{Y}}(T(t)u_0, \mathcal{P}_1) \overset{t \to +\infty}{\longrightarrow} 0, \ u_0 \in V.$$

Proof. Given Corollary 4.1.4 only Hölder's $\mathcal{Y}$ needs to be considered. Thus, let $\mu \in (0, \mu_0)$ be fixed and $\mathcal{Y} = \mathcal{X}^\mu$. According to Lemma 8.5.1 we then know that V, as a subset of $\mathcal{X}^\mu$, must be contained in $X^{\alpha(\mu)}_{p_0}$ with $\alpha(\mu)$ as in Lemma 8.5.7. From Corollary 4.1.4 there is a bounded and closed subset $\mathcal{P}$ of $X^{\alpha(\mu)}_{p_0}$ such that

$$dist_{X^{\alpha(\mu)}_{p_0}}(T(t)u_0, \mathcal{P}) \to 0 \text{ when } t \to +\infty. \tag{8.5.44}$$

Define

$$\mathcal{P}_1 = \bigcup_{u_0 \in V} \omega_{X^{\alpha(\mu)}_{p_0}}(u_0). \tag{8.5.45}$$

From Proposition 1.1.1 it is clear that $\mathcal{P}_1$ also attracts points of V with respect to $T(t) : X_{p_0}^{\alpha(\mu)} \longrightarrow X_{p_0}^{\alpha(\mu)}$. Moreover, as a sum of invariant sets $\omega_{X_{p_0}^{\alpha(\mu)}}(u_0)$, $\mathcal{P}_1$ is itself invariant, i.e.

$$T(t)\mathcal{P}_1 = \mathcal{P}_1, \ t \geq 0.$$

Hence, $\mathcal{P}_1 \subset \mathcal{P}$ and therefore $\mathcal{P}_1$ is bounded in $X_{p_0}^{\alpha(\mu)}$. Given Lemmas 8.5.7, 8.5.8, $\mathcal{P}_1$ is thus precompact and, of course, bounded in $\mathcal{X}^\mu$ and

$$\mathcal{P}_1 = \bigcup_{u_0 \in V} \omega_{\mathcal{X}^\mu}(u_0). \tag{8.5.46}$$

Since for each $u_0 \in V$ a positive orbit $\gamma^+(u_0)$ is bounded in $X_{p_0}^{\alpha(\mu)}$ (Thoerem 3.1.1), then by Lemma 8.5.7,

$$T(1)\gamma^+(u_0) = \{T(t)u_0 : t \geq 1\} \text{ is precompact in } \mathcal{X}^\mu. \tag{8.5.47}$$

However, from the linear estimate (8.4.19), $\{T(t)u_0 : t \in [0,1]\}$ is bounded in $\mathcal{X}^\mu$, so that, recalling (8.5.47), the positive orbit $\gamma^+(u_0)$ is bounded in $\mathcal{X}^\mu$. Referring, given Lemma 8.5.7, to Proposition 1.1.1, we observe that $\bigcup_{u_0 \in V} \omega_{\mathcal{X}^\mu}(u_0)$ attracts each point of V with respect to $T(t) : \mathcal{X}^\mu \longrightarrow \mathcal{X}^\mu$. The proof of Lemma 8.5.8 is finished. □

Theorem 8.4.1 follows directly from Lemmas 8.5.6, 8.5.7, 8.5.9. For the proof of Theorem 8.4.2 note that if (8.4.25) holds with some nonempty V, then V must contain a nonempty invariant and bounded subset of $\mathcal{Y}'$ (see (8.5.45), (8.5.46)), so that from the condition (8.5.43) of Lemma 8.5.7 and the result of Proposition 3.2.1 for each $\mathcal{Y} \in \mathcal{F}$ the set $V \cap \mathcal{Y}$ is nonempty. Moreover, $V \cap \mathcal{Y}$ is closed in $\mathcal{Y}$ whenever $\mathcal{Y} \subset \mathcal{Y}'$. Thus for $\mathcal{Y} \in \mathcal{F}$, $\mathcal{Y} \subset \mathcal{Y}'$ the existence of a global attractor $\mathcal{A}_{V\cap\mathcal{Y}}$ for $T(t)$ restricted to a metric space $V \cap \mathcal{Y}$ (i.e. in a topology of $\mathcal{Y}$) is a consequence of Theorem 8.4.1 and the results of Corollary 1.1.8.

In order to prove that all attractors coincide, let us note first that since $\mathcal{Y} \subset \mathcal{Y}'$ (see Lemma 8.5.1) $\mathcal{A}_{V\cap\mathcal{Y}}$ is bounded in $\mathcal{Y}'$ and therefore, since $\mathcal{A}_{V\cap\mathcal{Y}}$ is invariant,

$$\mathcal{A}_{V\cap\mathcal{Y}} \subset \mathcal{A}_{V\cap\mathcal{Y}'} \ \text{ for } \ \mathcal{Y} \subset \mathcal{Y}', \ \mathcal{Y} \in \mathcal{F}. \tag{8.5.48}$$

It is also clear that $\mathcal{A}_{V\cap\mathcal{Y}'}$ must be bounded in $X_{p_0}^{\alpha_0}$ (which is the largest space of the family $\mathcal{F}$), and hence the result of Proposition 3.2.1 together with the condition (8.5.43) shows that $\mathcal{A}_{V\cap\mathcal{Y}'}$ is bounded in $\mathcal{Y}$ for any $\mathcal{Y} \subset \mathcal{Y}'$, $\mathcal{Y} \in \mathcal{F}$. However, the latter implies the required inclusions

$$\mathcal{A}_{V\cap\mathcal{Y}'} \subset \mathcal{A}_{V\cap\mathcal{Y}} \ \text{ for } \ \mathcal{Y} \subset \mathcal{Y}', \ \mathcal{Y} \in \mathcal{F},$$

which in the light of (8.5.48) finishes the proof of Theorem 8.4.2.

Remark 8.5.2. In a special case when $V = Y = X_{p_0}^{\alpha}$ for some $\alpha \in [\alpha_0, 1)$, the proof of Theorems 8.4.1, 8.4.2 reduces in the main part to direct verification of all the assumptions of J. K. Hale theorem [HA 2, Cor. 3.9.5]:

Proposition 8.5.1. *Suppose that X_1, X_2 are Banach spaces, and*

(i) X_1 is compactly embedded in X_2, X_1 dense in X_2.

(ii) $T(t) = S(t) + U(t) : X_j \longrightarrow X_j$, $S(t) : X_j \longrightarrow X_j$ *is a linear semigroup with norm* $\leq K(t)$ *and* $K(t) \to 0$ *as* $t \to +\infty$, $j = 1, 2$.
(iii) $U(t) : X_2 \longrightarrow X_1$ *is continuous, and for any* $\tau \geq 0$, *if* B, $\{U(t)B : 0 \leq t \leq \tau\}$ *are bounded in* X_2, *then* $\{U(t)B : 0 \leq t \leq \tau\}$ *is bounded in* X_1.
(iv) $T(t)$ *is point dissipative in* X_2.

Then there is a global attractor $\mathcal{A}$ *for* $T(t)$ *in* X_2 *and* $T(t)$ *is bounded dissipative in* X_1.

If, in addition,

(v) $U(t)$ *is conditionally completely continuous in* X_j, $j = 1, 2$,

then $\mathcal{A} \subset X_1$ *and is a global attractor for* $T(t)$ *in* X_1.

Nevertheless, in applications V may be a proper subset of Y, so that a global attractor for $T(t)$ can then be obtained only on a metric subspace $V \cap \mathcal{Y}$ of $\mathcal{Y}$. For this purpose the proof of Theorems 8.4.1, 8.4.2 is slightly more complicated and, although the idea of *dissipativeness in two spaces* ([HA 2, Sec. 3.9]) is generally followed, in spite of the verification of the assumptions required in Proposition 8.5.1 (Lemmas 8.5.1-8.5.6 above) additional smoothing properties of the semigroup generated by (8.4.1) were also considered (see Lemmas 8.5.7-8.5.9).

Comparing Theorems 8.4.1, 8.4.2 of the present section with those of Chapter 4 (see Remark 8.4.3) it is easy to see that the above regularity results are applicable to the examples (in the case of bounded domains) listed in Chapter 6 with the limitations that Theorems 8.4.1, 8.4.2 have not been proved here for systems but only for a single equation and that for the Dirichlet boundary condition the implication (8.4.18) should also hold. Hence, only a very brief description of the celebrated *Cahn-Hilliard* equation will be given here.

Example 8.5.1. Consider the problem (see (6.4.1)-(6.4.3))

$$\textbf{(C-H)} \begin{cases} u_t = \Delta\left[-\varepsilon^2 \Delta u + f(u)\right], \ x \in \Omega, \ t > 0, \\ \frac{\partial u}{\partial N} = \frac{\partial(\Delta u)}{\partial N} = 0, \ x \in \partial\Omega, \ t > 0, \\ u(0, x) = u_0(x), \ x \in \Omega, \end{cases}$$

where Ω is a bounded domain in R^n, $n \leq 3$, having $C^{4+\text{Lipschitz}}$ regular boundary $\partial\Omega$ and the nonlinear term $f \in C^{2+\text{Lipschitz}}(R)$ satisfies the requirements (6.4.5)-(6.4.7).

Since the triple $(\Delta^2, \{\frac{\partial}{\partial N}, \frac{\partial(\Delta)}{\partial N}\}, \Omega)$ forms a regular elliptic boundary value problem, for the sample choice of parameters

$$\alpha_0 = \frac{5}{6}, \ p_0 = 5$$

(i.e. for $X_{p_0} := L^5(\Omega)$), the requirements of Assumptions 8.4.1, 8.4.2 are satisfied with $\mu_0 = \frac{1}{180}$. The existence of the Lyapunov functional for **(C-H)** and the property that the spatial average $\overline{u}(t) = |\Omega|^{-1} \int_\Omega u(t)dx$ of any solution of **(C-H)** is independent of $t \geq 0$ allow us to obtain in Chapter 6 (see (6.4.24)) an a priori

$H^2(\Omega)$ estimate of solutions of **(C-H)**:

$$\|u(t)\|_{H^2(\Omega)} \leq c_1(\|u_0\|_{X_5^{\frac{5}{8}}}),\ u_0 \in X_5^{\frac{5}{8}}, \tag{8.5.49}$$

with $c_1 : [0,+\infty) \to [0,+\infty)$ locally bounded. This estimate is a counterpart of (8.4.11) from Assumption 8.4.3 although to fulfill condition (8.4.12) we cannot directly use (6.4.26) because now **(C-H)** is considered in $L^5(\Omega)$ (not in $L^2(\Omega)$ as previously).

However, based on (1.2.36) and Remark 1.2.1, for $n \leq 3$, we have

$$\begin{cases} H^2(\Omega) \subset L^5(\Omega), \\ H^2(\Omega) \subset C(\overline{\Omega}), \\ \|u\|_{W^{2,5}(\Omega)} \leq const.\|u\|_{H^2(\Omega)}^{\frac{1}{4}}\|u\|_{W^{3,5}(\Omega)}^{\frac{3}{4}} \\ \qquad \leq const.'\|u\|_{H^2(\Omega)}^{\frac{1}{4}}\|u\|_{X_5^{\frac{5}{6}}}^{\frac{3}{4}}, \\ \|u\|_{W^{1,10}(\Omega)} \leq const.\|u\|_{H^2(\Omega)}^{\frac{5}{8}}\|u\|_{W^{3,5}(\Omega)}^{\frac{3}{8}} \\ \qquad \leq const.'\|u\|_{H^2(\Omega)}^{\frac{5}{8}}\|u\|_{X_5^{\frac{5}{6}}}^{\frac{3}{8}}. \end{cases} \tag{8.5.50}$$

Calculating as in (6.4.26) and using (8.5.50) we obtain further

(8.5.51)

$$\begin{aligned} &\|\Delta f(u(t)) + \delta u\|_{L^5(\Omega)} \\ &\leq \|f'(u(t))\|_{L^\infty(\Omega)}\|\Delta u\|_{L^5(\Omega)} + \|f''(u(t))\|_{L^\infty(\Omega)}\|\nabla u\|_{L^{10}(\Omega)}^2 + \|\delta u\|_{L^5(\Omega)} \\ &\leq \|f'(u(t))\|_{L^\infty(\Omega)} const.'\|u\|_{H^2(\Omega)}^{\frac{1}{4}}\|u\|_{X_5^{\frac{5}{6}}}^{\frac{3}{4}} \\ &\quad + \|f''(u(t))\|_{L^\infty(\Omega)} const.'^2\|u\|_{H^2(\Omega)}^{\frac{5}{4}}\|u\|_{X_5^{\frac{5}{6}}}^{\frac{3}{4}} + c\|\delta u\|_{H^2(\Omega)} \\ &\leq \sup_{|s| \leq c\|u(t)\|_{H^2(\Omega)}} \Big(const.'|f'(s)|\|u\|_{H^2(\Omega)}^{\frac{1}{4}} + const.'^2|f''(s)|\|u\|_{H^2(\Omega)}^{\frac{5}{4}} \\ &\quad + c\delta\|u(t)\|_{H^2(\Omega)}\Big)\Big(1 + \|u(t)\|_{X_5^{\frac{5}{6}}}^{\frac{3}{4}}\Big) \\ &=: g(\|u(t)\|_{H^2(\Omega)})\left(1 + \|u\|_{X_5^{\frac{5}{6}}}^{\frac{3}{4}}\right), \end{aligned}$$

with an increasing function $g : [0,+\infty) \to [0,+\infty)$.

In this example we thus have

$$\mathcal{F} = \left\{\mathcal{X}^\mu : \mu \in \left(0, \frac{1}{180}\right)\right\} \cup \left\{X_5^\alpha : \alpha \in \left[\frac{5}{6}, 1\right)\right\}$$

and, by Theorem 8.4.1:

Corollary 8.5.1. *For each $\mathcal{Y} \in \mathcal{F}$ a completely continuous C^0 semigroup of solutions of the Cahn-Hilliard problem is generated on $\mathcal{Y}$.*

Since each constant function is a stationary solution of **(C-H)**, a global attractor may be shown to exist only on a proper metric subspace of $\mathcal{Y}$. Thus, for any $\gamma > 0$, we define

$$V_\gamma := \{w \in X_5^{\frac{5}{8}} : |\overline{w}| \leq \gamma\} \tag{8.5.52}$$

(see (6.4.33)). Although in considerations in Chapter 4, we did not show explicitly that the introductory $H^2(\Omega)$ estimate (8.5.49) of solutions to **(C-H)** was asymptotically independent of $u_0 \in V_\gamma$ (point dissipativeness was proved based on the boundedness of the set of stationary solutions and properties of the Lyapunov function), nevertheless (8.5.49) must be asymptotically independent of $u_0 \in V_\gamma$ as a consequence of the existence of a global attractor for **(C-H)** in $H_\gamma^{\frac{1}{2}}$ (see Proposition 6.4.3). Since $T(t)V_\gamma \subset V_\gamma$ for $t \geq 0$ (see (6.4.9)) and V_γ is closed in $X_5^{\frac{5}{8}}$, we obtain in the light of Theorems 8.4.1, 8.4.2 that, for arbitrary $\gamma > 0$:

Corollary 8.5.2. *For each $\mathcal{Y} \in \mathcal{F}$ there is a global attractor $\mathcal{A}_\mathcal{Y}$ for the semigroup generated by the Cahn-Hilliard problem on a metric space $V_\gamma \cap \mathcal{Y}$ (in a topology of Y). Also, $\mathcal{A}_\mathcal{Y}$ is independent of the choice of $\mathcal{Y} \in \mathcal{F}$.*

8.6. Equations with monotone operators

In this section we would like to take a glance at degenerate parabolic equations. We are focused here on quasilinear problems which admit the form of the abstract evolutionary equation (8.6.2) with *monotone* principal part. A particular feature of such problems, in comparison with those of semilinear type, is that the dissipation properties of the nonlinear main part operator are often much stronger than is observed in the linear case. This may be easily seen if one considers e.g. the Dirichlet initial boundary value problem for the equation

$$u_t = div(|\nabla u|^{p-2}\nabla u) + \lambda u + g,\ t > 0,\ x \in \Omega \tag{8.6.1}$$

($p > 2$), for which the corresponding semigroup possesses (as shown below) a compact global attractor for all values of λ. Whereas, obviously, its linear counterpart

$$u_t = \Delta u + \lambda u + g,\ t > 0,\ x \in \Omega,$$

does not possesses this property, unless λ is strictly less then the first positive eigenvalue of the negative Laplacian with Dirichlet boundary conditions (see Remark 5.3.2).

8.6.1. Formulation of the problem. Let X be a reflexive Banach space, H be a Hilbert space such that X is dense in H and denote by X^* the topological dual of X with $\langle\cdot,\cdot\rangle_{X^*,X}$ indicating the pairing between elements of X^* and X.

Definition 8.6.1. *A single valued operator $\mathcal{M} : X \to X^*$ is called*

- ***hemicontinuous**, if*

$$R \ni s \to \langle\mathcal{M}(v + sw), z\rangle_{X^*,X} \in R$$

is a continuous function for all $v, w, z \in X$,

• ***monotone***, *if*

$$\langle \mathcal{M}(v) - \mathcal{M}(w), v - w \rangle_{X^*,X} \geq 0, \ v, w \in X,$$

• ***coercive***, *if*

$$\lim_{n \to +\infty} \frac{\langle \mathcal{M}(v_n), v_n \rangle_{X^*,X}}{\|v_n\|_X} = +\infty,$$

for all $\{v_n\} \subset X$ *such that* $\|v_n\|_X \to +\infty$.

Assumption 8.6.1. *Let*

(i) $X \subset H \subset X^*$ *(with both embeddings continuous; H being identified with its dual),*
(ii) $\mathcal{M} : X \to X^*$ *be a nonlinear, monotone, coercive and hemicontinuous operator defined everywhere on* X,
(iii) $\mathcal{N} : H \to H$ *be a globally Lipschitz map.*

From the results of [BR, p. 26] (see also [BAR, Theorem 1.3]) it is known that the operator $\mathcal{M}_H : H \to H$ defined by

$$\mathcal{M}_H(u) = \mathcal{M}(u) \ \text{ for } \ u \in D(\mathcal{M}_H) := \{v \in V : \mathcal{M}(v) \in H\}$$

is a *maximal monotone* operator in H.

Consider thus the abstract Cauchy problem in H:

$$\begin{cases} \frac{du}{dt}(t) + \mathcal{M}_H(u(t)) + \mathcal{N}(u(t)) = 0, \ t > 0, \\ u(0) = u_0. \end{cases} \tag{8.6.2}$$

Definition 8.6.2. *The following two notions of the solution are related to* (8.6.2) *(see* [BR, p. 64]*):*

• ***strong solution***, *which is a function* $u \in C([0,+\infty); H)$ *absolutely continuous on any compact subinterval of* $(0, +\infty)$ *and such that*

$$u(t) \in D(\mathcal{M}_H) \text{ for a. e. } t \in (0, +\infty),$$
$$\frac{du}{dt}(t) + \mathcal{M}_H(u(t)) + \mathcal{N}(u(t)) = 0 \ \text{ for a. e. } \ t \in (0, +\infty),$$
$$u(0) = u_0;$$

• ***weak solution***, *being a function* $u \in C([0,+\infty); H)$ *such that there is a sequence* $\{u_n\}$ *of strong solutions of* (8.6.2) *convergent to* u *in* $C([0,\tau]; H)$ *for each* $\tau > 0$.

Based on [BR, Theorem 3.17, Remark 3.14] we obtain the existence of a global weak solution $u(\cdot, u_0)$ to (8.6.2) for each $u_0 \in H$. Furthermore, $u(\cdot, u_0)$ is Lipschitz continuous on each $[0, \tau]$ (and therefore is a strong solution to (8.6.2)) whenever $u_0 \in D(\mathcal{M}_H)$.

Let us mention that $u(\cdot, u_0)$ is also continuous at u_0 uniformly for t varying in compact subintervals of $[0, +\infty)$. Indeed, if $u_0, u_1 \in D(\mathcal{M}_H)$ and $\tau > 0$, we obtain from (8.6.2) that

$$\begin{aligned}\frac{d}{dt}\|u(t,u_0) - u(t,u_1)\|_H^2 &= -\langle \mathcal{M}(u(t,u_0)) - \mathcal{M}(u(t,u_1)), u(t,u_0) - u(t,u_1)\rangle_H \\ &\quad - \langle \mathcal{N}(u(t,u_0)) - \mathcal{N}(u(t,u_1)), u(t,u_0) - u(t,u_1)\rangle_H \\ &\leq L_{\mathcal{N}}\|u(t,u_0) - u(t,u_1)\|_H^2 \quad \text{for a. e. } t > 0\end{aligned}$$

($L_{\mathcal{N}}$ being the Lipschitz constant of $\mathcal{N}$), and consequently, integrating over $[0, \tau]$ and using Lemma 1.2.9,

$$\sup_{t\in[0,\tau]} \|u(t,u_0) - u(t,u_1)\|_H \leq C(\tau)\ \|u_0 - u_1\|_H.$$

This justifies the following existence result.

Proposition 8.6.1. *Under Assumption* 8.6.1 *a C^0 semigroup $\{T(t)\}$ of weak solutions of* (8.6.2) *is defined on $cl_H(D(\mathcal{M}_H))$. Additionally, whenever $u_0 \in D(\mathcal{M}_H)$, $T(\cdot)u_0 = u(\cdot, u_0)$ is a strong solution of* (8.6.2) *which is also Lipschitz continuous on $[0, \tau]$ for each $\tau > 0$.*

Our further concern will be to prove the existence of a global attractor for $\{T(t)\}$ under some reasonable conditions on $\mathcal{M}$.

8.6.2. Complete continuity and dissipativeness in H.

Assumption 8.6.2. *Let Assumption* 8.6.1 *hold and $\{T(t)\}$ denote a C^0 semigroup of weak solutions $T(t)u_0 = u(t, u_0)$ of* (8.6.2) *resulting from Proposition* 8.6.1. *Assume additionally that*

(8.6.3) *the embedding $X \subset H$ is compact*

and that $\mathcal{M}$ satisfies the following two conditions:

(8.6.4)

$$\begin{aligned}&\exists_{\omega_1>0}\ \exists_{c_1\in R}\ \exists_{p>1}\ \forall_{u_0\in D(\mathcal{M}_H)} \\ &\langle \mathcal{M}(u(t,u_0)), u(t,u_0)\rangle_{X^*,X} \geq \omega_1\|u(t,u_0)\|_X^p + c_1 \quad \textit{for a. e. } t > 0\end{aligned}$$

and

(8.6.5)

$$\begin{aligned}&\exists_{T>0}\ \exists_{\theta>1}\ \exists_{C:[0,+\infty)\times[0,+\infty)\to R,\ \text{locally bounded}} \\ &\int_0^T \|\mathcal{M}u(s,u_0)\|_{X^*}^\theta ds \leq C(\|u_0\|_H, \tau).\end{aligned}$$

Theorem 8.6.1. *Under Assumption* 8.6.2 *and the additional restriction on the parameter p appearing in* (8.6.4),

(8.6.6) *either $p > 2$, or $p = 2$ and $L_{\mathcal{N}}E^2 < \omega_1$*

(E being the constant for the embedding $X \subset H$), the semigroup $\{T(t)\}$ has a global attractor in $cl_H(D(\mathcal{M}_H))$.

For the proof of this result we shall verify the conditions of Corollary 1.1.8 which will be done in the lemmas below.

Lemma 8.6.1. *Under Assumption* 8.6.2, *the semigroup* $\{T(t)\}$ *on* $cl_H D(\mathcal{M}_H)$ *is completely continuous.*

Proof. To prove compactness of the semigroup $\{T(t)\}$ it suffices, according to Corollary 3.3.4, to consider bounded subsets of H having the form $\mathcal{B} = B_H(r) \cap D(\mathcal{M}_H)$ ($B_H(r)$ being the ball in H with radius r centered at zero).

From equation (8.6.2) and inequality (8.6.4) it follows that

$$\begin{aligned}\frac{1}{2}\frac{d}{dt}\|u\|_H^2 &= -\langle \mathcal{M}(u), u\rangle_{X^*,X} - \langle B(u), u\rangle_H \\ &\le -\omega_1\|u\|_X^p - c_1 + L_{\mathcal{N}}\|u\|_H^2 + const.\|u\|_H \\ &\le -\frac{\omega_1}{2}\|u\|_X^p + const._1\|u\|_H^2 + \frac{1}{2}\end{aligned}$$

(note that $p > 2$). Integrating over $(0, \tau)$ we obtain

$$\|u(\tau)\|_H^2 + \omega_1 \int_0^\tau \|u(s)\|_X^p ds \le \|u_0\|_H^2 + 2const._1 \int_0^\tau \|u(s)\|_H^2 ds + \tau. \tag{8.6.7}$$

This shows in particular that (see e.g. Lemma 1.2.9)

$$\sup_{u_0\in\mathcal{B}} \sup_{t\in[0,\tau]} \|u(t, u_0)\|_H \le const._2(\mathcal{B}, \tau) \quad \text{for each} \quad \tau > 0 \tag{8.6.8}$$

and

$$\int_0^\tau \|u(s, u_0)\|_X^p ds \le C_1(\|u_0\|_H, \tau). \tag{8.6.9}$$

Now take $\tau > 0$ and $\theta > 1$ according to the assumption (8.6.5). Using equation (8.6.2) again we find

$$\begin{aligned}\left\|\frac{du}{dt}\right\|_{X^*}^\theta &\le 2^\theta\left(\|\mathcal{M}(u)\|_{X^*}^\theta + \|\mathcal{N}(u)\|_{X^*}^\theta\right) \\ &\le const._3\left(1 + \|\mathcal{M}(u)\|_{X^*}^\theta + \|u\|_H^\theta\right),\end{aligned}$$

which in the light of (8.6.9), (8.6.8) ensures that

$$\int_0^\tau \left\|\frac{du}{dt}(s)\right\|_{X^*}^\theta ds \le C_2(\|u_0\|_H, \tau) \tag{8.6.10}$$

with a locally bounded function $C_2 : [0, +\infty) \times [0, +\infty) \to [0, +\infty)$.

Define further the subset $\tilde{\mathcal{B}} \subset C([0, \tau]; H)$,

$$\tilde{\mathcal{B}} := \{T(\cdot)u_0 : u_0 \in \mathcal{B}\},$$

where $u(\cdot) = T(\cdot)u_0 \in C([0,\infty); H)$ denotes a weak solution of (8.6.2) resulting from Proposition 8.6.1, and consider a Banach space W,

$$W := \left\{ v \in L^p(0,\tau; X) : \frac{dv}{dt} \in L^\theta(0,\tau; X^*) \right\}$$

endowed with the norm

$$\|v\|_W := \|v\|_{L^p(0,\tau;X)} + \left\| \frac{dv}{dt} \right\|_{L^\theta(0,\tau;X^*)}.$$

As a consequence of (8.6.9), (8.6.10) the set $\tilde{\mathcal{B}}$ is bounded in the norm of W. Therefore, from [LI, Theorem 5.1, Chapter 1],

(8.6.11) $\qquad \tilde{\mathcal{B}}$ is precompact in $L^p(0,\tau; H)$.

Take any sequence $\{u_n\} \subset \mathcal{B}$ and consider the sequence $\{T(\cdot)u_n\} \subset \tilde{\mathcal{B}}$. From (8.6.11) there are a subsequence $\{T(\cdot)u_{n_k}\}$ of $\{T(\cdot)u_n\}$ and $v_0 \in L^p(0,\tau; H)$ such that

$$\left(\int_0^\tau \|T(s)u_{n_k} - v_0(s)\|_H^p ds \right)^{\frac{1}{p}} \to 0 \text{ when } k \to \infty. \tag{8.6.12}$$

Hence the sequence $\{\|T(\cdot)u_{n_k} - v_0(\cdot)\|_H\}$ of real functions

$$\|T(\cdot)u_{n_k} - v_0(\cdot)\|_H : (0,T) \to R$$

converges to zero in $L^p(0,\tau; R)$ and, in particular, there is a subsequence $\{\|T(\cdot)u_{n_{k_l}} - v_0(\cdot)\|_H\}$ such that

$$\|T(\cdot)u_{n_{k_l}} - v_0(\cdot)\|_H \to 0 \text{ a. e. on } (0,\tau). \tag{8.6.13}$$

Using (8.6.13) we have

$$\forall_{t>0}\ \exists_{s\in(0,t)}\ T(s)u_{n_{k_l}} \to v_0(s) \text{ in } H.$$

Therefore

$$T(t)u_{n_{k_l}} = T(t-s)T(s)u_{n_{k_l}} \to T(t-s)v_0(\tau),$$

which proves that the sequence $\{T(t)u_n\}$ has a convergent subsequence. Therefore the semigroup $\{T(t)\}$ is compact, which together with (8.6.8) completes the proof of Lemma 8.6.1. □

Lemma 8.6.2. *Assume* (8.6.6) *and the condition* (8.6.4). *Then the semigroup* $\{T(t)\}$ *on* $cl_H D(\mathcal{M}_H)$ *is bounded dissipative.*

Proof. Consider initial data from $D(\mathcal{M}_H)$ lying in a bounded subset of H. From equation (8.6.2), inequality (8.6.4), Lipschitz continuity of $\mathcal{N}$ and the embedding $X \subset H$ it follows that

$$\begin{aligned}\frac{1}{2}\frac{d}{dt}\|u\|_H^2 &= -\langle \mathcal{M}(u), u\rangle_{V^*,V} - \langle \mathcal{N}(u), u\rangle_H \\ &\le -\omega_1\|u\|_X^p - c_1 + const.(\|u\|_H^2 + 1) \\ &\le -\frac{\omega_1}{2}E^{-p}\|u\|_H^p + const'.,\end{aligned}$$

i.e. the function $y(t) := \|u(t)\|_H^2$ satisfies the Bernoulli type differential inequality (1.2.4). Using a version of Lemma 1.2.4 we obtain that

$$\|u(t,u_0)\|_H^2 \le \max\left\{\|u_0\|_H^2, \left(\frac{2const'.}{\omega_1 E^{-p}}\right)^{\frac{2}{p}}\right\}$$

and, asymptotically,

$$\limsup_{t\to+\infty}\|u(t,u_0)\|_H^2 \le E^2\left(\frac{2const'.}{\omega_1}\right)^{\frac{2}{p}}.$$

The proof is complete. □

Theorem 8.6.1 is now an immediate consequence of Lemmas 8.6.1, 8.6.2 and Corollary 1.1.8.

Remark 8.6.1. The requirements (8.6.4), (8.6.5) are generalizations of the conditions, widely exploited in the literature,

$$\langle \mathcal{M}v, v\rangle_{X^*,X} \ge \omega_1\|v\|_X^p + c_1,\ v \in X, \tag{8.6.14}$$

$$\|\mathcal{M}v\|_{X^*} \le \omega_2(1 + \|v\|_X^{p-1}),\ v \in X \tag{8.6.15}$$

(see [LI], [BAR], [TE 1]). Inequalities (8.6.14), (8.6.15) imply in particular that $D(\mathcal{M}_H)$ is dense in H (see [BAR, p. 141]), although it may be known in advance in many examples. Lemma 8.6.2 is based on [TE 1]; however, sufficiency of (8.6.14), (8.6.15) for the existence of the global attractor was shown in [C-C-D 2] thanks to Lemma 8.6.2. Note that the condition $p > 1$ is sufficient for proving that the semigroup is completely continuous. Also the assumption (8.6.5) is essentially weaker than (8.6.15); this may be seen e.g. in Example 8.6.1 below for which (8.6.15) fails.

Remark 8.6.2. When $p > 1$, the condition (8.6.14) is stronger than coerciveness. Monotonicity and hemicontinuity of $\mathcal{M}$ are connected with the existence of a Gateaux differentiable functional $\mathcal{J} : X \to R$ satisfying $\mathcal{J}'(v) = \mathcal{M}(v)$ for $v \in X$. Recall that the Gateaux derivative $\mathcal{J}'(v)$ is an element of X^* such that

$$\lim_{h\to 0}\frac{\mathcal{J}(v+hw) - \mathcal{J}(v)}{h} = \mathcal{J}'(v)w,\ w \in X.$$

As shown in [LI, Chapter 2], if $\mathcal{J}'(v)$ exists for all $v \in X$, and $\mathcal{J} : X \to R$ is convex, then $\mathcal{J}' : X \to X^*$ is monotone and hemicontinuous.

8.6.3. Degenerate parabolic equations.

Example 8.6.1. Consider the quasilinear parabolic problem (see [B-V 2, pp. 34, 127])

$$(8.6.16)\qquad \begin{cases} u_t = div(|\nabla u|^{p-2}\nabla u) - |u|^{\rho-1}u + f(u),\ t > 0,\ x \in \Omega, \\ u_{|\partial\Omega} = 0, \\ u(0,x) = u_0(x), x \in \Omega, \end{cases}$$

with $p > 2$, $\rho > 1$ and $f : R \to R$ a globally Lipschitz function. Let $H = L^2(\Omega)$, $X = W_0^{1,p}(\Omega) \cap L^{\rho+1}(\Omega)$ and Ω be an open, bounded subdomain of R^n with Lipschitz boundary $\partial\Omega$.

Lemma 8.6.3. *The Banach space X normed by $\|\cdot\|_X = \|\cdot\|_{W_0^{1,p}(\Omega)} + \|\cdot\|_{L^{\rho+1}(\Omega)}$ is reflexive.*

Proof. By the Eberlein-Shmulyan theorem (see [YO 1, p. 141]) it suffices to prove that the unit ball $B_X(0,1)$ is locally sequentially weakly compact. If $\{v_n\} \subset B_X(0,1)$ then there is a subsequence $\{v_{n'}\}$ convergent weakly both in $W_0^{1,p}(\Omega)$ and in $L^{\rho+1}(\Omega)$ to some $v_0 \in L^2(\Omega)$. Since from the characterization given in [G-G-Z] we have

$$X^* = W^{-1,p'}(\Omega) \oplus L^{\frac{\rho+1}{\rho}}(\Omega)\ (\text{ direct sum}),$$

$v_{n'}$ must converge weakly to v_0 in X. The proof is complete. □

Consider further the nonlinear operator $\mathcal{M} : X \to X^*$,

$$(8.6.17)\quad \langle \mathcal{M}(v), w\rangle_{X^*,X} = \int_\Omega |\nabla v|^{p-2}\nabla v \nabla w dx + \int_\Omega |v|^{\rho-1} v w dx,\ w \in V,$$

and denote by $\mathcal{N}$ a substitution operator on H corresponding to $-f$. It may be easily seen that $\mathcal{M}(v) = \mathcal{J}'(v)$ for $v \in X$, where $\mathcal{J}'(v)$ is the Gateaux derivative of the convex functional

$$X \ni v \to \mathcal{J}(v) = \frac{1}{p}\int_\Omega |\nabla v|^p dx + \frac{1}{\rho+1}\int_\Omega |v|^{\rho+1} dx \in R.$$

Moreover, from (8.6.17) and the Young inequality, we have

$$\langle \mathcal{M}(v), v\rangle_{X^*,X} = \|\nabla v\|^p_{L^p(\Omega)} + \|v\|^{\rho+1}_{L^{\rho+1}(\Omega)} \geq \omega_1 \|v\|^\eta_X + c_1,\ v \in X,$$

where $\eta = \min\{p, \rho+1\}$ and c_1 is negative.

The above considerations show that Assumption 8.6.1 is satisfied. Moreover, $D(\mathcal{M}_H)$ is dense in X and the inclusion $X \subset H$ is compact (see (1.2.36)).

Looking next at Assumption 8.6.2 it remains to prove (8.6.5). As a result of Proposition 8.6.1 the solution $u = u(\cdot, u_0)$ corresponding to $u_0 \in D(\mathcal{M}_H)$ satisfies

$$(8.6.18)\qquad \begin{aligned} &\int_\Omega \frac{du}{dt} v dx + \int_\Omega |\nabla u|^{p-2}\nabla u \nabla v dx \\ &+ \int_\Omega |u|^{\rho+1} u v dx - \int_\Omega f(u) v dx = 0,\ v \in X. \end{aligned}$$

Therefore we obtain

$$\frac{1}{2}\frac{d}{dt}\|u\|^2_{L^2(\Omega)} + \|\nabla u\|^p_{L^p(\Omega)} + \|u\|^{\rho+1}_{L^{\rho+1}(\Omega)} \leq L_{\mathcal{N}}\|u\|^2_{L^2(\Omega)} + const.,$$

which ensures the estimate

$$\begin{aligned}(8.6.19)\qquad &\int_0^\tau \left(\|\nabla u(s,u_0)\|^p_{L^p(\Omega)} + \|u(s,u_0)\|^{\rho+1}_{L^{\rho+1}(\Omega)}\right)ds \\ &\leq \mathcal{C}\left(\|u_0\|_{L^2(\Omega)},\tau\right),\ u_0 \in D(\mathcal{M}_H),\ \tau > 0,\end{aligned}$$

with a bounded function $\mathcal{C} : [0+\infty) \times [0,+\infty) \to R$. Next, using (8.6.17), we get the bound for $\|\mathcal{M}(u)\|_{X^*}$,

$$(8.6.20)\qquad \|\mathcal{M}(v)\|_{X^*} \leq \|\nabla v\|^p_{L^p(\Omega)} + \|v\|^\rho_{L^{\rho+1}(\Omega)},\ v \in X,$$

which together with (8.6.19) gives

$$\begin{aligned}\int_0^\tau \|\mathcal{M}(u,u_0)\|^\theta_{X^*} ds &\leq 2^\theta \int_0^\tau \left(\|\nabla u(s,u_0)\|^{\theta(p-1)}_{L^p(\Omega)} + \|u(s,u_0)\|^{\theta\rho}_{L^{\rho+1}(\Omega)}\right)ds \\ &\leq 2^\theta \mathcal{C}(\|u_0\|_{L^2(\Omega)},\tau),\ u_0 \in D(\mathcal{M}_H),\ \tau > 0.\end{aligned}$$

Theorem 8.6.1 is thus applicable and we conclude that:

Theorem 8.6.2. *The semigroup generated by* (8.6.16) *on* $L^2(\Omega)$ *has a global attractor.*

Example 8.6.2. Again let $f : R \to R$ be globally Lipschitz continuous, $\Omega \subset R^n$ be a bounded domain with Lipschitz boundary $\partial\Omega$ and $p > 2$. Consider the problem (see [LI, Chapter 2])

$$(8.6.21)\qquad \begin{cases} u_t = div(|u|^{p-2}\nabla u) + f(u),\ t > 0,\ x \in \Omega, \\ u_{|\partial\Omega} = 0, \\ u(0,x) = u_0(x),\ x \in \Omega, \end{cases}$$

choosing $X = L^p(\Omega)$ and $H = H^{-1}(\Omega)$. Here $H^{-1}(\Omega)$ is endowed with the inner product being an extension of the bilinear form

$$\langle v, w\rangle_H = \langle v, (-\Delta)^{-1}w\rangle_{H^{-1}(\Omega),H^1_0(\Omega)},$$

where $(-\Delta)^{-1}w =: \psi_w$ stands for the $H^1_0(\Omega)$ solution of the Dirichlet boundary value problem

$$\begin{cases} -\Delta\psi_w = w \ \text{ in } \ \Omega, \\ \psi_{w|\partial\Omega} = 0. \end{cases}$$

Let $-f$ generate the Lipschitz operator $\mathcal{N} : H \to H$ and $\mathcal{M} : X \to X^*$ be defined by

$$(8.6.22)\qquad \langle \mathcal{M}(v), w\rangle_{X^*,X} = \frac{1}{p-1}\int_\Omega |v|^{p-2}vw\,dx,\ v,w \in L^p(\Omega).$$

Now $\mathcal{M}(v)$ is seen to coincide with the Gateaux derivative of the functional

$$\mathcal{J}(v) = \frac{1}{p(p-1)} \int_\Omega |v|^p dx;$$

$\mathcal{M}$ is thus monotone and hemicontinuous.

As a consequence of (8.6.22)

$$\langle \mathcal{M}(v), v \rangle_{X^*,X} = \frac{1}{p-1} \|v\|_X^p, \ v \in X,$$

$$\|\mathcal{M}(v)\|_{X^*} \le \frac{1}{p-1} \|v\|_X^{p-1}, \ v \in X.$$

Also $L^p(\Omega) \subset H^{-1}(\Omega)$ is compact (see [LI-MA, Theorem 16.1]) and $D(\mathcal{M}_H)$ is dense in H.

It thus follows from Theorem 8.6.1 that:

Theorem 8.6.3. *The problem* (8.6.21) *defines in* $H^{-1}(\Omega)$ *a* C^0 *semigroup which has a global attractor.*

Bibliographical notes.

- **Section 8.1.** The approach of Section 8.1 to the problem with non-Lipschitz nonlinearities is mostly based on the results of [L-M] and [MAT]. We mention also an interesting paper [PAZ 1] in which a number of aspects appearing in this section are nicely described.
- **Sections 8.2 and 8.3.** The abstract theorem in Section 8.2 follows [HE 1]. Its application to the n-dimensional Navier-Stokes system is new but with the essential use of the results of [GI 2] and [GI-MI].
- **Sections 8.4 and 8.5.** These sections are based on [C-D 5] and refer to the concept of *dissipativeness in two spaces* which comes from [HA 2]. We are grateful to a referee for making us realize the characterization of the space $\mathcal{X}^\mu$ given in (8.4.15).
- **Section 8.6.** This closing section exploits the results of [C-C-D 2], partially [TE 1], and is essentially connected with the theory of monotone operators developed in [LI], [BR], [BAR].

CHAPTER 9

Appendix

9.1. Notation, definitions and conventions

For convenience we collect here the basic assumptions appearing in the main body of the book. These are Assumption 2.1.1 and conditions (A_2), (A_2') of Chapter 3.

Assumption 2.1.1. *Let X be a Banach space, $A : D(A) \to X$ a sectorial and positive operator in X and for some $\alpha \in [0,1)$, $F : X^\alpha \to X$ be Lipschitz continuous on bounded subsets of X^α.*

Condition (A_2). *It is possible to choose*

- *a Banach space Y, with $D(A) \subset Y$,*
- *a locally bounded function $c : [0,+\infty) \to [0,+\infty)$,*
- *a nondecreasing function $g : [0,+\infty) \longrightarrow [0,+\infty)$,*
- *a certain number $\theta \in [0,1)$,*

such that, for each $u_0 \in X^\alpha$, both conditions

$$\|u(t,u_0)\|_Y \leq c(\|u_0\|_{X^\alpha}), \ t \in (0,\tau_{u_0}),$$

and

$$\|F(u(t,u_0))\|_X \leq g(\|u(t,u_0)\|_Y)(1 + \|u(t,u_0)\|^\theta_{X^\alpha}), \ t \in (0,\tau_{u_0}),$$

hold.

Condition (A_2'). *There are given*

- *a Banach space Y, with $D(A) \subset Y$,*
- *a function $c : [0,+\infty) \times (0,+\infty) \to R$ for which $c(\cdot,s)$ is locally bounded for each fixed $s > 0$,*
- *a nondecreasing function $g : [0,+\infty) \longrightarrow [0,+\infty)$,*
- *a number $\theta \in [0,1)$,*

such that, for each $u_0 \in X^\alpha$, both

$$\|u(t,u_0)\|_Y \leq c(\|u_0\|_{X^\alpha}, s), \ 0 < s \leq t < \tau_{u_0},$$

and

$$\|F(u(t,u_0))\|_X \le g(\|u(t,u_0)\|_Y)(1+\|u(t,u_0)\|_{X^\alpha}^{\theta}),\ t\in(0,\tau_{u_0}),$$

hold.

Notation. The notation used in this book is standard.

Let Ω be a domain (open, connected subset) of R^n. By $|\Omega|$ we denote its n-dimensional Lebesgue measure.

Usually in this book we study real Banach spaces. But sometimes the techniques force us to work with the complexification of the space (e.g. when the resolvent operators or *complex interpolation method* are considered). We will not mention this evident procedure directly in the text.

For any real $p\in[1,+\infty)$ we denote by $L^p(\Omega)$ a Banach space of (classes of) all measurable functions $\phi:\Omega\to R$ normed by

$$\|\phi\|_{L^p(\Omega)} = \left(\int_\Omega |\phi(x)|^p dx\right)^{\frac{1}{p}}.$$

Two functions equal a.e. are identified.

When $p=+\infty$ a space $L^\infty(\Omega)$ consists of all measurable functions $\phi:\Omega\to R$ that are *essentially bounded*; i.e.

$$\exists_{K>0}\ |\phi(x)|\le K \ \text{ a. e. on } \Omega.$$

The greatest lower bound of such constants K is denoted by $\operatorname{ess\,sup}_{x\in\Omega}|\phi(x)|$. Again, we identify the functions equal a. e. on Ω.

Recall (see [AD, p. 25]) that when $|\Omega|<+\infty$ and $\phi\in L^\infty(\Omega)$, we have

$$\lim_{p\to+\infty}\|\phi\|_{L^p(\Omega)} = \|\phi\|_{L^\infty(\Omega)}.$$

We also consider the *weighted Lebesgue spaces* $L^p(\Omega;\rho)$, $p\in[1,+\infty)$,

$$L^p(\Omega;\rho) := \left\{\phi\in L^p_{loc}(\Omega) : \int_\Omega \rho(x)|\phi(x)|^p dx<\infty\right\}$$

where $\rho:\Omega\to R$ is a continuous, bounded and positive function. This space is equipped with the norm

$$\|\phi\|_{L^p(\Omega;\rho)} = \left(\int_\Omega \rho(x)|\phi(x)|^p dx\right)^{\frac{1}{p}}.$$

When $\alpha=(\alpha_1,\dots,\alpha_n)$ is an n-tuple of natural numbers N (we admit $0\in N$), we call it a *multi-index* and define $|\alpha|:=\alpha_1+\ldots+\alpha_n$. For a smooth function $\phi:\Omega\to R$ its partial derivative of order α is denoted by

$$D^\alpha\phi := \frac{\partial^{|\alpha|}\phi}{\partial x_1^{\alpha_1}\ldots\partial x_n^{\alpha_n}},$$

where $x=(x_1,\dots,x_n)\in\Omega$.

$C_0^\infty(\Omega)$ stands for the set of all smooth functions with compact support in Ω, having (continuous in Ω) partial derivatives of arbitrary order.

For $m \in N$ and $p \in [1, +\infty]$ we define

$$W^{m,p}(\Omega) := \{\phi \in L^p(\Omega) : D^\alpha\phi \in L^p(\Omega) \text{ for } 0 \leq |\alpha| \leq m\}, \tag{9.1.1}$$

Here $D^\alpha\phi$ denotes the *distributional partial derivative* of ϕ; i.e. $D^\alpha\phi$ is a locally integrable function in Ω (determined uniquely up to sets of measure zero) such that

$$\int_\Omega \phi(x)D^\alpha\psi(x)dx = (-1)^{|\alpha|}\int_\Omega \psi(x)D^\alpha\phi(x)dx$$

for each $\psi \in C_0^\infty(\Omega)$.

The equality (9.1.1) defines the Sobolev space $W^{m,p}(\Omega)$ of natural order m, where the element $\phi \in W^{m,p}(\Omega)$ is normed by

$$\|\phi\|_{W^{m,p}(\Omega)} = \sum_{|\alpha|\leq m} \|D^\alpha\phi\|_{L^p(\Omega)}.$$

For basic properties of these spaces we refer to [AD].

When $p = 2$ the $W^{m,2}(\Omega)$ spaces become Hilbert spaces. We use an equivalent notation $W^{m,2}(\Omega) = H^m(\Omega)$.

The completion in the topology of $W^{m,p}(\Omega)$ of the set $C_0^\infty(\Omega)$ is denoted by $W_0^{m,p}(\Omega)$. Usually there is only a proper inclusion $W_0^{m,p}(\Omega) \subset W^{m,p}(\Omega)$, but $W_0^{m,p}(R^n) = W^{m,p}(R^n)$ (see [AD, p. 56]).

We define the spaces $W^{-m,p'}(\Omega)$, where $\frac{1}{p} + \frac{1}{p'} = 1$, as dual to $W_0^{m,p}(\Omega)$. We thus have

$$W^{-m,p'}(\Omega) := (W_0^{m,p}(\Omega))^*.$$

Also, for each element $L \in W^{-m,p'}(\Omega)$ there are $\psi_\alpha \in L^{p'}(\Omega)$, $|\alpha| \leq m$, such that $L(\phi) = \sup_{|\alpha|\leq m}\int_\Omega D^\alpha\phi(x)\psi_\alpha(x)dx$ (see [AD, Chapter III]).

The definition of *fractional order Sobolev spaces* is given in Section 1.2.2.

In Chapter 5, studying the *regular elliptic boundary value problems*, we consider spaces $W_{\{B_j\}}^{2m,p}(\Omega)$ (see formula (1.2.64)) which are closed subspaces of the corresponding Sobolev spaces $W^{2m,p}(\Omega)$ consisting of functions satisfying boundary conditions B_j, $j = 0, \dots, m-1$, in the sense of *traces* (see [LI-MA], [AD]).

Introducing fractional order Sobolev spaces $W^{s,p}(R^n)$ with real $s \geq 0$ we use the space $H_p^s(R^n)$, $-\infty < s < +\infty$, $1 < p < +\infty$, of *Bessel potentials* defined as ([TR, 2.3.3])

$$H_p^s(R^n) := \{f \in S'(R^n) : \|f\|_{H_p^s(R^n)} = \|F^{-1}(1+|x|^2)^{\frac{s}{2}}Ff\|_{L^p(R^n)} < +\infty\},$$

where F, F^{-1} denote the Fourier transform and its inverse respectively and S' denotes the space of *tempered distributions* [SZ]. For the definition of the *Besov spaces* $B_{p,p}^s(R^n)$, $0 < s < +\infty$, $1 < p < +\infty$, we refer to [TR, §2.3.1, §2.3.2].

For $k \in N$ by $C^k(\overline{\Omega})$ we denote the space of functions having all partial derivatives of order less than or equal to k continuous in $\overline{\Omega}$. This space is considered with the *supremum* norm:

$$\|\phi\|_{C^k(\overline{\Omega})} = \sum_{|\alpha|\leq k} \sup_{x\in\Omega} |D^\alpha \phi(x)|.$$

Usually we write $C(\overline{\Omega})$ instead of $C^0(\overline{\Omega})$. If $\Omega \subset R^n$ is bounded, the *Ascoli-Arzelà compactness criterion* ensures that a bounded set $K \subset C(\overline{\Omega})$ is precompact in $C(\overline{\Omega})$ whenever the family of functions $\{\phi\}_{\phi\in K}$ is equicontinuous in Ω.

When $\mu \in (0,1)$, we introduce a *Hölder semi-norm*

$$H_\mu^\Omega(\phi) := \sup_{\substack{x,y\in\Omega \\ x\neq y}} \frac{|\phi(x)-\phi(y)|}{|x-y|^\mu}.$$

We set (see [FR 2, p. 61])

$$C^{k+\mu}(\overline{\Omega}) := \{\phi \in C^k(\overline{\Omega}) : \forall_{|\alpha|=k}\ H_\mu^\Omega(D^\alpha\phi) < +\infty\},$$

and norm this Banach space with

$$\|\phi\|_{C^{k+\mu}(\overline{\Omega})} = \|\phi\|_{C^k(\overline{\Omega})} + \sum_{|\alpha|=k} H_\mu^\Omega(D^\alpha\phi).$$

Solutions of evolutionary problems studied in the book are functions of variables $(t,x) \in [0,+\infty) \times \Omega$. We will call variable t the *time variable* and x the *space variable.* In the abstract setting such solutions are considered as the mappings from $[0,T)$ into X, where X denotes a certain Banach space. Usually we do not mark explicitly the dependence of a solution on the space variable x; hence $u(t)$ stands for a function depending on the (t,x) variable.

By $C^k((0,T),X)$, $k \in N$, we denote the space of k times continuously differentiable functions $v : (0,T) \to X$. The derivative $\dot{v}$ of a function $v \in C^1((0,T),X)$ is understood here as

$$\dot{v}(t) = \lim_{h\to 0} \frac{v(t+h)-v(t)}{h},$$

where the limit is taken with respect to the norm of X.

When X and Y are normed linear spaces, $\mathcal{L}(X,Y)$ stands for a normed linear space of all bounded, linear operators from X into Y. The element $T \in \mathcal{L}(X,Y)$ is normed by

$$\|T\|_{\mathcal{L}(X,Y)} = \sup_{\|v\|_X=1} \|Tv\|_Y.$$

The space X is said to be *continuously embedded* in Y (which is written as $X \subset Y$), if there is defined a continuous identity operator I from X into Y. In this case there exists *const.* > 0 such that

$$\|Iv\|_Y \leq const. \|v\|_X \ \text{ for each } \ v \in X.$$

If I is in addition a *compact map* (i.e. I takes bounded sets of X into precompact subsets of Y), then we say that the embedding $X \subset Y$ is *compact.*

Extension property. Most of the theorems describing the relations between different Sobolev spaces are proved initially for $\Omega = R^n$ and then generalized for domains $\Omega \subset R^n$ (bounded or not) having an *extension property.* This property was studied first by Calderón and Stein and may be described using the notion of the *extension operator for* Ω as follows (see [HE 2]).

Let Ω be a domain in R^n such that there is a linear map E_Ω (an extension operator for Ω) possessing the property that if $\phi_{|\Omega}$ is the restriction to Ω of any function $\phi \in C_0^\infty(R^n)$, then

- $E_\Omega(\phi) \in C_0^\infty(R^n)$ *and* $E_\Omega(\phi) = \phi$ *on* Ω,
- $\|\phi\|_{W^{m,p}(\Omega)} \leq \|E_\Omega(\phi)\|_{W^{m,p}(R^n)} \leq K\|\phi\|_{W^{m,p}(\Omega)}$,
- $\|\phi\|_{C^{k+\mu}(\Omega)} \leq \|E_\Omega(\phi)\|_{C^{k+\mu}(R^n)} \leq K'\|\phi\|_{C^{k+\mu}(\Omega)}$,

for constants K, K' depending only on Ω, m, p, k, μ.

There are few sufficient conditions guaranteeing that Ω has an *extension property.*

(i) The results of Calderón and Stein (see [ST]) says that if Ω is an open set lying on one side of its boundary $\partial\Omega$ and $\partial\Omega$ is a Lipschitz continuous graph over a portion of a hyperplane (with some additional requirements when $\partial\Omega$ is not compact), then Ω has an extension property.

(ii) If Ω is a bounded domain in R^n satisfying the *cone condition* of Definition 9.1.1 below, then it has an extension property (see [TR, Theorem 4.2.3], [AD, Theorem 4.32]).

Definition 9.1.1. *A bounded domain $\Omega \subset R^n$ has a* **cone property** *if there are a collection of domains $U_1, \ldots, U_M$ and a collection of cones $C_1, \ldots, C_M$ (C_i are rotations of a standard cone $C = \{x \in R^n : x = (x', x_n) \in R^n, 0 < x_n < h,\ |x'| < ax\}$, a, h are positive constants), such that $\partial\Omega \subset \bigcup_{i=1}^M U_i$ and $(U_i \cap \Omega) + C_i \subset \Omega$ for $i = 1, \ldots, M$.*

(iii) If Ω is a bounded domain in R^n with $\partial\Omega$ of the class C^m, then Ω has an extension property (see [FR 1, Lemma 5.2], [AD, Theorem 4.26]). Recall (see [FR 1]) that $\partial\Omega$ *is of class* C^m if for each point $x \in \partial\Omega$ there are a ball $B_{R^n}(x)$ centered at x and a C^m smooth real function f such that $\partial\Omega \cap B_{R^n}(x)$ may be represented for some $i = 1, 2, \ldots, n$ in the form

$$x_i = f(x_1, \ldots, x_{i-1}, x_{i+1}, \ldots, x_n).$$

As mentioned in [FR 1, p. 22] if $\partial\Omega$ is of class C^1 then the cone condition is satisfied. A similar result holds if Ω is convex (see [FR 1, p. 22]) or has a Lipschitz boundary (see [AD, p. 67]). Therefore, the result of (ii) is more general than those of (i) and (iii).

Nemytskiĭ operator. A large portion of the book strongly involves the idea of a *substitution operator* corresponding to a nonlinear function. This leads to the notion of the *Nemytskiĭ operator* that will be recalled below (see [F-K] and references therein).

Consider a real function $h = h(x, y)$ defined for $(x, y) \in \Omega \times R^{m_0}$, where Ω is a domain in R^n. Denote by $\mathcal{H}$ a map defined on any sequence consisting of m_0 real functions $\phi_i : \Omega \to R$, $i = 1, \dots, m_0$, according to the formula

$$\mathcal{H}(\phi_1, \dots, \phi_{m_0})(x) = h(x, \phi_1(x), \dots, \phi_{m_0}(x)), \; x \in \Omega.$$

Definition 9.1.2. *$\mathcal{H}$ is called the* **Nemytskiĭ operator** *corresponding to a function h.*

It is obvious that, if h is continuous, then $\mathcal{H}$ takes a sequence of continuous functions into a continuous function. For the purpose of measurability, less restrictive assumptions on $h : \Omega \times R^{m_0} \to R$ are sufficient. That is, the function $h : \Omega \times R^{m_0} \to R$ is required to satisfy the *Carathéodory conditions* below:

- for each fixed $y \in R^{m_0}$, the function $\Omega \ni x \to h_y(x) = h(x, y) \in R$ is measurable,
- for almost all $x \in \Omega$, $R^{m_0} \ni y \to h_x(y) = h(x, y) \in R$ is a continuous function.

Under the Carathéodory conditions, the following well known result holds.

Proposition 9.1.1. *$\mathcal{H}(\phi_1, \dots, \phi_{m_0})$ is a measurable function on Ω, whenever ϕ_i, $i = 1, \dots, m_0$, are measurable on Ω.*

To ensure further that $\mathcal{H}(\phi_1, \dots, \phi_{m_0})$ is an element of $L^q(\Omega)$ it is necessary to control the growth of h. For $q \geq 1$ and $p_i \geq 1$, $i = 1, \dots, m_0$, we introduce the *growth restriction*

$$(9.1.2) \qquad |h(x, y_1, \dots, y_{m_0})| \leq \psi(x) + const. \sum_{i=1}^{m_0} |y_i|^{\frac{p_i}{q}}, \; y \in R^{m_0},$$

where $\psi \in L^q(\Omega)$ and $const. > 0$.

The basic result concerning integrability of $\mathcal{H}$ as well as its continuity in Lebesgue type spaces reads (see [F-K, Theorem 12.10]):

Proposition 9.1.2. *Suppose that h satisfies the Carathéodory conditions. Then $\mathcal{H}(\phi_1, \dots, \phi_{m_0}) \in L^q(\Omega)$ for $\phi_i \in L^{p_i}(\Omega)$, $i = 1, \dots, m_0$, if and only if* (9.1.2) *holds. Moreover, the growth restriction* (9.1.2) *is sufficient for continuity of $\mathcal{H}$ acting from $L^{p_1}(\Omega) \times \ldots \times L^{p_{m_0}}(\Omega)$ into $L^q(\Omega)$.*

Generalized Poincaré inequality. Further extensions of the Poincaré inequality covering, in particular, the L^p theory instead of the L^2 theory in (5.2.2) can be found in [ZI, p. 182]. We recall here this result briefly for completeness.

Let $0 \leq k < m$ be integers, $p > 1$ and Ω be a bounded, connected domain in R^n having the *extension property* (see above). The symbol $\mathcal{P}^k(R^n)$ denotes the set of all polynomials in R^n of degree k. In [ZI, p. 179] we find the following *projection lemma.*

Lemma 9.1.1. *Let k, m, p and Ω be as above. Consider a functional $T \in (W^{m-k,p}(\Omega))^*$ having the property that $T(\chi_\Omega) \neq 0$. Then there is a projection $L : W^{m,p}(\Omega) \to \mathcal{P}^k(R^n)$ such that for each $u \in W^{m,p}(\Omega)$ and all $|\alpha| \leq k$*

$$T(D^\alpha u) = T(D^\alpha P), \tag{9.1.3}$$

where $P = L(u)$. Such an L has the form

$$L(u)(x) = \sum_{|\alpha|\leq k} T(P^\alpha(Du))x^\alpha, \tag{9.1.4}$$

where $P^\alpha \in \mathcal{P}^k(R^n), Du = \nabla u$, and

$$\|L\| \leq c(k,p,\Omega)\left(\frac{\|T\|}{T(\chi_\Omega)}\right)^{k+1}, \tag{9.1.5}$$

χ_Ω being the characteristic function of Ω.

We next have

Proposition 9.1.3. *If the assumptions of Lemma 9.1.1 are satisfied and $L : W^{m,p}(\Omega) \longrightarrow \mathcal{P}(R^n)$ is associated with T, then*

$$\|u - L(u)\|_{W^{k,p}(\Omega)} \leq c(k,p,\Omega)\left(\frac{\|T\|}{T(\chi_\Omega)}\right)^{k+1} \|D^{k+1}u\|_{W^{m-(k+1),p}(\Omega)}. \tag{9.1.6}$$

Usually we will use the estimate (9.1.6) with $k = m-1$, that means the particular version

$$\|u - L(u)\|_{W^{m-1,p}(\Omega)} \leq c(k,p,\Omega)\left(\frac{\|T\|}{T(\chi_\Omega)}\right)^{m} \|D^m u\|_{L^p(\Omega)}.$$

With the above estimate, we can immediately find an estimate of the $W^{m,p}(\Omega)$ norm of u from above. We have

(9.1.7)
$$\begin{aligned}
\|u\|_{W^{m,p}(\Omega)} &= \|u\|_{W^{m-1,p}(\Omega)} + \|D^m u\|_{L^p(\Omega)} \\
&\leq \|u - L(u)\|_{W^{m-1,p}(\Omega)} + \|Lu\|_{W^{m-1,p}(\Omega)} + \|D^m u\|_{L^p(\Omega)} \\
&\leq const.\|D^m u\|_{L^p(\Omega)} + \|Lu\|_{W^{m-1,p}(\Omega)},
\end{aligned}$$

where $Lu = P$. The final question is connected with determining sufficient conditions under which $L(u) = 0$. Such conditions will be presented in examples below.

All classical versions of the Poincaré inequality will be derived from the above estimates (see [ZI, Chapter 4.4]), provided we choose properly the functional L.

Fix real $p \geq 1$ and integer $m \geq 1$.

Example 9.1.1. Let Ω be a bounded domain in R^n and $u \in C_0^\infty(\Omega)$. Such an Ω is contained in a sufficiently large ball $B(0,r) \subset R^n$. Extend u with zero onto $B(0,2r) \setminus \Omega$, set $\Omega' = B(0,2r)$ and define

$$T(w) = \int_{\Omega'} (\chi_{B(0,2r)} - \chi_{B(0,r)}) w dx, \ w \in W^{1,p}(\Omega'). \tag{9.1.8}$$

Since $supp\, u \subset B(0,r)$, we have

$$T(D^\alpha u) = 0 \ \text{ for } \ 0 \le |\alpha| \le m-1, \tag{9.1.9}$$

which, by Lemma 9.1.1, proves that $L(u) = 0$. Next, (9.1.7) provides the required estimate:

$$\|u\|_{W^{m,p}(\Omega)} \le const. \|D^m u\|_{L^p(\Omega)} \ \text{ for } \ u \in W_0^{m,p}(\Omega). \tag{9.1.10}$$

Example 9.1.2. Let $\Omega \subset R^n$ be a bounded domain having the extension property. Assume that $u \in W^{m,p}(\Omega)$ has the property, that

$$\int_E D^\alpha u dx = 0 \ \text{ for } \ 0 \le |\alpha| \le m-1, \tag{9.1.11}$$

where $E \subset \Omega$ is a set of positive Lebesgue measure. Then defining a functional $T \in (W^{1,p}(\Omega))^*$ as

$$T(w) = \int_E w dx, w \in W^{1,p}(\Omega), \tag{9.1.12}$$

we see that $T(\chi_\Omega) \ne 0$ and

$$T(D^\alpha u) = 0 \ \text{ for } \ 0 \le |\alpha| \le m-1. \tag{9.1.13}$$

Consequently, $Lu = 0$ and, according to (9.1.7), we claim an estimate:

$$\|u\|_{W^{m,p}(\Omega)} \le const. \|D^m u\|_{L^p(\Omega)}. \tag{9.1.14}$$

Example 9.1.3. Let $\Omega \subset R^n$ be a bounded domain having the extension property and let $u \in W^{1,p}(\Omega), p \ge 1$. Let A and B be subsets of Ω of positive Lebesgue measures and such that $u > 0$ on A and $u < 0$ on B. Then

$$\|u\|_{L^p(\Omega)} \le c(p, n, |A|, |B|) \|\nabla u\|_{L^p(\Omega)}. \tag{9.1.15}$$

In this example we define $T \in (W^{1,p}(\Omega))^*$ as

$$T(w) = \int_\Omega \left(\frac{1}{\alpha} \chi_A - \frac{1}{\beta} \chi_B \right) w dx, \ w \in W^{1,p}(\Omega), \tag{9.1.16}$$

where

$$\alpha = \int_A u dx, \ \beta = \int_B u dx \tag{9.1.17}$$

(see [ZI, Chapter 4.4] for details).

9.2. Abstract version of the maximum principle

Following [HE 1] we will recall below an abstract version of the *maximum principle* which expresses one of the most specific features of solutions of many second order parabolic equations.

Let $X_{(\geq)}$ denote further a Banach space with an *order relation* $\geq$ that satisfies the conditions below:

(i) $\forall_{x \in X_{(\geq)}}$ $x \geq x$,

(ii) $\forall_{x,y \in X_{(\geq)}}$ $x \geq y,\ y \geq z \Rightarrow x \geq z$,

(iii) $\forall_{x,y,z \in X_{(\geq)}}$ $\forall_{\alpha \in [0,+\infty)}$ $x \geq y \Rightarrow x + z \geq y + z,\ \alpha x \geq \alpha y$,

(iv) the *nonnegative cone* $C^+ = \{x \in X_{(\geq)} : x \geq 0\}$ is closed.

A typical example of an ordered Banach space is $X_{(\geq)} = L^p(\Omega), 1 \leq p \leq +\infty$. In this case $x \geq y$ means that $x(t) \geq y(t)$ a. e. in Ω, whereas the nonnegative cone C^+ is a collection of all (classes of) functions that are nonnegative a. e. in Ω.

A function $g : X_{(\geq)} \to X_{(\geq)}$ is *increasing* with respect to the relation $\geq$ provided that

$$g(x) \geq g(y) \text{ for } x \geq y.$$

If g is linear the latter is equivalent to saying that g is a *nonnegative function*, i.e.

$$g(x) \geq 0 \text{ for } x \geq 0.$$

Returning to the standard notation we quote an abstract version of the *maximum principle* as formulated in [HE 1, p. 61] (see also [AL 1]).

Lemma 9.2.1. *Let A be a sectorial, positive operator in an ordered Banach space $X := X_{(\geq)}$ for which $(A + \lambda I)^{-1}$ is a nonnegative function for all $\lambda > 0$. Suppose that $F : [0, t_1) \times X^\alpha \to X$ is Lipschitz continuous on bounded sets and for any bounded set $B_b^+ = \{u \in X^\alpha : u \geq 0, \|u\|_{X^\alpha} \leq b\}$ let there exist a real constant $\beta = \beta(B_b^+)$ such that*

$$F(t,u) + \beta u \geq 0 \quad \textit{for} \quad u \in B_b^+,\ t \in [0, t_1). \tag{9.2.1}$$

If $u_0 \in X^\alpha$ and $u_0 \geq 0$, then the corresponding X^α solution of the problem

$$\dot{u} + Au = F(t,u),\ u(0) = u_0, \tag{9.2.2}$$

satisfies the condition $u(t) \geq 0$ for each $t \in [0, \min\{t_1, \tau_{u_0}\})$ (τ_{u_0} being the lifetime of $u(\cdot, u_0)$).

Proof. Let $u_0 \geq 0$, $b_{u_0} := \|u_0\|_{X^\alpha} + 1$ and β_{u_0} be a constant corresponding to $B^+_{b_{u_0}}$. For $\tilde{\beta} = \max\{\beta_{u_0}, 0\}$ consider translations $\tilde{A} = A + \tilde{\beta} I$, $\tilde{F}(t,u) = F(t,u) + \tilde{\beta} u$ and the problem

$$\dot{u} + \tilde{A} = \tilde{F}(t,u),\ t > 0,\ u(0) = u_0, \tag{9.2.3}$$

which has the same solution as the original problem (9.2.2).

The operator $\tilde{A}$ is sectorial and positive, $(\tilde{A}+\lambda I)^{-1}$ is nonnegative for all $\lambda > 0$ and, by the exponential formula (see [PAZ 2, p. 33]),

$$(9.2.4) \qquad e^{-\tilde{A}t}u = \lim_{n\to+\infty}\left(I + \frac{t}{n}\tilde{A}\right)^{-n} u.$$

Iterating the nonnegative operators $(I + \frac{t}{n}\tilde{A})^{-1}$ it is easy to see that $(I + \frac{t}{n}\tilde{A})^{-n}$ is nonnegative and that the limit passage preserves this property. Therefore, $e^{-\tilde{A}t}$ is nonnegative for each $t \geq 0$.

By our assumptions, there exists a unique X^α solution $u(\cdot, u_0)$ to (9.2.2). Since $u(\cdot, u_0)$ is also a solution of (9.2.3), $u(\cdot, u_0)$ is in particular a limit of the sequence of iterates $\tilde{\Psi}^n(u_0)$ of the operator

$$v \to \tilde{\Psi}(v)(t) = e^{-\tilde{A}t}u_0 + \int_0^t e^{-\tilde{A}(t-s)}\tilde{F}(s, v(s))ds$$

(see the proof of Theorem 2.1.1). If $0 < r \leq 1$ and v is an element of the closed ball $B_{r,\delta} = B_{C([0,\delta],X^\alpha)}(u_0, r)$, then $\|v(t)\|_{X^\alpha} \leq \|u_0\|_{X^\alpha} + 1$ for all $t \in [0,\delta]$. If, in addition, $\delta < t_1$ the latter implies that $F(t, v(t)) \geq 0$ for all $t \in [0,\delta]$, $v \in B_{r,\delta}$. As shown in Chapter 2, when $\delta \in (0, \delta_0]$ (see (2.3.6)) the operator $\tilde{\Psi}$ transforms $B_{r,\delta}$ into itself. This and the form of $\tilde{\Psi}$ ensure that, for $u_0 \geq 0$, the values of all iterates $\tilde{\Psi}^n(u_0)$, and hence also the values of $u(\cdot, u_0)$, belong to the nonnegative cone C^+ for each $0 \leq t \leq \delta < \min\{\delta_0, t_1\}$.

Since, by the previous analysis, the solution corresponding to a nonnegative initial condition exists and remains nonnegative on some nondegenerate interval $[0,\delta]$, the set $\mathcal{I}^+_{u_0} := \{\delta \in (0, \min\{t_1, \tau_{u_0}\}) : \forall_{t\in[0,\delta]}\ u(t) \geq 0\}$ is nonempty and $\sup \mathcal{I}^+_{u_0}$ cannot belong to $\mathcal{I}^+_{u_0}$. Therefore the solution is nonnegative on each closed subinterval of $[0, \min\{t_1, \tau_{u_0}\})$. The proof is complete. □

Remark 9.2.1. Application of the above theorem is limited, mostly, to second order parabolic problems. The reason for this lies in the assumption that $(A+\lambda I)^{-1}$ is a nonnegative function for all $\lambda \geq 0$.

Example 9.2.1. To see a typical application of the above theorem consider a simple second order parabolic problem in $X_{(\geq)} = L^p(\Omega)$,

$$(9.2.5) \qquad \begin{cases} \frac{\partial u}{\partial t} = \Delta u + f(t, x, u) \text{ in } (0, t_1) \times \Omega, \\ u = 0 \text{ on } \partial\Omega, \\ u(0, x) = u_0(x) \text{ in } \Omega, \end{cases}$$

under the usual smoothness assumptions on the data guaranteeing existence of local X^α solutions. We assume additionally that the condition (9.2.1) is satisfied. For F corresponding to a real function $f : [0, t_1) \times \bar{\Omega} \times R \to R$ this condition will be satisfied provided that

$$(9.2.6) \qquad \forall_{t\in[0,t_1)}\ \forall_{x\in\bar{\Omega}}\ \forall_{0\leq u\leq b}\ f(t, x, u) \geq -\beta u$$

with a real constant $\beta = \beta(t_1, b)$ (not necessarily positive).

It is then known, based on the properties of the *Green's function* of $(-\Delta, I, \Omega)$ (see [P-W, pp. 85-88]), that the relation $(-\Delta+\lambda I)^{-1} \geq 0$ holds for every $\lambda > 0$. Thus, by Lemma 9.2.1, the solution of (9.2.5) originating in C^+ remains in C^+ until the time $\tau = \min\{t_1, \tau_{u_0}\}$.

For the application of the result of Lemma 9.2.1 to second order parabolic systems we refer further to [AL 1].

9.3. $L^\infty(\Omega)$ estimate for second order problems

We shall present here in detail the Moser-Alikakos iteration technique which allows us to obtain $L^\infty(\Omega)$ estimates for solutions of a single second order parabolic equation.

Iteration procedure. Consider the problem

$$(9.3.1)\quad u_t = \sum_{i,j=1}^n \frac{\partial}{\partial x_j}\left(a_{ij}(x)\frac{\partial u}{\partial x_i}\right) + \sum_{i=1}^n b_i(x)\frac{\partial u}{\partial x_i} + f(x,u),\ (t,x) \in R^+ \times \Omega,$$

$$(9.3.2)\qquad u(0,x) = u_0(x),\ x \in \Omega,$$

$$(9.3.3)\qquad u_{|\partial\Omega} = 0, \text{ or } u_{N_{|\partial\Omega}} = \left(\sum_{i,j=1}^n a_{ij}(x)\frac{\partial u}{\partial x_i}\cos(N, x_j)\right)_{|\partial\Omega} = 0,$$

where $\Omega \subset R^n$ is a bounded domain with C^2 boundary $\partial\Omega$, $a_{ij}, b_i \in C^1(\overline{\Omega})$, and $f \in C(\overline{\Omega} \times R)$. We assume the ellipticity condition

$$(9.3.4)\qquad \exists_{a_0>0}\ \forall_{x\in\Omega}\ \forall_{\xi\in R^n}\ \sum_{i,j=1}^n a_{ij}(x)\xi_i\xi_j \geq a_0|\xi|^2,$$

and the growth restriction

$$(9.3.5)\qquad \exists_{C,D>0}\ \forall_{x\in\Omega}\ \forall_{v\in R}\quad vf(x,v) \leq Cv^2 + D.$$

Following [AL 1], [DL 4], we shall show how the $L^1(\Omega)$ estimate for the solutions of (9.3.1)-(9.3.3) implies the $L^\infty(\Omega)$ estimate.

Remark 9.3.1. Since the result stated below is purely technical we shall omit here the precise definition of what kind of solutions of (9.3.1)-(9.3.3) is actually considered. Indeed, as seen in the formulation of the following lemma, it suffices to study any *sufficiently smooth solutions* for which the further calculations presented are sensible. In particular, such a result holds for X^α solutions of (9.3.1)-(9.3.3).

Lemma 9.3.1. *All hypothetical sufficiently smooth solutions of* (9.3.1)-(9.3.3) *fulfill the estimate*

$$\sup_{t\geq 0} \|u(t,u_0)\|_{L^\infty(\Omega)} \leq 2^{n+1}c' \max\left\{\sup_{t\geq 0}\|u(t,u_0)\|_{L^1(\Omega)}, 1\right\},$$

where $c' > 0$ is defined in (9.3.11) *below.*

Proof. Multiplying (9.3.1) by u^{2^k-1}, $k = 1, 2, \ldots$, and integrating by parts we get

(9.3.6)
$$\frac{1}{2^k}\frac{d}{dt}\int_\Omega u^{2^k}dx = -(2^k-1)2^{2-2k}\int_\Omega \sum_{i,j=1}^n a_{ij}(x)\frac{\partial(u^{2^{k-1}})}{\partial x_i}\frac{\partial(u^{2^{k-1}})}{\partial x_j}dx$$
$$+2^{1-k}\int_\Omega \sum_{i=1}^n b_i(x)\frac{\partial(u^{2^{k-1}})}{\partial x_i}u^{2^{k-1}}dx + \int_\Omega f(x,u)u^{2^k-1}dx.$$

Now, using the ellipticity condition (9.3.4) and the easy inequality $u^{2^k-2} \leq u^{2^k}+1$ we get

(9.3.7)
$$\frac{d}{dt}\int_\Omega u^{2^k}dx \leq -(2^k-1)2^{2-k}a_0\int_\Omega \sum_{i=1}^n\left[\frac{\partial(u^{2^{k-1}})}{\partial x_i}\right]^2 dx$$
$$+2B\left(\int_\Omega \sum_{i=1}^n\left[\frac{\partial(u^{2^{k-1}})}{\partial x_i}\right]^2 dx\right)^{\frac{1}{2}}\left(n\int_\Omega u^{2^k}dx\right)^{\frac{1}{2}}$$
$$+2^k(C+D)\int_\Omega u^{2^k}dx + 2^kD|\Omega|,$$

where $B := \max_{i=1,\ldots,n}\sup_{x\in\Omega}|b_i(x)|$. Consider the following consequence of the Nirenberg-Gagliardo inequality (1.2.46) (with $j=0$, $p=2$, $m=1$, $r=2$, $q=1$, $\theta = \frac{n}{n+2}$ obtained with the use of the Young inequality with $m = \frac{n+2}{n}$):

(9.3.8) $\forall_{\varepsilon\in(0,1)}\ \exists_{C>0}\ \forall_{v\in H^1(\Omega)}\ (1-\varepsilon)\|v\|^2_{L^2(\Omega)} \leq \varepsilon\|v_x\|^2_{L^2(\Omega)} + C_\varepsilon\|v\|^2_{L^1(\Omega)}$,

where C_ε is proportional to $const.\varepsilon^{-\frac{n}{2}}$.

From the Cauchy inequality (with $\overline{\varepsilon} = \frac{a_0}{B}$) and condition (9.3.8) we get further

(9.3.9)
$$\frac{d}{dt}\int_\Omega u^{2^k}dx \leq \left(-(2^k-1)2^{2-k}a_0 + B\overline{\varepsilon}\right)\int_\Omega \sum_{i=1}^n\left[\frac{\partial(u^{2^{k-1}})}{\partial x_i}\right]^2 dx$$
$$+\left(2^k(C+D)+\frac{Bn}{\overline{\varepsilon}}\right)\int_\Omega u^{2^k}dx + 2^kD|\Omega|$$
$$\leq -a_0\frac{1-\varepsilon}{\varepsilon}\int_\Omega u^{2^k}dx + \frac{C_\varepsilon}{\varepsilon}a_0\left(\int_\Omega u^{2^{k-1}}dx\right)^2$$
$$+\left(2^k(C+D)+\frac{B^2n}{a_0}\right)\int_\Omega u^{2^k}dx + 2^kD|\Omega|,$$

noting that for all $k \in N$, $2 \leq \frac{2^k-1}{2^{k-2}} \leq 4$. Then, for fixed $\varepsilon = \varepsilon_k$, $k = 1, 2, \ldots$, such that

$$-a_0\frac{(1-\varepsilon_k)}{\varepsilon_k} + 2^k(C+D) + \frac{B^2n}{a_0} \leq -2^k$$

and $\varepsilon_k = const.2^{-k}$, we arrive at the estimate

$$\frac{d}{dt}\int_\Omega u^{2^k}dx \le -2^k\int_\Omega u^{2^k}dx + const.(2^k)^{\frac{n}{2}+1}\left(\int_\Omega u^{2^{k-1}}dx\right)^2 + 2^k D|\Omega|.$$

This, with the use of Lemma 1.2.5, gives

$$\int_\Omega u^{2^k}(t,x)dx \le \max\left\{\int_\Omega u_0^{2^k}(t,x)dx, const.2^{k\frac{n}{2}}m_{k-1}^{2^k} + D|\Omega|\right\},$$

where $m_{k-1} := \sup_{t\ge 0}\left(\int_\Omega u^{2^{k-1}}(t,x)dx\right)^{2^{-k+1}}$. Taking the 2^k-th roots of both sides of the last estimate and supremum on the left side we arrive at

$$m_k \le \max\{\|u_0\|_{L^{2^k}(\Omega)}, (const.2^{k\frac{n}{2}}m_{k-1}^{2^k} + D|\Omega|)^{\frac{1}{2^k}}\}. \tag{9.3.10}$$

Now, since $\|u_0\|_{L^{2^k}(\Omega)} \le \max\{1,|\Omega|\}\|u_0\|_{L^\infty(\Omega)} =: K$, the first term in (9.3.10) is estimated uniformly in $k \in N$ by K. Enlarging *const.* to the value

$$c' := \max\{const., 1, D|\Omega|, K^2\} \tag{9.3.11}$$

and enlarging m_0 (defined as $\sup_{t\ge 0}\|u(t,\cdot)\|_{L^1(\Omega)}$) to the value $x_0 := \max\{m_0, 1\}$ it is easy to see that all the numbers m_k, $k = 0,1,\dots$, are dominated by the corresponding numbers x_k satisfying the recurrence

$$x_k = \max\left\{K, (c'2^{k\frac{n}{2}}x_{k-1}^{2^k} + c')^{\frac{1}{2^k}}\right\},\ k = 1,2,\dots.$$

Now since $c' \ge 1$ and $x_0 \ge 1$, it is clear that the sequence $\{x_k\}$ is increasing. Moreover, for $k = 1$ the second term is always dominated. In turn, the sequence $\{x_k\}$ is dominated by the sequence $\{x'_k\}$ given by

$$x'_0 = x_0,\ x'_k = (2c'2^{k\frac{n}{2}})^{\frac{1}{2^k}}x'_{k-1}$$

($x_k \ge 1$), or through the passage to the limit

$$\sup_{t\ge 0}\|u(t,\cdot)\|_{L^\infty(\Omega)} \le x'_\infty := \lim_{k\to\infty} x'_k = x'_0\prod_{k=1}^\infty(2c'2^{k\frac{n}{2}})^{\frac{1}{2^k}} = x_0 2c'P^{\frac{n}{2}}, \tag{9.3.12}$$

where $P = \prod_{k=1}^\infty 2^{\frac{k}{2^k}} = 4$. Finally,

$$\sup_{t\ge 0}\|u(t,\cdot)\|_{L^\infty(\Omega)} \le 2^{n+1}c'\max\left\{\sup_{t\ge 0}\|u(t,\cdot)\|_{L^1(\Omega)}, 1\right\}. \tag{9.3.13}$$

The proof is complete. □

Iteration proof of the maximum principle. Below we use the iteration technique to prove a version of the maximum principle for an autonomous parabolic problem in divergence form:

(9.3.14)
$$\begin{cases} u_t = \sum_{i,j=1}^n \frac{\partial}{\partial x_i}\left(a_{ij}(x)\frac{\partial u}{\partial x_j}\right) + a(x,u,\nabla u)u + f(x), \ (t,x) \in R^+ \times \Omega, \\ u(0,x) = u_0(x) \ \text{in} \ \Omega, \quad u = 0 \ \text{on} \ \partial\Omega. \end{cases}$$

As in Lemma 9.3.1, $\Omega \subset R^n$ is a bounded domain with a regular boundary $\partial\Omega$, $a_{ij} \in C^1(\overline{\Omega})$, $f \in L^\infty(\Omega)$, $a(x,u,\nabla u) \leq A_0$, $A_0 \geq 0$, and we assume the ellipticity condition (9.3.4).

Following [DL 4] we show (see Remark 9.3.1)

Lemma 9.3.2. *All hypothetical sufficiently smooth solutions of* (9.3.14) *satisfy the estimate*

$$\|u(t)\|_{L^\infty(\Omega)} \leq (\|u_0\|_{L^\infty(\Omega)} + \|f\|_{L^\infty(\Omega)}t)e^{A_0 t}. \tag{9.3.15}$$

Proof. It suffices to consider the case $A_0 = 0$, elsewhere we can use the transformation $v(t) = u(t)e^{-A_0 t}$ and study the equation for v.

Since $a(x,u,\nabla u) \leq 0$, multiplying (9.3.14) by u^{2^k-1}, integrating and using the Hölder inequality (with $p = \frac{2^k}{2^k-1}, q = 2^k$), we obtain

$$2^{-k}\frac{d}{dt}\int_\Omega u^{2^k}dx \leq -\int_\Omega \sum_{i,j=1}^n a_{ij}(x)\frac{\partial u}{\partial x_j}\frac{\partial(u^{2^k-1})}{\partial x_i}dx + \left(\int_\Omega f^{2^k}dx\right)^{2^{-k}}\left(\int_\Omega u^{2^k}dx\right)^{1-2^{-k}}. \tag{9.3.16}$$

For $w = u^{2^{k-1}}$, using (9.3.4) and the Poincaré inequality, we get

(9.3.17)
$$\begin{aligned} \frac{d}{dt}\int_\Omega w^2 dx &\leq -a_0(2^k-1)2^{2-k}\int_\Omega \sum_{i=1}^n w_{x_i}^2 dx + \|f\|_{L^\infty(\Omega)}2^k|\Omega|^{2^{-k}}\left(\int_\Omega w^2 dx\right)^{1-2^{-k}} \\ &\leq -\alpha_k\|w\|^2_{L^2(\Omega)} + \beta_k 2^k(\|w\|^2_{L^2(\Omega)})^{1-2^{-k}}, \end{aligned}$$

where $\alpha_k = \frac{a_0}{const.}\frac{2^k-1}{2^{k-2}}$, $\beta_k = \|f\|_{L^\infty(\Omega)}|\Omega|^{2^{-k}}$. Solving (9.3.17), we obtain

$$\|w(t)\|^2_{L^2(\Omega)} \leq \left[\|w(0)\|^{2^{1-k}}_{L^2(\Omega)} + \int_0^t \beta_k e^{\alpha_k 2^{-k} z}dz\right]^{2^k} e^{-\alpha_k t}, \tag{9.3.18}$$

which shows that

$$\|u(t)\|_{L^{2^k}(\Omega)} \leq \left[\|u_0\|_{L^{2^k}(\Omega)} + \beta_k \frac{2^k}{\alpha_k}\left(e^{\frac{\alpha_k t}{2^k}} - 1\right)\right] e^{-\frac{\alpha_k t}{2^k}}.$$

Finally, letting k tend to $+\infty$, we get the estimate

$$\|u(t)\|_{L^\infty(\Omega)} \leq \|u_0\|_{L^\infty(\Omega)} + \|f\|_{L^\infty(\Omega)}t. \tag{9.3.19}$$

The proof is complete. □

The above lemma, with a different proof, can be found in [I-K-O]. Further versions of Lemma 9.3.2 can be found in [DL 4] and also in [TU].

9.4. Comparison of X^α solution with other types of solutions

Through the last 40 years a number of different classes of solutions of parabolic problems has been studied by various authors. Usually the definition and properties of such solutions were determined by the admitted method of proving their existence. Working with various basic monographs (like [L-S-U], [FR 1], [TE 1], [HE 1]), one is often forced to compare various types of solutions or use the existing results obtained within one approach to solution defined in another way. We discuss briefly below the relations between the X^α solutions and other leading classes of solutions known in the literature.

We will study an initial boundary value problem for autonomous semilinear $2m$-th order parabolic equation (8.4.1) (see Chapters 5, 8)

$$\begin{cases} u_t = -\sum_{|\alpha|\le 2m} a_\alpha(x)D^\alpha u + f(x, d^{m_0}u), \ t>0, \ x\in\Omega\subset R^n, \\ \sum_{|\beta|\le m_j} b_{j\beta}(x)D^\beta u = 0, \ m_j \le 2m-1, \ j=1,\dots,m, \ t>0, \ x\in\partial\Omega, \\ u(0,x)=u_0(x), \ x\in\Omega. \end{cases}$$

Here Ω is a bounded domain in R^n with C^{2m} boundary $\partial\Omega$, the operator in the main part of (8.4.1) is uniformly strongly elliptic in Ω, and $d^{m_0}u$ denotes a vector of partial derivatives with respect to x of order not exceeding $m_0 \le 2m-1$. Moreover, we shall assume that

(9.4.1)

the triple $(A, \{B_j\}, \Omega)$ forms a strongly regular initial boundary value problem (see Definition 1.2.1) and the nonlinear term $f : \overline{\Omega} \times R^{d_0} \to R$ is continuous and locally Lipschitz continuous with respect to each functional argument separately, uniformly with respect to $x \in \overline{\Omega}$.

We may write (8.4.1) in the form (2.1.1),

$$\begin{cases} \dot u + Au = F(u), \ t>0, \\ u(0)=u_0, \end{cases}$$

where A is a sectorial, positive operator in $X = L^p(\Omega)$ ($p>1$, $D(A) := X^1 = W^{2m,p}_{\{B_j\}}(\Omega)$) and, for some $\alpha \in [0,1)$, $F : X^\alpha \to X$ is Lipschitz continuous on bounded sets.

We will discuss some relations among the following types of solutions to (8.4.1):

(s_1) $C^{1+\frac{\epsilon}{2m}, 2m+\epsilon}([0,T]\times\overline{\Omega})$ solution (*Hölder solution*),
(s_2) $W^{1,2m}_p((0,\tau)\times\Omega)$ solution (*Sobolev solution*),
(s_3) $C([0,\tau), X^\alpha)\cap C^1((0,\tau), X^\gamma)\cap C((0,\tau), X^1)$ solution, $\gamma\in[0,1)$ (*X^α solution*),

(s_4) *classical solution* to (8.4.1) such that:

(*i*) u is continuous on $[0,T] \times \overline{\Omega}$,

(*ii*) there exist classical derivatives $\frac{\partial u}{\partial t}$ and $D^\alpha u$, $|\alpha| \le 2m$, in $(0,T] \times \overline{\Omega}$,

(s_5) $C([0,\tau), X^{\frac{1}{2}}) \cap L^2((0,\tau), X^1)$ solution with $\dot{u} \in L^2((0,\tau), X)$ (*Galerkin solution*).

The Hölder and Sobolev solutions in (s_1), (s_2) were described in the classical monograph [L-S-U].

The X^α solution in (s_3) comes back to [HE 1]. According to Definition 2.1.1 and Corollary 2.3.1, the X^α solution satisfies the following smoothness requirements:

$$u \in C([0,\tau), X^\alpha) \cap C((0,\tau), X^1), \; \dot{u} \in C((0,\tau), X^\gamma), \; \gamma \in [0,1).$$

To justify (s_3) we need to observe its additional property

$$u \in C^1((0,\tau), X^\gamma), \; \gamma \in [0,1). \tag{9.4.2}$$

The last condition means that $\dot{u}$, for positive t, is in fact the X^α limit of the difference quotient and not only the X limit as previously reported. To prove (9.4.2) we fix $t \in (0,\tau)$, $\gamma \in (0,1)$, and write the relation

$$\frac{u(t+h) - u(t)}{h} = \frac{1}{h} \int_t^{t+h} \dot{u}(s) ds. \tag{9.4.3}$$

Next, by closeness of A^γ, we get

$$\frac{A^\gamma u(t+h) - A^\gamma u(t)}{h} = \frac{1}{h} \int_t^{t+h} A^\gamma \dot{u}(s) ds. \tag{9.4.4}$$

Since $\dot{u} \in C((0,\tau), X^\gamma)$, it is easy to see that $\frac{1}{h} \int_t^{t+h} A^\gamma \dot{u}(s) ds$ tends to $A^\gamma \dot{u}(t) = \frac{d}{dt} A^\gamma u(t)$. The condition (9.4.2) is thus justified.

Requirements of (s_4) concerning clasical solutions are very basic; they simply say that all the derivatives and all relations appearing in (8.4.1) are understood in the classical sense. Note that the classical solutions studied in Chapter 7 had more regularity; now we do not require continuity of the highest order space derivatives with respect to time variable.

The discussion of (s_5) will be limited to the case when X is a Hilbert space. Such an assumption, although inconvenient in general, is an unquestionable advantage in many applications. As shown in [TE 1], in a number of problems originating in applied sciences A may be viewed as an operator corresponding to certain symmetric, continuous, coercive, bilinear form on a Hilbert space V densely embedded in X. In that case one may follow the Faedo-Galerkin method to construct the solutions of (8.4.1) (see [TE 1, Chapter II] for details). Since A viewed as an operator in H will be self-adjoint and positive definite, the approach of [HE 1] is also applicable. Proposition 9.4.4 below ensures that the X^α solution (s_3) is in fact Galerkin solution (s_5).

X^α solution being Sobolev solution. We first compare the X^α solution with the Sobolev solution.

Proposition 9.4.1. *Let*

$$2m\alpha - \frac{n}{p} > m_0, \ \alpha > 1 - \frac{1}{p}, \tag{9.4.5}$$

and u be an X^α solution to (2.1.1) defined on a maximal interval of existence $[0, \tau_{u_0})$. Then both

$$u \in W_p^{1,2m}((0,\tau) \times \Omega), \ \tau \in (0, \tau_{u_0}), \tag{9.4.6}$$

and

$$\|u\|_{W_p^{1,2m}((0,\tau)\times\Omega)} \leq const.(\tau,\alpha,p)\Big(1 + \|F(u)\|_{L^p((0,\tau)\times\Omega)}\Big), \ \tau \in (0,\tau_{u_0}), \tag{9.4.7}$$

hold with const.(τ,α,p) increasing with respect to τ.

Proof. Following [L-S-U], recall that $W_p^{1,2m}((0,\tau)\times\Omega)$ is a space of (classes of) measurable functions $\psi : (0,\tau)\times\Omega \to R$ with distributional derivatives $\frac{\partial\psi}{\partial t}$, $D^\alpha\psi = \frac{\partial^{|\alpha|}\psi}{\partial x_1^{\alpha_1}\ldots\partial x_n^{\alpha_n}}$, $|\alpha| \leq 2m$ in $L^p((0,\tau)\times\Omega)$. This space is normed by

$$\|\psi\|_{W_p^{1,2m}((0,\tau)\times\Omega)} := \left\|\frac{\partial\psi}{\partial t}\right\|_{L^p((0,\tau)\times\Omega)} + \sum_{|\alpha|\leq 2m} \|D^\alpha\psi\|_{L^p((0,\tau)\times\Omega)}.$$

As a consequence of (s_3), u belongs to $C((0,\tau), X^1)$. Hence there exist distributional derivatives $D^\alpha u$, $|\alpha| \leq 2m$ in $(0,\tau)\times\Omega$. Since $u \in C^1((0,\tau), X^\gamma)$ and $X^\gamma \subset C(\overline{\Omega})$ for γ close to 1, $\dot{u}$ is a continuous classical derivative for $t \in (0, \tau_{u_0})$ and $x \in \overline{\Omega}$. To justify its integrability we recall the estimate of [HE 1, Theorem 3.5.2]:

$$\|\dot{u}(t)\|_{X^\gamma} \leq Ct^{\alpha-\gamma-1}, \ 0 < t \leq \tau < \tau_{u_0}, \ \gamma \in [0,1). \tag{9.4.8}$$

The above inequality with $\gamma = 0$ implies that

$$\|\dot{u}\|_{L^p((0,\tau)\times\Omega)} \leq \frac{C}{[(\alpha-1)p+1]^{\frac{1}{p}}}\tau^{\alpha-1+\frac{1}{p}}, \ \tau \in (0,\tau_{u_0}). \tag{9.4.9}$$

Next, since F is Lipschitz continuous on bounded sets and $2m\alpha - \frac{n}{p} > m_0$, it is easy to verify that

$$F(u) \in L^p((0,\tau)\times\Omega), \ \tau \in (0,\tau_{u_0}). \tag{9.4.10}$$

From (2.1.1) we thus have

$$\int_0^\tau \|u\|^p_{W_p^{2m}(\Omega)}dt \leq c^p \int_0^\tau \|Au\|^p_{L^p(\Omega)}dt = c^p \int_0^\tau \|-\dot{u} + F(u)\|^p_{L^p(\Omega)}dt. \tag{9.4.11}$$

Connecting (9.4.9)-(9.4.11) we finally obtain the estimate

$$\|u\|_{W_p^{1,2m}((0,\tau)\times\Omega)} \leq \frac{C}{[(1-\alpha)p+1]^{\frac{1}{p}}}(1+c)\tau^{\alpha-1+\frac{1}{p}} + c\|F(u)\|_{L^p((0,\tau)\times\Omega)},$$

with $\tau \in (0,\tau_{u_0})$, which completes the proof. □

Remark 9.4.1. The above $W_p^{1,2m}((0,\tau)\times\Omega)$ estimate based on the properties of X^α solutions is independent of the linear theory of [L-S-U]. Of course, since it was shown that X^α solution is the Sobolev solution, one may use the result of [L-S-U, Chapter VII, §10, Theorem 10.4] to write directly the estimate

$$\|u\|_{W_p^{1,2m}((0,\tau)\times\Omega)} \leq const. \left(\|f\|_{L^p((0,\tau)\times\Omega)} + \|u_0\|_{B_p^{2m-\frac{2m}{p}}(\Omega)} \right).$$

Application of the Sobolev embedding in R^{n+1} now shows that $u \in C^{\mu,\mu}([0,\tau]\times\overline{\Omega})$, $\mu \in (0, 1-\frac{n+1}{p})$, whenever $p > n+1$.

X^α solution being classical solution. Our next concern is the relation between X^α solutions and the classical solutions. Assume that

- $\partial\Omega \in C^{2m+\nu}$,
- $f = f(x, d^{m_0}u)$ is locally Lipschitz continuous with respect to each variable separately,
- the coefficients of the operator A are of class $C^\nu(\overline{\Omega})$ and those of B_j, $j = 1,\dots,m$, are of class $C^{2m-m_j+\nu}(\partial\Omega)$,

where $\nu \in (0,1)$ is fixed from now on.

Proposition 9.4.2. *Let the above assumptions hold. Let also $2m\alpha - \frac{n}{p} > m_0$ and u be an X^α solution to (2.1.1) defined on $[0, \tau_{u_0})$. Then u is a classical solution to (8.4.1) on $[0,\tau]\times\overline{\Omega}$ for any $\tau \in (0, \tau_{u_0})$.*

Proof. We shall show that (s_4) holds. Condition (i) of (s_4) as well as continuity of $\frac{\partial u}{\partial t}$, $D^\alpha u$ ($|\alpha| \leq m_0$) are immediate consequences of the properties in (s_3). Furthermore, by our assumptions on f, there is certain $\mu \in (0,\nu)$ such that $F(u(t))$ as a function of x satisfies the condition

$$F(u(t)) \in C^\mu(\overline{\Omega}),\ t \in [0,\tau]. \tag{9.4.12}$$

Additionally, it is seen from (s_3) that the number μ may be chosen in such a way that

$$u(t), \dot{u}(t) \in C^\mu(\overline{\Omega}),\ t \in (0,\tau]. \tag{9.4.13}$$

The following Schauder type estimate is reported in [L-S-W, Theorem 5.2]:

$$\|\phi\|_{C^{2m+\mu}(\overline{\Omega})} \leq const.'(\|A\phi\|_{C^\mu(\overline{\Omega})} + \|\phi\|_{C(\overline{\Omega})}) \tag{9.4.14}$$

for

$$\phi \in \mathcal{D} = \left\{ \phi \in \bigcap_{p>1} W^{2m,p}_{\{B_j\}}(\Omega) : A\phi \in C^\mu(\overline{\Omega}) \right\}.$$

Since (8.4.1) is well posed for any $p > 1$ as a sectorial problem in $L^p(\Omega)$ and the X^α solution is unique, it is easy to see that $u \in \mathcal{D}$. Connecting next (9.4.12)-(9.4.14) we justify that $u \in C^{2m+\mu}(\overline{\Omega})$. Thus, all the derivatives appearing in (8.4.1) are classical derivatives. However, the space derivatives $D^\alpha u$ with $m_0 + 1 \leq |\alpha| \leq 2m$ are not necessarily continuous with respect to time variable t. This continuity may be shown based on the results of [L-S-U] (see also considerations in [CHO]). The proof is complete. □

X^α solution being Hölder solution. Recall that $C^{1+\frac{\mu}{2m},2m+\mu}([0,\tau]\times\overline{\Omega})$ denotes the space of functions $\psi:[0,\tau]\times\overline{\Omega}\to R$ having derivatives $\frac{\partial\psi}{\partial t}$, $D^\alpha\psi$, $|\alpha|\le 2m$ uniformly Hölder continuous in $[0,\tau]\times\overline{\Omega}$ (see [L-S-U, Chapter VII, §10]).

The relation between X^α solutions and Hölder solutions is described in the following proposition.

Proposition 9.4.3. *Suppose that the assumptions above Proposition* 9.4.2 *hold and* $2m\alpha-\frac{n}{p}>m_0$, $p>n+1$. *Let u be an X^α solution to* (2.1.1) *defined on* $[0,\tau_{u_0})$. *Then, for a certain* $\mu\in[0,\nu]$ *and any* $\tau\in(0,\tau_{u_0})$, *we have*

$$u\in C^{1+\frac{\mu}{2m},2m+\mu}([0,\tau]\times\overline{\Omega}) \tag{9.4.15}$$

and

$$\begin{aligned}&\|u\|_{C^{1+\frac{\mu}{2m},2m+\mu}([0,\tau]\times\overline{\Omega})}\\&\le c(\tau)\left(\|F(u)\|_{C^{\frac{\mu}{2m},\mu}([0,\tau]\times\overline{\Omega})}+\|u_0\|_{C^{2m+\mu}(\overline{\Omega})}\right),\end{aligned} \tag{9.4.16}$$

provided that $u_0\in C^{2m+\mu}(\overline{\Omega})$ *and the necessary compatibility conditions are satisfied (see Remark 8.4.2).*

Proof. Following the proof of Lemma 8.5.3 (see also [C-D 5]) we obtain the condition

$$u,F(u)\in C^{\mu,\mu}([0,\tau]\times\overline{\Omega}). \tag{9.4.17}$$

Here $C^{\mu,\mu}([0,\tau]\times\overline{\Omega})$ denotes the space of functions $\psi:[0,\tau]\times\overline{\Omega}\to R$ being μ-Hölder continuous with respect to t and x separately, equipped with the norm

$$\begin{aligned}&\|\psi\|_{C^{\mu,\mu}([0,\tau]\times\overline{\Omega})}=\|\psi\|_{C([0,\tau]\times\overline{\Omega})}\\&+\sup_{(t,x),(t,\overline{x})\in[0,\tau]\times\overline{\Omega},\,x\ne\overline{x}}\frac{|\psi(t,x)-\psi(t,\overline{x})|}{|x-\overline{x}|^\mu}+\sup_{(t,x),(\overline{t},x)\in[0,\tau]\times\overline{\Omega},\,t\ne\overline{t}}\frac{|\psi(t,x)-\psi(\overline{t},x)|}{|t-\overline{t}|^\mu}.\end{aligned}$$

Having (9.4.17), one may use the estimate (8.4.19) of the linear theory to get (9.4.16) (see [L-S-W, Theorem 4.1]; also [L-S-U, Chapter VII, §10, Theorem 10.1]). The proof is complete. □

X^α solution in a Hilbert space being Galerkin solution. Suppose that $p=2$, $X=L^2(\Omega)$ and the operator A is self-adjoint and positive definite in X. If $\alpha=\frac{1}{2}$ and $F:X^{\frac{1}{2}}\to X$ is Lipschitz continuous on bounded sets, then Proposition 9.4.1 does not determine whether $\|\dot{u}\|_X$ and $\|Au\|_X$ are square integrable over $(0,\tau)$, $\tau<\tau_{u_0}$. In this case we have

Proposition 9.4.4. *Under the above assumptions the following estimate holds*

$$\|Au\|^2_{L^2((0,\tau)\times\Omega)}\le\|F(u)\|^2_{L^2((0,\tau)\times\Omega)}-\|u\|^2_{X^{\frac{1}{2}}}+\|u_0\|^2_{X^{\frac{1}{2}}},\ \tau\in(0,\tau_{u_0}). \tag{9.4.18}$$

Proof. Multiplying (2.1.1) by Au in $L^2(\Omega)$ and integrating the result over $(0,\tau)$, we obtain

$$\int_0^\tau \langle \dot{u}, Au\rangle_{L^2(\Omega)} ds + \|Au\|^2_{L^2((0,\tau)\times\Omega)} \le \int_0^\tau \|F(u)\|_{L^2(\Omega)}\|Au\|_{L^2(\Omega)} ds$$
$$\le \frac{1}{2}\|F(u)\|^2_{L^2((0,\tau)\times\Omega)} + \frac{1}{2}\|Au\|^2_{L^2((0,\tau)\times\Omega)}$$

and consequently

$$\int_0^\tau \langle \dot{u}, Au\rangle_{L^2(\Omega)} ds + \frac{1}{2}\|Au\|^2_{L^2((0,\tau)\times\Omega)} \le \frac{1}{2}\|F(u)\|^2_{L^2((0,\tau)\times\Omega)}. \tag{9.4.19}$$

Since A is self-adjoint and positive definite, it is easy to see that

(9.4.20)
$$\int_0^\tau \langle \dot{u}, Au\rangle_{L^2(\Omega)} ds = \lim_{\varepsilon\to 0}\int_\varepsilon^\tau \langle \dot{u}, Au\rangle_{L^2(\Omega)} ds$$
$$= \lim_{\varepsilon\to 0}\int_\varepsilon^\tau \langle A^{\frac{1}{2}}\dot{u}, A^{\frac{1}{2}}u\rangle_{L^2(\Omega)} ds = \frac{1}{2}\lim_{\varepsilon\to 0}\int_\varepsilon^\tau \frac{d}{ds}\|A^{\frac{1}{2}}u\|^2_{L^2(\Omega)} ds$$
$$= \frac{1}{2}\|A^{\frac{1}{2}}u\|^2_{L^2(\Omega)} - \frac{1}{2}\|A^{\frac{1}{2}}u_0\|^2_{L^2(\Omega)}.$$

Estimates (9.4.19) and (9.4.20) lead to (9.4.18). Note also that the equality $\dot{u} = -Au + F(u)$ implies that $\dot{u} \in L^2((0,\tau)\times\Omega)$. The proof is complete. □

9.5. Final remarks

As was mentioned in the Preface, this book was devoted rather to the introductory studies of the global attractors related to abstract parabolic problems than to the detailed properties of such objects. In the light of the existing huge literature devoted to specific properties of the global attractors we shall, however, list below a few further directions of studies with necessary indications of literature.

The first direction contains the studies of the *topological dimension (Hausdorff or fractal)* of the attractor. It has been known since the paper of R. Mañé [MAN] (see also J. Mallet-Paret [MA]) that under suitable conditions on the semigroup any compact and positively invariant set has finite Hausdorff dimension. Moreover, this dimension will be estimated from above. This fact is important in the eventual applications since finite dimensionality of the attractor ensures that, at least asymptotically, the flow considered is finite dimensional. Since, in general, finite dimensional flows are simpler to study, finite dimensionality of the global attractors was one of the main reasons for introducing such a notion. For studies of the dimension of attractors we note the articles of B. Nicolaenko, B. Scheurer, R. Temam [N-S-T] and the monographs [HA 2], [TE 1].

A large number of further investigations are devoted to the description of the *content* of the global attractor. While any equilibrium point or periodic (in time)

orbit belongs to the global attractor, there are also other types of elements belonging to this set. The pioneering studies of the content of an attractor for second order semilinear parabolic problem are due to N. Chafee and E. Infante [C-I].

There is a great number of later publications devoted to the content of attractors. We recall here the results of P. Brunovsky and B. Fiedler [B-F] describing possible *connections* between various stationary solutions, again for second order semilinear parabolic equations. The structure of attractors generated by higher order parabolic equations, in particular by the Cahn-Hilliard problem, was studied by S. Zheng [ZH] (in one space dimension), also by B. Nicolaenko, B. Scheurer and R. Temam (see [N-S-T]).

The third large group of results is devoted to generalizations of the notion of a global attractor to similar objects having, however, better structure or shape than the attractor. We shall recall here the idea of an *inertial manifold* introduced by C. Foias, G. R. Sell and R. Temam [F-S-T]. The inertial manifold is a *'Lipschitz manifold containing the global attractor, attracting exponentially fast all solutions, it is also stable with respect to perturbations'*. Again, this object also has finite topological dimension. Later the last concept has been extended to a notion of the *approximate inertial manifold* by A. Debussche and M. Marion [D-M], since the original inertial manifolds appear to be difficult objects for computational methods.

Another way of generalizing the notion of the global attractor was proposed by A. Eden, C. Foias, B. Nicolaenko and R. Temam [E-F-N-T] (see also [E-M-N]) by introducing the notion of *inertial set* (which is a less specified object than the inertial manifold). This set has finite fractal dimension and attracts all solutions at an exponential rate. Existence of inertial sets was studied further for both first and second order in time nonlinear differential equations.

The fourth direction in which the idea of the global attractor has been recently generalized deals with studies of the nonautonomous processes corresponding to parabolic equations with time dependent coefficients or boundary data. The counterpart of the global attractor for dynamical systems generated for such processes was studied by M. I. Vishik [VI] (see also [C-V 1]). Here mention should be made of the *trajectory attractors* that have been constructed for various systems governed by partial differential equations by V. V. Chepyzhov and M. I. Vishik (see [C-V 2], [C-V 3], [C-V 4]).

We should next point to the large literature devoted to global attractors for problems having less regular behavior than those given by parabolic equations. We mean here in particular ordinary differential equations with delay and hyperbolic partial differential equations. We are unable to recall all important results in these directions. Without having strong smoothing properties of parabolic semigroups, compactness of trajectories and existence of the corresponding ω-limit sets are hard introductory questions in these considerations (see e.g. O. Ladyzenskaya [LA 3], J. Arrieta, A. N. Carvalho and J. K. Hale [A-C-H] or E. Feireisl [FE 2]).

We also recall the recent survey article by J. K. Hale [HA 4] in which detailed questions related to dynamics of scalar second order parabolic equations have been notified. Special properties of solutions and techniques specific to second order parabolic problems allow one to get the most complete results just for such types of problems.

Once having sufficient knowledge of the dynamics and a global attractor for a separate problem one can try to compare the flows and the attractors for different systems (see e.g. [F-H]). There is, at the present, significant literature (see [HA 2], [H-R 2], [H-R 3]) in which the dependence of attractors on parameters appearing in differential equations, or the domain in which the space variable varies, has been studied using the notion of the *upper semicontinuity* of the attractors. Asymptotic equivalence of the flow on the whole phase space and on a corresponding global attractor called the *shadow property* (see [NIS]) is another interesting and popular field of studies through recent years (see [HA 3], [HA-SA], [K-M-N-O-T]).

Bibliography

[AB] F. Abergel, Existence and finite dimensionality of the global attractor for evolution equations on unbounded domains, *J. Diff. Equations* 83 (1990), 85-108.

[A-T] P. Acquistapace, B. Terreni, Hölder classes with boundary conditions as interpolation spaces, *Math. Z.* 195 (1987), 451-471.

[AD] R. A. Adams, *Sobolev Spaces*, Academic Press, New York, 1975.

[AG] S. Agmon, *Lectures on Elliptic Boundary Value Problems*, Mathematical Studies 2, Van Nostrand, New York, 1965.

[A-D-N 1] S. Agmon, A. Douglis, L. Nirenberg, Estimates near the boundary for solutions of elliptic partial differential equations satisfying general boundary conditions, I, *Comm. Pure Appl. Math.* 12 (1959), 623-727.

[A-D-N 2] S. Agmon, A. Douglis, L. Nirenberg, Estimates near the boundary for solutions of elliptic partial differential equations satisfying general boundary conditions, II, *Comm. Pure Appl. Math.* 7 (1964), 35-92.

[A-N] S. Agmon, L. Nirenberg, Lower bounds and uniqueness theorems for solutions of differential equations in a Hilbert space, *Comm. Pure Appl. Math.* 20 (1967), 207-229.

[AL 1] N. D. Alikakos, An application of the invariance principle to reaction-diffusion equations, *J. Diff. Equations* 33 (1979), 201-225.

[AL 2] N. D. Alikakos, Quantitative maximum principles and strongly coupled gradient-like reaction-diffusion systems, *Proc. Roy. Soc. Edinburgh* 94A (1983), 265-286.

[A-B-C] N. D. Alikakos, P. W. Bates, X. Chen, Convergence of the Cahn-Hilliard equation to the Hele-Shaw model, *Arch. Rational Mech. Anal.* 128 (1994), 165-205.

[A-F] N. D. Alikakos, G. Fusco, Slow dynamics for the Cahn-Hilliard equation in higher space dimensions. Part I: Spectral estimates, *Comm. Partial Diff. Equations* 19 (1994), 1397-1447.

[A-M] N. D. Alikakos, W. R. McKinney, Remarks on the equilibrium theory for the Cahn-Hilliard equation in one space dimension, in *Reaction-Diffusion Equations* (Eds. K. J. Brown and A. A. Lacey), Oxford University Press, Oxford, 1990, 75-94.

[AM 1] H. Amann, Global existence for semilinear parabolic systems, *J. Reine Angew. Math.* 360 (1985), 47-83.

[AM 2] H. Amann, Quasilinear evolution equations and parabolic systems, *Trans. Amer. Math. Soc.* 293 (1986), 191-227.

[AM 3] H. Amann, Existence and regularity for semilinear parabolic equations, *Ann. Sc. Norm. Sup. Pisa* 11 (1984), 593-676.

[AM 4] H. Amann, Parabolic evolution equations and nonlinear boundary conditions, *J. Diff. Equations* 72 (1988), 201-269.

[AM 5] H. Amann, *Linear and Quasilinear Parabolic Problems*, Birkhäuser, Basel, 1995.

[A-H-S] H. Amann, M. Hieber, G. Simonett, Bounded H_∞-calculus for elliptic operators, *Diff. Int. Equations* 3 (1994), 613-653.

[A-C-H] J. Arrieta, A. N. de Carvalho, J. K. Hale, A damped hyperbolic equation with critical exponent, *Comm. Partial Diff. Equations* 17 (1992), 841-866.

[B-S] A. V. Babin, G. R. Sell, Attractors of non-autonomous parabolic equations and their symmetry properties, *J. Diff. Equations* 160 (2000), 1-50.

[B-V 1] A. V. Babin, M. I. Vishik, Attractors of partial differential evolution equations in unbounded domain, *Proc. Roy. Soc. Edinburgh* 116A (1990), 221-243.

[B-V 2] A. V. Babin, M. I. Vishik, *Attractors of Evolution Equations*, North-Holland, Amsterdam, 1992.

[B-V 3] A. V. Babin, M. I. Vishik, *Properties of Global Attractors of Partial Differential Equations*, AMS, RI, 1992.

[BAL] J. M. Ball, Remarks on blow-up and nonexistence theorems for nonlinear evolution equations, *Quart. J. Math. Oxford* 28 (1977), 473-486.

[BAR] V. Barbu, *Nonlinear Semigroups and Differential Equations in Banach Spaces*, Noordhoff, Leiden, 1976.

[BA-FI] P. W. Bates, P. C. Fife, Spectral comparison principles for the Cahn-Hilliard and phase-field equations, and time scales for coarsening, *Phys. D* 43 (1990), 335-348.

[BA-FU] P. W. Bates, G. Fusco, Equilibria with many nuclei for the Cahn-Hilliard equation, *J. Diff. Equations* 160 (2000), 283-356.

[B-E] J. Bebernes, D. Eberly, *Mathematical Problems from Combustion Theory*, Springer, New York, 1989.

[B-K-W] P. Biler, G. Karch, W. A. Woyczyński, Asymptotics for multifractal conservation laws, *Studia Math.* 135, (1999), 231-252.

[B-L] J. E. Billotti, J. P. LaSalle, Periodic dissipative processes, *Bull. Amer. Math. Soc.* 6 (1971), 1082-1089.

[BR] H. Brezis, *Opérateurs Maximaux Monotones et Semi-groupes de Contractions dans les Espaces de Hilbert,* North-Holland, Amsterdam, 1973.

[B-F] P. Brunovsky, B. Fiedler, Connecting orbits in scalar reaction diffusion equations, in *Dynamics Reported,* Teubner, Stuttgart, 1988, 57-89.

[B-P] P. Brunovsky, P. Poláčik, The Morse-Smale structure of a generic reaction-diffusion equation in higher space dimension, *J. Diff. Equations* 135 (1997), 129-181.

[CA-HI] J. W. Cahn, J. E. Hilliard, Free energy of a nonuniform system - I. Interfacial free energy, *J. Chem. Phys.* 28 (1958), 258-267.

[CAR] A. N. de Carvalho, Contracting sets and dissipation, *Proc. Roy. Soc. Edinburgh* 125A (1995), 1305-1329.

[C-C] A. N. de Carvalho, J. W. Cholewa, Strongly damped wave equation with critical nonlinearities I; Case $\theta = \frac{1}{2}$, *Notas do ICMC*, Saõ Carlos, 1999.

[C-C-D 1] A. N. de Carvalho, J. W. Cholewa, T. Dlotko, Examples of global attractors in parabolic problems, *Hokkaido Math. J.* 27 (1998), 77-103.

[C-C-D 2] A. N. de Carvalho, J. W. Cholewa, T. Dlotko, Global attractors for problems with monotone operators, *Boll. Un. Mat. Ital.* (8) 2-B (1999), 693-706.

[C-R] A. N. de Carvalho, J. G. Ruas-Filho, Global attractors for parabolic problems in fractional power spaces, *SIAM J. Math. Anal.* 26 (1995), 415-427.

[CAT] L. Cattabriga, Sur un problema al contorno relativo al sistema di equazioni di Stokes, *Rend. Sem. Mat. Univ. Padova* 31 (1961), 308-340.

[C-L] T. Cazenave, P. L. Lions, Solutions globales d'équations de la chaleur semi linéaires, *Comm. Partial Diff. Equations* 9 (1984), 955-978.

[CHA] J. Chabrowski, *Variational Methods for Potential Operator Equations*, Studies in Mathematics 24, De Gruyter, Berlin, 1997.

[CHF] N. Chafee, Asymptotic behavior for solutions of a one-dimensional parabolic equation with homogeneous Neumann boundary conditions, *J. Diff. Equations* 18

(1975), 111-134.

[C-I] N. Chafee, E. F. Infante, A bifurcation problem for a nonlinear partial differential equation of parabolic type, *Appl. Anal.* 4 (1974), 17-37.

[C-V 1] V. V. Chepyzhov, M. I. Vishik, Attractors of non-autonomous partial differential equations and their dimension, Equadiff 8 (Bratislava, 1993), *Tatra Mt. Math. Publ.* 4 (1994), 221-234.

[C-V 2] V. V. Chepyzhov, M. I. Vishik, Trajectory attractors for the 2D Navier-Stokes system and some generalizations, *Topol. Methods Nonlinear Anal.* 8 (1996), 217-243.

[C-V 3] V. V. Chepyzhov, M. I. Vishik, Trajectory attractors for evolution equations, *C. R. Acad. Sci. Paris* 321 (1995), 1309-1314.

[C-V 4] V. V. Chepyzhov, M. I. Vishik, Evolution equations and their trajectory attractors, *J. Math. Pures Appl.* 76 (1997), 913-964.

[CHO] J. W. Cholewa, Local existence of solutions of $2m$-th order semilinear parabolic equations, *Demonstratio Math.* 28 (1995), 929-944.

[C-D 1] J. W. Cholewa, T. Dlotko, Global solutions via partial information and the Cahn-Hilliard equation, in *Singularities and Differential Equations*, (Eds. S. Janeczko, W. M. Zajączkowski and B. Ziemian), Banach Center Publications 33, PWN, Warsaw, 1996, 39-50.

[C-D 2] J. W. Cholewa, T. Dlotko, Global attractor for the Cahn-Hilliard system, *Bull. Austral. Math. Soc.* 49 (1994), 277-293.

[C-D 3] J. W. Cholewa, T. Dlotko, Global attractor for sectorial evolutionary equation, *J. Diff. Equations* 125 (1996), 27-39.

[C-D 4] J. W. Cholewa, T. Dlotko, Cauchy problem with subcritical nonlinearity, *J. Math. Anal. Appl.* 210 (1997), 531-548.

[C-D 5] J. W. Cholewa, T. Dlotko, Global attractors for parabolic p.d.e.'s in Hölder spaces, *Tsukuba J. Math.* 21 (1997), 263-283.

[C-D 6] J. W. Cholewa, T. Dlotko, Local attractor for n-D Navier-Stokes system, *Hiroshima Math. J.* 28 (1998), 309-319.

[CH-HA] J. W. Cholewa, J. K. Hale, Some counterexamples in dissipative systems, *Dynam. Contin. Discrete Impuls. Systems* 7 (2000), 159-176.

[C-C-S] K. Chueh, C. Conley, J. Smoller, Positively invariant regions for systems of nonlinear diffusion equations, *Indiana Univ. Math. J.* 26 (1977), 373-392.

[CO] C. Cosner, Pointwise a priori bounds for strongly coupled semi-linear systems of parabolic partial differential equations, *Indiana Univ. Math. J.* 30 (1981), 607-620.

[C-P-T] M. G. Crandall, A. Pazy, L. Tartar, Remarks on generators of analytic semigroups, *Israel J. Math.* 32 (1979), 363-374.

[D-M] A. Debussche, M. Marion, On the construction of families of approximate inertial manifolds, *J. Diff. Equations* 100 (1992), 173-201.

[DL 1] T. Dlotko, Global attractor for the Cahn-Hilliard equation in H^2 and H^3, *J. Diff. Equations* 113 (1994), 381-393.

[DL 2] T. Dlotko, The two-dimensional Burgers' turbulence model, *J. Math. Kyoto Univ.* 21 (1981), 809-823.

[DL 3] T. Dlotko, Global solutions of reaction-diffusion equations, *Funkcial. Ekvac.* 30 (1987), 31-43.

[DL 4] T. Dłotko, *Parabolic Equations in Divergence Form. Theory of Global Solutions,* Wydawnictwo Uniwersytetu Śląskiego, Katowice, 1987.

[DL 5] T. Dlotko, Examples of parabolic problems with blowing-up derivatives, *J. Math. Anal. Appl.* 154 (1991), 226-237.

[DL 6] T. Dlotko, Parabolic equation modelling diffusion with strong absorption, *Atti Sem. Mat. Fis. Univ. Modena* 38 (1990), 61-70.

[D-S] N. Dunford, J. T. Schwartz, *Linear Operators*, Vol. I, Interscience, New York, 1958.

[DU] L. Dung, Hölder regularity for certain strongly coupled parabolic systems, *J. Diff. Equations* 151 (1999), 313-344.

[E-F-N-T] A. Eden, C. Foias, B. Nicolaenko, R. Temam, Ensembles inertiels pour des équations d'évolution dissipatives, *C. R. Acad. Sci. Paris* 310 (1990), 559-562.

[E-M-N] A. Eden, A. J. Milani, B. Nicolaenko, Finite dimensional exponential attractors for semilinear wave equations with damping, *J. Math. Anal. Appl.* 169 (1992), 408-419.

[EY] D. J. Eyre, Systems of Cahn-Hilliard equations, *SIAM J. Appl. Math.* 53 (1993), 1686-1712.

[F-H] H. T. Fan, J. K. Hale, Attractors in inhomogeneous conservation laws and parabolic regularizations, *Trans. Amer. Math. Soc.* 347 (1995), 1239-1254.

[FA] H. O. Fattorini, *The Cauchy Problem,* Addison-Wesley, London, 1983.

[FE 1] E. Feireisl, Bounded, locally compact global attractors for semilinear damped wave equations on $\mathbf{R}^N$. *Diff. Int. Equations* 9 (1996), 1147-1156.

[FE 2] E. Feireisl, Strong decay for wave equations with nonlinear nonmonotone damping, *Nonlinear Anal.* 21 (1993), 49-63

[F-S] E. Feireisl, F. Simondon, Convergence for degenerate parabolic equations, *J. Diff. Equations* 152 (1999), 439-466.

[F-G] C. Foias, C. Guillopé, On the behavior of the solutions of the Navier-Stokes equations lying on invariant manifolds, *J. Diff. Equations* 61 (1986), 128-148.

[F-P] C. Foias, G. Prodi, Sur le comportement global des solutions non stationnaires des équations de Navier-Stokes en dimension 2, *Rend. Sem. Mat. Univ. Padova* 39 (1967), 1-34.

[FO-SA] C. Foias, J. C. Saut, Asymptotic behavior as $t \to +\infty$ of solutions of Navier-Stokes equations and nonlinear spectral manifolds, *Indiana Univ. Math. J.* 3 (1984), 459-477.

[F-S-T] C. Foias, G. R. Sell, R. Temam, Inertial manifolds for nonlinear evolutionary equations, *J. Diff. Equations* 73 (1988), 309-353.

[FR 1] A. Friedman, *Partial Differential Equations,* Holt, Rinehart and Winston, New York, 1969.

[FR 2] A. Friedman, *Partial Differential Equations of Parabolic Type,* Prentice-Hall, Englewood Cliffs, NJ, 1964.

[F-H] A. Friedman, M. A. Herrero, Extinction properties of semilinear heat equations with strong absorption, *J. Math. Anal. Appl.*, 124 (1987), 530-546.

[F-M] A. Friedman, B. McLeod, Blow-up of positive solutions of semilinear heat equations, *Indiana Univ. Math. J.* 34 (1985), 425-447.

[F-K] S. Fučik, A. Kufner, *Nonlinear Differential Equations,* Elsevier, Amsterdam, 1980.

[FU-MO] D. Fujiwara, H. Morimoto, An L_r-theorem of the Helmholtz decomposition of vector fields, *J. Fac. Sci. Univ. Tokyo* 24 (1977), 685-700.

[G-G-Z] H. Gajewski, K. Gröger, K. Zacharias, *Nichtlineare Operatorgleichungen und Operatordifferentialgleichungen,* Akademie, Berlin, 1974.

[GH-MA] J. M. Ghidaglia, A. Marzocchi, Exact decay estimates for solutions to semilinear parabolic equations, *Appl. Anal.* 42 (1991), 69-81.

[G-G-S] M. Giga, Y. Giga, H. Sohr, L^p estimates for the Stokes system, in *Functional Analysis and Related Topics* (Kyoto, 1991), (Ed. H. Komatsu), Lecture Notes in Mathematics 1540, Springer, Berlin, 1993, 55-67.

[GI 1] Y. Giga, Analyticity of the semigroup generated by the Stokes operator in L^r spaces, *Math. Z.* 178 (1981), 297-329.

[GI 2] Y. Giga, Domains of fractional powers of the Stokes operator in L_r spaces, *Arch. Rational Mech. Anal.* 89 (1985), 251-265.

[GI 3] Y. Giga, A bound for global solutions of semilinear heat equations, *Comm. Math. Phys.* 103 (1986), 415-421.

[GI-MI] Y. Giga, T. Miyakawa, Solutions in L_r of the Navier-Stokes initial value problem, *Arch. Rational Mech. Anal.* 89 (1985), 267-281.

[GI-SO] Y. Giga, H. Sohr, Abstract L^p estimates for the Cauchy problem with applications to the Navier-Stokes equations in exterior domains, *J. Functional Anal.* 102 (1991), 72-94.

[GI-TR] D. Gilbarg, N. Trudinger, *Elliptic Partial Differential Equations of Second Order,* Springer, Berlin, 1983.

[GO-SA] M. Gobbino, M. Sardella, On the connectedness of attractors for dynamical systems, *J. Diff. Equations* 133 (1997), 1-14.

[GR] P. Grisvard, Caractérisation de quelques espaces d'interpolation, *Arch. Rational Mech. Anal.* 25 (1967), 40-63.

[GU 1] D. Guidetti, On interpolation with boundary conditions, *Math. Z.* 207 (1991), 439-460.

[GU 2] D. Guidetti, On boundary value problems for parabolic equations of higher order in time, *J. Diff. Equations* 124 (1996), 1-26.

[GUI] C. Guillopé, Comportement à l'infini des solutions des équations de Navier-Stokes et propriété des ensembles fonctionnels invariants (ou attracteurs), *Ann. Inst. Fourier* 3 (1982), 1-37.

[HA 1] J. K. Hale, Large diffusivity and asymptotic behavior in parabolic systems, *J. Math. Anal. Appl.* 118 (1986), 455-466.

[HA 2] J. K. Hale, *Asymptotic Behavior of Dissipative Systems,* AMS, Providence, RI, 1988.

[HA 3] J. K. Hale, Shadow systems for evolutionary equations, in *Differential Equations* (Xanthi, 1987), Lecture Notes in Pure and Appl. Math. 118, Dekker, New York, 1989, 303-310.

[HA 4] J. K. Hale, Dynamics of a scalar parabolic equation, *Canadian Appl. Math. Quart.* 5 (1997), 209-305.

[H-R 1] J. K. Hale, G. Raugel, Attractors for dissipative evolutionary equations, in *International Conference on Differential Equations* (Barcelona, 1991), World Sci., River Edge, NJ, 1993, 3-22.

[H-R 2] J. K. Hale, G. Raugel, Limits of semigroups depending on parameters, *Resenhas* 1 (1993), 1-45.

[H-R 3] J. K. Hale, G. Raugel, A reaction-diffusion equation on a thin L-shaped domain. *Proc. Roy. Soc. Edinburgh* Sect. A 125 (1995), 283-327.

[HA-SA] J. K. Hale, K. Sakamoto, Shadow systems and attractors in reaction-diffusion equations, *Appl. Anal.* 32 (1989), 287-303.

[HA-SC] J. K. Hale, J. Scheurle, Smoothness of bounded solutions of nonlinear evolution equations, *J. Diff. Equations* 56 (1985), 142-163.

[HA-WA] J. K. Hale, P. Waltman, Persistence in infinite-dimensional systems, *SIAM J. Math. Anal.* 20 (1989), 388-395.

[HE 1] D. Henry, *Geometric Theory of Semilinear Parabolic Equations,* Lecture Notes in Mathematics 840, Springer, Berlin, 1981.

[HE 2] D. Henry, How to remember the Sobolev inequalities, in *Differential Equations* (Eds. D. G. de Figueiredo and C. S. Hönig), Lecture Notes in Mathematics 957, Springer, New York 1982, 97-109.

[H-P] E. Hille, R. S. Phillips, *Functional Analysis and Semi-groups,* AMS, Providence, RI, 1957.

[H-Y] H. Hoshino, Y. Yamada, Solvability and smoothing effect for semilinear parabolic equations, *Funkcial. Ekvac.* 34 (1991), 475-494.

[I-K-O] A. M. Il'in, A. S. Kalahsnikov, O. A. Olejnik, Linear second order equations of parabolic type, *Uspehi Mat. Nauk* 17 (1962), 3-146 (in Russian).

[KAL] A. S. Kalashnikov, The propagation of disturbances in problems of nonlinear heat conduction with absorption, *U.S.S.R. Computational Math. Math. Phys.* 14 (1974), 70-85.

[KA] T. Kato, Strong L^p-solutions of the Navier-Stokes equation in R^m, with applications to weak solutions, *Math. Z.* 187 (1984), 471-480.

[KI] H. Kielhöfer, Global solutions of semilinear evolution equations satisfying an energy inequality, *J. Diff. Equations* 36 (1980), 188-222.

[KL] S. Klainerman, Global existence for nonlinear wave equations, *Comm. Pure Appl. Math.* 33 (1980), 43-101.

[K-M-N-O-T] H. Kokubu, K. Mischaikow, Y. Nishiura, H. Oka, T. Takaishi, Connecting orbit structure of monotone solutions in the shadow system, *J. Diff. Equations* 140 (1997), 309-364.

[KO 1] H. Komatsu, Fractional powers of operators, *Pacific J. Math.* 19 (1966), 285-346.

[KO 2] H. Komatsu, Fractional powers of operators. II. Interpolation spaces, *Pacific J. Math.* 21 (1967), 89-111.

[K-J-F] A. Kufner, O. John, S. Fučik, *Function Spaces,* Academia, Prague, 1997.

[LA 1] O. A. Ladyženskaya, A dynamical system generated by the Navier-Stokes equation, *Zap. Naucn. Sem. Leningrad. Otdel. Mat. Inst. Steklov. (LOMI)* 27 (1972), 91-115 (in Russian).

[LA 2] O. A. Ladyženskaya, Some recent mathematical results concerning the Navier-Stokes equations, *Arch. Mech.* 30 (1978), 217-224.

[LA 3] O. A. Ladyženskaya, *Attractors for Semigroups and Evolution Equations,* Cambridge University Press, Cambridge, 1991.

[L-S-U] O. A. Ladyženskaja, V. A. Solonnikov, N. N. Ural'ceva, *Linear and quasilinear equations of parabolic type*, Translations of Mathematical Monographs, AMS, Providence, RI, 1967.

[L-U] O. A. Ladyženskaya, N. N. Ural'ceva, *Linear and Quasilinear Elliptic Equations,* Academic Press, New York, 1968.

[LA-MA] A. Lasota, M. C. Mackey, *Probabilistic properties of deterministic systems,* Cambridge University Press, Cambridge, 1985.

[L-Z] De-Sheng Li, Chen-Kui Zhong, Global attractor for the Cahn-Hilliard system with fast growing nonlinearity, *J. Diff. Equations* 149 (1998), 191-210.

[L-M] J. H. Lightbourne, R. H. Martin, Relatively continuous nonlinear perturbations of analytic semigroups, *Nonlinear Anal.* (1977), 277-292.

[LI] J. L. Lions, *Quelques Méthodes de Résolutions des Problèmes aux Limites non Linéaires,* Dunod, Paris, 1969.

[LI-MA] J. L. Lions, E. Magenes, *Problèmes aux Limites non Homogènes et Applications,* Vol. I, Dunod, Paris 1968.

[LU 1] A. Lunardi, *Analytic Semigroup and Optimal Regularity in Parabolic Problems,* Birkhäuser, Berlin, 1995.

[LU 2] A. Lunardi, Interpolation spaces between domains of elliptic operators and spaces of continuous functions with applications to nonlinear parabolic equations, *Math. Nachr.* 121 (1985), 295-318.

[L-S-W] A. Lunardi, E. Sinestrari, W. von Wahl, A semigroup approach to the time dependent parabolic initial boundary value problem, *Diff. Int. Equations* 6 (1992), 1275-1306.

[MA] J. Mallet-Paret, Negatively invariant sets of compact maps and an extension of a theorem of Cartwright, *J. Diff. Equations* 22 (1976), 331-348.

[MAN] R. Mañé, On the dimension of the compact invariant sets of certain non-linear maps, in *Dynamical Systems and Turbulence,* (Eds. D. A. Rand and L.-S. Young), Lecture Notes in Mathematics 898, Springer, Berlin, 1981, 230-242.

[MAR] J. T. Marti, *Introduction to Sobolev Spaces and Finite Element Solution of Elliptic Boundary Value Problems,* Academic Press, London, 1986.

[MAT] R. H. Martin, Invariant sets and a mathematical model involving semilinear differential equations, in *Nonlinear Equations in Abstract spaces,* Academic Press, New York, 1978, 135-148.

[MAZ] V. G. Maz'ja, *Sobolev Spaces,* Springer, Berlin, 1985.

[MIK] M. Miklavčič, Stability for semilinear equations with noninvertible linear operator, *Pacific J. Math.* 118 (1985), 199-214.

[MIY 1] T. Miyakawa, The L^p approach to the Navier-Stokes equations with Neumann boundary conditions, *Hiroshima Math. J.* 10 (1980), 517-537.

[MIY 2] T. Miyakawa, On the initial value problem for the Navier-Stokes equations in L^p spaces, *Hiroshima Math. J.* 11 (1981), 9-20.

[MIY 3] T. Miyakawa, Application of Hardy space techniques to the time-decay problem for incompressible Navier-Stokes flows in R^n, *Funkcial. Ekvac.* 41 (1998), 383-434.

[M-Y] N. Mizoguchi, E. Yanagida, Critical exponents for the blow-up of solutions with sign changes in a semilinear parabolic equation, *Math. Ann.* 307 (1997), 663-675.

[ML] W. Mlak, *Hilbert Spaces and Operator Theory,* Kluwer (PWN), Dordrecht, Holland, 1991.

[MO] C. B. Morrey, Jr, *Multiple Integrals in the Calculus of Variations*, Springer, New York, 1966.

[NE] J. Nečas, *Introduction to the Theory of Nonlinear Elliptic Equations,* Teubner, Leipzig, 1983.

[N-S-T] B. Nicolaenko, B. Scheurer, R. Temam, Some global dynamical properties of a class of pattern formation equations, *Comm. Partial Diff. Equations* 14 (1989), 245-297.

[NI 1] L. Nirenberg, On elliptic partial differential equations, *Ann. Sc. Norm. Sup. Pisa* 13 (1959), 123-131.

[NI 2] L. Nirenberg, An extended interpolation inequality, *Ann. Sc. Norm. Sup. Pisa* 20 (1966), 733-737.

[NIS] Y. Nishiura, Global structure of bifurcating solutions of some reaction-diffusion systems, *SIAM J. Math. Anal.* 13 (1982), 555-593.

[PA] L. E. Payne, Some general remarks on improperly posed problems for partial differential equations, in *Symposium on Non-Well-Posed Problems and Logaritmic Convexity* (Ed. R. J. Knops), Lecture Notes in Mathematics 316, Springer, Berlin, 1973, 1-30.

[PAZ 1] A. Pazy, A class of semi-linear equations of evolution, *Israel J. Math.* 20 (1975), 23-36.

[PAZ 2] A. Pazy, *Semigroups of Linear Operators and Applications to Partial Differential Equations,* Springer, Berlin, 1983.

[PR] K. Promislow, Time analyticity and Gevrey regularity for solutions of a class of dissipative partial differential equations, *Nonlinear Anal.* (1991), 959-980.

[P-W] M. H. Protter, H. F. Weinberger, *Maximum Principles in Differential Equations,* Springer, New York, 1984.

[P-S] J. Prüss, H. Sohr, Imaginary powers of elliptic second order differential operators in L^p-spaces, *Hiroshima Math. J.* 23 (1993), 161-192.

[RA] R. Racke, Global solutions to semilinear parabolic systems for small data, *J. Diff. Equations* 76 (1988), 312-338.

[RB 1] A. Rodriguez-Bernal, Existence, uniqueness and regularity of solutions of evolution equations in extended scales of Hilbert spaces, *CDSNS GaTech Report Series,* Atlanta, 1991.

[RB 2] A. Rodriguez-Bernal, *Semilinear Evolutionary Equations,* Lecture Notes Departamento de Matemática Aplicada, Universidad Complutense de Madrid, Madrid, 2000.

[RB 3] A. Rodriguez-Bernal, Initial value problem and asymptotic low dimensional behavior in the Kuramoto-Velardc equation, *Nonlinear Anal.* 19 (1992), 643-685.

[RO] F. Rothe, *Global Solutions of Reaction-Diffusion Systems,* Lecture Notes in Mathematics 1072, Springer, Berlin, 1984.

[RY 1] K. P. Rybakowski, *The Homotopy Index and Partial Differential Equations,* Springer, Berlin, 1987.

[RY 2] K. P. Rybakowski, Realization of arbitrary vector fields on center manifolds of parabolic Dirichlet BVPs, *J. Diff. Equations* 114 (1994), 199-221.

[SEE] R. Seeley, Interpolation in L^p with boundary conditions, *Studia Math.* 44 (1972), 47-60.

[SEG] I. Segal, Non-linear semi-groups, *Ann. Math.* 78 (1963), 339-364.

[SEL] G. R. Sell, Global attractors for the three-dimensional Navier-Stokes equations, *J. Dynam. Diff. Equations* 8 (1996), 1-33.

[SIN] E. Sinestrari, On the abstract Cauchy problem of parabolic type in spaces of continuous functions, *J. Math. Anal. Appl.* 107 (1985), 16-66.

[SM] J. Smoller, *Shock Waves and Reaction-Diffusion Equations,* Springer, New York, 1988.

[SOB] P. E. Sobolevski, On equations of parabolic type in a Banach space, *Trudy Moskov. Mat. Obšč.* 10 (1961), 297-340 (in Russian).

[SOL] V. A. Solonnikov, On L_p estimates of solutions of elliptic and parabolic systems, *Trudy Mat. Inst. Steklov.* 102 (1967), 137-160 (in Russian).

[S-T] H. Sohr, G. Thäter, Imaginary powers of second order differential operators and L^q-Helmholtz decomposition in the infinite cylinder, *Math. Ann.* 311 (1998), 577-602.

[S-W] H. Sohr, W. von Wahl, On the singular set and the uniqueness of weak solutions of the Navier-Stokes equations, *Manuscripta Math.* 49 (1984), 27-59.

[ST] E. Stein, *Singular Integrals and Differentiability Properties of Functions,* Princeton University Press, Princeton, NJ, 1970.

[SZ] Z. Szmydt, *Fourier Transformation and Linear Differential Equations,* Reidel (PWN), Dordrecht, Holland, 1977.

[TA 1] H. Tanabe, *Equations of Evolution*, Pitman, London, 1979.

[TA 2] H. Tanabe, *Functional Analytic Methods for Partial Differential Equations,* Dekker, New York, 1997.

[TE 1] R. Temam, *Infinite-Dimensional Dynamical Systems in Mechanics and Physics,* Springer, New York, 1988.

[TE 2] R. Temam, *Navier-Stokes Equations, Theory and Numerical Analysis,* North-Holland, Amsterdam, 1979.

[TR] H. Triebel, *Interpolation Theory, Function Spaces, Differential Operators,* Veb Deutscher, Berlin, 1978.

[TU] A. W. Turski, A modern proof of the Maximum Principle, *Ann. Polon. Math.* 53 (1991), 147-152.

[VA] M. M. Vainberg, *Variational Method and Method of Monotone Operators,* Wiley, New York, 1973.

[VI] M. I. Vishik, *Asymptotic Behaviour of Solutions of Evolutionary Equations,* Cambridge University Press, Cambridge, 1992.

[VR] I. Vrabie, *Compactness Methods for Nonlinear Evolutions,* Pitman Monographs and Surveys in Pure and Applied Mathematics 32, Longman, Burnt Mill, 1987.

[WA 1] W. von Wahl, Global solutions to evolution equations of parabolic type, in *Differential Equations in Banach Spaces,* Proceedings, 1985 (Eds. A. Favini and E. Obrecht), Lecture Notes in Mathematics 1223, Springer, Berlin, 1986, 254-266.

[WA 2] W. von Wahl, Semilinear elliptic and parabolic equations of arbitrary order, *Proc. Roy. Soc. Edinburgh* 78A (1978), 193-207.

[WA 3] W. von Wahl, Über die stationären Gleichungen von Navier-Stokes semilineare elliptische und parabolische Gleichungen, *Jahresber. Deutsch. Math.-Verein.* 80 (1978), 129-149.

[WA 4] W. von Wahl, Die stationären Gleichungen von Navier-Stokes und semilineare elliptische Systeme, *Amer. J. Math.* 100 (1977), 1173-1184.

[WA 5] W. von Wahl, *Equations of Navier-Stokes and Abstract Parabolic Equations*, Vieweg, Braunschweig/Wiesbaden, 1985.

[WAL] W. Walter, *Differential and Integral Inequalities*, Springer, Berlin, 1970.

[WAT] J. Watanabe, On some properties of fractional powers of linear operators, *Proc. Japan Acad.* 37 (1961), 273-275.

[WEI] H. F. Weinberger, An example of blowup produced by equal diffusions, *J. Diff. Equations* 154 (1999), 225-237.

[WES] F. Weissler, Local existence and nonexistence for semilinear parabolic equations in L^p, *Indiana Univ. Math. J.* 29 (1980), 79-102.

[YO 1] K. Yosida, *Functional Analysis*, Springer, Berlin, 1978.

[YO 2] K. Yosida, Fractional powers of linear operators, *Proc. Japan Acad.* 36 (1960), 94-96.

[ZE 1] T. I. Zelenyak, The stability of solutions of mixed problems for a particular quasi-linear equation, *Diff. Equations* 3 (1967), 9-13.

[ZE 2] T. I. Zelenyak, Stabilization of solutions of boundary value problems for a second order parabolic equation with one space variable, *Diff. Equations* 4 (1968), 17-22.

[ZE 3] T. I. Zelenyak, *Qualitative Theory of Boundary Value Problems for Second Order Quasilinear Equations of Parabolic Type*, Novosibirsk. Gos. Univ., Novosibirsk, 1972 (in Russian).

[ZH] S. Zheng, Asymptotic behavior of solutions to the Cahn-Hilliard equation, *Appl. Anal.* 23 (1986), 165-184.

[ZI] W. P. Ziemer, *Weakly Differentiable Functions, Sobolev Spaces and Functions of Bounded Variation*, Springer, New York, 1989.

Index

Printed in the United States
By Bookmasters